UBC Civil

604 822 2775

About Island Press

Island Press is the only nonprofit organization in the United States whose principal purpose is the publication of books on environmental issues and natural resource management. We provide solutions-oriented information to professionals, public officials, business and community leaders, and concerned citizens who are shaping responses to environmental problems.

In 2004, Island Press celebrates its twentieth anniversary as the leading provider of timely and practical books that take a multidisciplinary approach to critical environmental concerns. Our growing list of titles reflects our commitment to bringing the best of an expanding body of literature to the environmental community throughout North America and the world.

Support for Island Press is provided by the Agua Fund, Brainerd Foundation, Geraldine R. Dodge Foundation, Doris Duke Charitable Foundation, Educational Foundation of America, The Ford Foundation, The George Gund Foundation, The William and Flora Hewlett Foundation, Henry Luce Foundation, The John D. and Catherine T. MacArthur Foundation, The Andrew W. Mellon Foundation, The Curtis and Edith Munson Foundation, National Environmental Trust, The New-Land Foundation, Oak Foundation, The Overbrook Foundation, The David and Lucile Packard Foundation, The Pew Charitable Trusts, The Rockefeller Foundation, The Winslow Foundation, and other generous donors.

The opinions expressed in this book are those of the author(s) and do not necessarily reflect the views of these foundations.reflect the views of these foundations.

About the Pacific Institute for Studies in Development, Environment, and Security

The Pacific Institute for Studies in Development, Environment, and Security, in Oakland, California, is an independent, nonprofit organization created in 1987 to conduct research and policy analysis in the areas of environmental protection, sustainable development, and international security. Underlying all of the Institute's work is the recognition that the urgent problems of environmental degradation, regional and global poverty, and political tension and conflict are fundamentally interrelated, and that long-term solutions dictate an interdisciplinary approach. Since 1987, we have produced more than sixty research studies, organized roundtable discussions, and held widespread briefings for policymakers and the public. The Institute has formulated a new vision for long-term water planning in California and internationally, developed a new approach for valuing well-being in local communities, worked on transborder environment and trade issues in North America, analyzed ISO 14000's role in global environmental protection, clarified key concepts and criteria for sustainable water use in the lower Colorado basin, offered recommendations for reducing conflicts over water in the Middle East and elsewhere, assessed the impacts of global warming on freshwater resources, and created programs to address environmental justice concerns in poor communities and communities of color.

For detailed information about the Institute's activities, visit www.pacinst.org, www.worldwater.org, and www.globalchange.org.

THE WORLD'S WATER
2004-2005

THE WORLD'S WATER

2004-2005

The Biennial Report on Freshwater Resources

Peter H. Gleick

with Nicholas L. Cain, Dana Haasz, Christine Henges-Jeck,
Michael Kiparsky, Catherine Hunt, Marcus Moench,
Meena Palaniappan, Veena Srinivasan, and Gary H. Wolff

Pacific Institute for Studies in Development,
Environment, and Security
Oakland, California

ISLAND PRESS

Washington • Covelo • London

Library of Congress Card Catalog Number 98–24877
ISBN 1–55963–812–5
ISBN 1–55963–536–3
ISSN 1528–7165

Printed on recycled, acid-free paper ♲
Design and production: Artech, LLC, Baltimore, MD

Manufactured in the United States of America.
10 9 8 7 6 5 4 3 2 1

To Nicki and the boys.

Contents

Foreword

Every day, the world's population grows and the amount of water available to each person decreases.

Every fortnight, somewhere in the world, nature delivers reminders of its might and of increasing climate variability: floods, storms, droughts, forest fires, and heat waves.

Every year, the world feeds itself by pumping more water from underground sources than will be replenished by rain, snow melt, and seepage in the year ahead. Globally, the deficit may be as high as 20 percent year after year.

Every month, water pollution increases, as does the demand for energy and food, particularly in the form of beef burgers, chicken curries, and fish stews. Some estimate that the earth has enough fresh water to reliably provide food for 14 billion vegetarians, but given current diets and water-use patterns, we have only enough water to sustainably feed 5 billion. Our population already exceeds 6 billion.

Every few days, meetings are held to assess the situation and talk about water policy.

In far too many places, every morning, women walk a little further to find water to keep their families alive.

And every evening, thousands of families weep for another child lost to a water-related disease.

Are there common threads that link these disparate elements of the water problem? Are there policy solutions that can actually make a difference to the women walking to find water or to the municipal water delivery systems of New Delhi and San Diego? Are big dams, bottled water, desalination, or conservation the answers? Can the private sector make things happen when the public sector system has failed to meet the basic water needs of billions of people?

Government ministers come together to talk about issues that would never have been on their tables a few decades ago. To mark the millennium, they pledged to achieve together goals that—if realized—would revolutionize the world for the poorest populations providing safe drinking water, sanitation, universal education, housing, improved nutrition, fewer child deaths, and reduced poverty. Progress in these directions continues but, by and large, we are working with inadequate commitments and resources. And many of these goals require more and better water. Some meetings of experts and policymakers seem to give real impetus to the struggle. Others, like the

Kyoto conference, seem to deliver less than their excellent preparation would have presaged. How can we learn the best from each and try to forge real commitments to reach our goals?

The jury is out, but there are positive signs. The nightly news often features water stories. Countries are paying attention—at least at the level of rhetoric. They have pledged to integrate their policies and to bring together energy, agricultural, transport, fisheries, and environmental interests so that water policy does not simply reflect the needs of single sectors. New water laws are passed; new responsibilities for water are formulated. But compliance and implementation remain difficult.

Until the end of the twentieth century, it was relatively easy to bring on more water supply; build another dam, divert a stream or river, drain a lake, and bore deeper into an aquifer. But we have reached ecological and economic limits. Major rivers no longer reach the sea, groundwater levels are dropping, and half of our wetlands are gone. Our impact on the environment raises moral and survival issues.

Demand-side measures pose their own challenges. They involve fixing irrigation systems and providing incentives to farmers to get more crop per drop in the choice of seed, planting time, and farming method. They may involve local policy decisions against growing water-intensive crops in water-poor regions and, instead, importing these products from rain-rich areas. Demand-side measures involve digging up city pipes and keeping them in good shape. Demand-side methods involve understanding how and where water is used through monitoring and metering—there has apparently never been a system in which meters were installed and demand failed to decline. But we must learn how to make the public welcome these approaches and create institutions that can manage them.

The task of creating an informed public is probably the single most important task in improving water management. Where people are aware of the magnitude of problems and the choices available to them, ownership, compliance, and—who knows?—maybe even conservation will occur. When the facts on bottled water are set out with well-researched, empirical data, it could be the beginning of a wake-up call. With enough evidence on the effects of climate on our water supply, even those who cherish their blindness to these phenomena will eventually be moved or replaced by others with greater awareness.

The World's Water series is helping to create an informed public throughout the world and is a vital part of the solution. This fourth invaluable volume in the set offers new updates, insights, and reasons for optimism. Peter Gleick and his colleagues perform two essential and signal services in their analyses. In the first place, their work underlines that where water is concerned there is *one world* of problems, with different countries having differing abilities to cope. If California, Singapore, or South Africa can conserve and become more efficient, there are real lessons for the rest of the world— and real hope. Second, Gleick and Co. insist that the debate of private versus public in water services must be subservient to the existence of, and compliance with principles and standards for water services, whether public or private. This simple, elegant, and essential thesis may yet move this debate to productive ground.

An informed, involved public together with transparent institutional arrangements may lead us down a pathway to water for all. Our path may even catch up with the women as they walk day after day after day.

Margaret Catley-Carlson
New York, May 2004

Introduction

Water is an amazing substance—just a simple mix of two fundamental elements found scattered throughout the galaxy. The world was recently mesmerized by the search for water on Mars by smart machines because we understand that the presence of water may mean the presence of life. Here on our own planet, water means far more than simple chemistry. It is infused with cultural, political, environmental, and religious importance. If we understand these complexities, there is hope that we can move forward to solve our water-related problems.

Welcome to the fourth volume of *The World's Water: The Biennial Report on Freshwater Resources*. As with each of the preceding volumes, this book offers information on issues of topical importance and the data and insights into water challenges facing the public, policymakers, and scientists. In the volume, with new authors and new data, we hope to add to the light, rather than the heat, surrounding the world's water problems. Each of the volumes of *The World's Water* builds on and adds to the previous ones. The chapters are an evolving mix of new and updated discussions, information, and raw data. In this volume, for the first time, we include a complete Table of Contents for all four volumes and an integrated index that permits readers to find information across the different books.

Now that we are well into the twenty-first century, concerns about Y2K and the millennium bug are fading into the forgotten past. The importance of water, however, continues to grow. Water continues to dominate the international environmental agenda, together with the related problem of climate change. In 2003, thousands of water experts met at the 3rd World Water Forum in Kyoto, Japan to argue, debate, and discuss water issues. Among the top priorities laid out at that forum is meeting the Millennium Development Goals (MDGs) set by the United Nations to address the woes of the world's poorest populations. Two of these MDGs are aimed at reducing unmet basic water supply and sanitation needs—a regular theme in previous volumes of *The World's Water* and continued here. Progress is being made, but for those of us who work on water every day, it seems excruciatingly—and unnecessarily—slow. Chapter 1 reviews the Millennium Development Goals for water and highlights the inadequate resources devoted to meeting them. Indeed, total overseas development assistance for water and sanitation projects, an important part of the international effort to address unmet needs, has actually decreased in recent years and remains at an inadequate level.

A new issue addressed in this volume is bottled water. In the past decade, the use of bottled water has grown at an unexpectedly quick rate and in some places little plastic water bottles have become ubiquitous. Looked at in isolation, bottled water may be innocuous; a convenient and reliable source of water. But as Chapter 2 notes, there are some serious public issues raised by growing bottled water use. Among the most important is the cost—bottled water is thousands of time more expensive than reliable, high-quality municipal water. Other issues include the question of commodification of basic drinking water, the challenge of truly understanding water-quality risks and benefits, inadequate and inconsistent bottled water regulations around the world, and even the environmental effects of producing and disposing of plastic bottles. This chapter includes some of the first survey data on bottled water costs from a few diverse places around the world. I worry that the availability of bottled water as an alternative to reliable and clean municipal water will reduce the push to provide clean water for all. I hope that this chapter begins to shed some light on the implications of the rapidly growing use of bottled water. Both the chapter and the Water Data section include new comprehensive data tables on bottled water.

Among the most controversial issues at the Kyoto meeting was water privatization, viewed by some as panacea and by others as serious threat to human rights and economic equity. The previous volume of this series explored water privatization and globalization in detail, offering principles to protect the clear public interests in water, whether publicly or privately managed. In an effort to shed some more light on these principles, Chapter 3 provides a summary of a longer set of case studies published by the Pacific Institute, exploring places around the world where one or more of the Pacific Institute principles are being met. The good news, we believe, is that these principles are appropriate and can successfully be applied. The bad news is that water managers—public or private—rarely adopt all of the principles we think necessary to protect the public interest. No doubt this debate will continue.

The World's Water has never given adequate attention to the vital issue of groundwater until now. Groundwater is often a hidden resource: inadequately monitored, insufficiently regulated, and often overpumped and overused. Chapter 4, by groundwater expert Marcus Moench, lays out these issues with both a global look at groundwater resources and regional insights into groundwater use, abuse, and management. Examples from India, California, and elsewhere are included to both show the similarities in problems facing water users from very different regions and to provide some flavor of the differences.

The volume also continues our exploration of the concept of a soft path for water, looking at various aspects of the changing paradigm for water management, development, and planning. One part that has received inadequate attention from water planners is the potential for improving the efficiency of water use without dramatic changes in technology. As populations grow and water demands increase, water managers are increasingly faced with the option of either developing new supplies at high economic and environmental costs, or figuring out how to improve the efficiency of existing uses. Indeed, in the United States and many other countries, total demand for water has not grown in more than two decades despite dramatic increases in population and gross domestic product. The change is the result of improved efficiency of water use and fluctuations in the nature of our economies. Both permit our existing supplies to go farther and do more.

But what is the true potential for improving conservation and efficiency? Few water managers, districts, or regions know because the methods for evaluating conservation potential are imperfect or key data are missing. Chapter 5 (co-authored with Dana Haasz and Gary Wolff) and Chapter 6 (co-authored with Veena Srinivasan, Christine Henges-Jeck, and Gary Wolff) report on the results of a new set of analyses of the potential for improving the efficiency of urban uses for the State of California. They conclude that total commercial, industrial, residential, and institutional water use could be cut by as much as 30 percent, cost-effectively, with existing off-the-shelf technologies. And this improvement can be obtained more quickly and cleanly than any new supply project being considered. Similar studies around the world could be very revealing and lead to dramatic changes in water planning for the future.

The connections between climate change and the sustainable management of freshwater resources are strong. The first and third volumes of *The World's Water* looked at the science of these connections and at the potential impacts for one especially vulnerable resource—the fresh water available to island nations. The likelihood that climate change will affect both the natural hydrologic cycle and the infrastructure we build to manage water is greater every day. Indeed, the scientific evidence that global warming is already affecting water resources is increasingly compelling. As a result, water managers around the world are starting to take notice. What changes are most likely? What indicators should water managers start to watch? And what, if anything, should they do to protect their systems and their customers? Chapter 7, by Kiparsky and Gleick, explores these questions in the context of more than a decade of effort in the State of California to understand the climate-water interface and to integrate science and policy in a rational way. While water managers and planners are increasingly paying attention to climate change, this issue still fails to rise to the level of attention and response that many scientists feel is necessary.

One of the fun parts of preparing this biennial report is producing the Water Briefs section, which includes shorter summaries of issues of recent importance or updates of regular features of interest to many readers. In the current volume, we provide both. New briefs have been written to summarize the successes and failures of the Kyoto water conference in March 2003, including both the official ministerial statement together with the statement of non-governmental organizations to show the differences in perceptions and priorities generated by the forum. A new brief has also been written summarizing a new tool available on the Internet for those interested in the connections between water and climate change: The Water and Climate Bibliography—a searchable database of over 3,000 references on this issue (www.pacinst.org/resources).

Another Water Brief in this new report updates progress in the area of the human right to water—covered in a full chapter in the 2000–2001 volume. While some still consider a human right to water to be a debatable legal issue, most in the legal and professional water community have come to accept the arguments made in support of such a right. Indeed, the United Nations General Comment 15 in fall 2002 clearly supports a legal right to water. This brief brings readers up to date on this issue and provides the full text of the United Nations General Comment.

The last Water Brief is an update of the enormously popular Water Conflict Chronology—published regularly since the first *World's Water* was released in 1998. Researchers and readers from around the world have consistently commented on its value as an analytical and educational tool and on its historical interest. Many new entries have been added and we will continue to maintain this as a regular feature. It is also available at www.worldwater.org.

Finally, as always, we are delighted to provide an extensive section of water-related data. Data Tables 1 and 2 on water availability and use by country remain consistently the most sought-after data by researchers, media, and the public, and we update them again here. Other data are also updated from prior volumes, but new data on bottled water use, among others, are also provided. All four volumes now offer unique data sets as well as some consistently updated data (take a look at the comprehensive Table of Contents and Index to get a listing of the data tables from all the volumes).

I acknowledge with great thanks the financial support of the Horace W. Goldsmith, William and Flora Hewlett, and Henry Luce Foundations.

A special note of thanks is owed to all of my colleagues in the water world who provide information, feedback, and encouragement for the effort. While these books are a labor of love, the labor is made far easier by the support of all those working on these challenges.

Peter H. Gleick
Oakland, California

The Millennium Development Goals for Water: Crucial Objectives, Inadequate Commitments

Peter H. Gleick

We recognize that, in addition to our separate responsibilities to our individual societies, we have a collective responsibility to uphold the principles of human dignity, equality, and equity at the global level. As leaders we have a duty therefore to all the world's people, especially the most vulnerable and, in particular, the children of the world, to whom the future belongs. UN MILLENNIUM DECLARATION

As has been chronicled in earlier volumes of *The World's Water*, the failure to provide safe drinking water and adequate sanitation services to all people is perhaps the greatest development failure of the twentieth century (Gleick 1998, 2000). The most egregious consequence of this failure is the high rate of sickness and mortality among young children from preventable water-related diseases. This chapter discusses recent efforts to address the gap over access to water supply and sanitation by setting targets, explores the likelihood of achieving those targets, and highlights the consequences of success or failure.

In the year 2000, the United Nations and the international water community announced explicit goals—the Millennium Development Goals (MDGs)—for human development over the next several decades. One of these explicitly addressed water by setting the goal of reducing in half the proportion of people unable to reach or afford safe drinking water by 2015. A comparable MDG for sanitation was announced in 2002 at the World Summit on Sustainable Development (WSSD) in Johannesburg, South Africa. Achieving these goals is a laudable objective, but even if they can be achieved, hundreds of millions of people will still lack basic water services two decades from now. Moreover, given current levels of financial and institutional commitment, there now seems to be little chance that the MDGs will be met. This problem must be considered one of the most serious public health crises facing us, yet it has received little attention and inadequate resources.

Setting Water and Sanitation Goals

The United Nations, in collaboration with individual nations, regularly monitors access to water and sanitation. The World Health Organization (WHO) published the most recent assessment in 2000, providing information for 89 percent of the world's population (WHO 2000). According to the WHO, 1.1 billion people around the world lacked access to "improved water supply" and more than 2.4 billion, or roughly 40 percent of the world's population lacked access to "improved sanitation" in 2000.[1]

In response, an increasing number of nations, international water conferences, and aid organizations have announced efforts to improve global access to fresh water and water-related services. For example, the recent ministerial statement from the 2nd World Water Forum in 2000 in The Hague (Hague 2000) called for efforts to guarantee:

> "that every person has access to enough safe water at an affordable cost to lead a healthy and productive life and that the vulnerable are protected from the risks of water-related hazards..."

In September of the same year, the United Nations General Assembly adopted the Millennium Development Goals (United Nations 2000a), including one for water. A complete list of the MDGs, together with the indicators that are to be used to measure progress toward these goals, is provided in Box 1.1. The MDG objective for water (Goal 7, Target 10) is to:

> "Halve by 2015 the proportion of people without sustainable access to safe drinking water" (UNDP 2003).

In August 2002 at the WSSD, the international community added a new goal of halving by the year 2015 the proportion of people without access to basic sanitation.

> "The provision of clean drinking water and adequate sanitation is necessary to protect human health and the environment. In this respect, we agree to halve, by the year 2015, the proportion of people who are unable to reach or to afford safe drinking water (as outlined in the Millennium Declaration) and the proportion of people who do not have access to basic sanitation..." (United Nations 2002).

Efforts to meet these targets will lead to a decrease in the total population at risk from water-related diseases.

Commitments to Achieving the MDGs for Water

Unfortunately, the United Nations water goals and solutions to other water problems are unlikely to be achieved given current levels of financial and political commitments.

---

1. Previous assessments were released in 1994, 1990, and during the International Safe Drinking Water Supply and Sanitation Decade of the 1980s. Data from these assessments are available in the 2002–2003 volume of *The World's Water* (Gleick et al. 2002). While each of these assessments offers a picture of the populations without access to water services, different rates of response to surveys, inconsistent and changing definitions of "access" and "adequate," and poor data availability make it difficult—and ill-advised—to try to draw conclusions about trends over time. In a recent attempt, Lomborg (2001) drew optimistic conclusions from incorrect interpretations and misunderstandings of these data sets (Gleick 2002a). At the same time, despite problems with the data, it is evident that limited resources, misguided priorities, and rapidly growing populations have made it difficult to provide comprehensive and complete water coverage.

Box 1.1　The Millennium Development Goals, Targets, and Indicators

Goals and Targets	Indicators
Goal 1　Eradicate extreme poverty and hunger	
Target 1: Halve, between 1990 and 2015, the proportion of people whose income is less than $1 a day	• 1a. Proportion of population below $1 a day[1] • 1b. National poverty headcount ratio • 2. Poverty gap ratio at $1 a day (incidence x depth of poverty) • 3. Share of poorest quintile in national consumption
Target 2: Halve, between 1990 and 2015, the proportion of people who suffer from hunger	• 4. Prevalence of underweight in children (under five years of age) • 5. Proportion of population below minimum level of dietary energy consumption
Goal 2　Achieve universal primary education	
Target 3: Ensure that, by 2015, children everywhere, boys and girls alike, will be able to complete a full course of primary schooling	• 6. Net enrollment ratio in primary education • 7a. Proportion of pupils starting grade 1 who reach grade 5 • 7b. Primary completion rate[2] • 8. Literacy rate of 15– to 24–year-olds
Goal 3　Promote gender equality and empower women	
Target 4: Eliminate gender disparity in primary and secondary education preferably by 2005 and in all levels of education no later than 2015	• 9. Ratio of girls to boys in primary, secondary, and tertiary education • 10. Ratio of literate females to males among 15– to 24–year-olds • 11. Share of women in wage employment in the nonagricultural sector • 12. Proportion of seats held by women in national parliament
Goal 4　Reduce child mortality	
Target 5: Reduce by two-thirds, between 1990 and 2015, the under-five mortality rate	• 13. Under-five mortality rate • 14. Infant mortality rate • 15. Proportion of one-year-old children immunized against measles
Goal 5　Improve maternal health	
Target 6: Reduce by three-quarters, between 1990 and 2015, the maternal mortality ratio	• 16. Maternal mortality ratio • 17. Proportion of births attended by skilled health personnel
Goal 6　Combat HIV/AIDS, malaria, and other diseases	
Target 7: Have halted by 2015 and begun to reverse the spread of HIV/AIDS	• 18. HIV prevalence among 15– to 24– year-old pregnant women • 19. Condom use rate of the contraceptive prevalence rate[3,2] • 19a. Condom use at last high-risk sex[2] • 19b. Percentage of population aged 15–24 with comprehensive correct knowledge of HIV/AIDS[4,2] • 19c. Contraceptive prevalence rate[3] • 20. Ratio of school attendance of orphans to school attendance of non-orphans aged 10–14

continues

Box 1.1 *continued*

Goals and Targets	Indicators
Goal 6 *continued* Target 8: Have halted by 2015 and begun to reverse the incidence of malaria and other major diseases	• 21. Prevalence and death rates associated with malaria • 22. Proportion of population in malaria-risk areas using effective malaria prevention and treatment measures[5] • 23. Prevalence and death rates associated with tuberculosis • 24. Proportion of tuberculosis cases detected and cured under directly observed treatment short course
Goal 7 Ensure environmental sustainability Target 9: Integrate the principles of sustainable development into country policies and programs and reverse the loss of environmental	• 25. Proportion of land area covered by forest • 26. Ratio of area protected to maintain biological diversity to surface area • 27. Energy use per unit of GDP • 28. Carbon dioxide emissions (per capita) and consumption of ozone-depleting chlorofluorocarbons • 29. Proportion of population using solid fuels[2]
Target 10: Halve, by 2015, the proportion of people without sustainable access to safe drinking water and basic sanitation	• 30. Proportion of population with sustainable access to an improved water source, urban and rural • 31. Proportion of population with access to improved sanitation
Target 11: Have achieved, by 2020, a significant improvement in the lives of at least 100 million slum dwellers	• 32. Proportion of households with access to secure tenure
Goal 8 Develop a global partnership for development Target 12: Develop further an open, rule-based, predictable, nondiscriminatory trading and financial system (includes a commitment to good governance, development, and poverty reduction—both nationally and internationally) Target 13: Address the special needs of the least developed countries (includes tariff-and quota-free access for exports, enhanced program of debt relief for HIPC and cancellation of official bilateral debt, and more generous ODA for countries committed to poverty reduction)	Some of the indicators listed below will be monitored separately for the least developed countries, Africa, landlocked countries, and small island developing states. **Official development assistance** • 33. Net ODA total and to least developed countries, as a percentage of OECD/DAC donors' gross income • 34. Proportion of bilateral, sector-allocable ODA of OECD/DAC donors for basic social services (basic education, primary health care, nutrition, safe water, and sanitation) • 35. Proportion of bilateral ODA of OECD/DAC donors that is untied • 36. ODA received in landlocked countries as proportion of their GNI • 37. ODA received in small island developing states as proportion of their GNI

continues

Box 1.1 *continued*

Goals and Targets	Indicators
Goal 8 *continued*	
Target 14: Address the special needs of landlocked countries and small island developing states (through the Program of Action for the Sustainable Development of Small Island Developing States and 22nd General Assembly provisions)	**Market access** • 38. Proportion of total developed country imports (excluding arms) from developing countries and least developed countries admitted free of duties • 39. Average tariffs imposed by developed countries on agricultural products and clothing from developing countries • 40. Agricultural support estimate for OECD countries as a percentage of their GDP • 41. Proportion of ODA provided to help build trade capacity
Target 15: Deal comprehensively with the debt problems of developing countries through national and international measures in order to make debt sustainable in the long term	**Debt sustainability** • 42. Total number of countries that have reached their HIPC decision points and completion points (cumulative) • 43. Debt relief committed under HIPC initiative, US$ • 44. Debt service as a percentage of exports of goods and services
Target 16: In cooperation with developing countries, develop and implement strategies for decent and productive work for youth Target 17: In cooperation with pharmaceutical companies, provide access to affordable, essential drugs in developing countries Target 18: In cooperation with the private sector, make available the benefits of new technologies, especially information and communications	**Other** • 45. Unemployment rate of 15– to 24–year-olds, male and female and total[6] • 46. Proportion of population with access to affordable, essential drugs on a sustainable basis[7] • 47. Telephone lines and cellular subscribers per 100 population • 48a. Personal computers in use per 100 population • 48b. Internet users per 100 population

Notes:

1. For monitoring country poverty trends, indicators based on national poverty lines should be used, where available.

2. These indicators are proposed as additional MDG indicators, but have not yet been adopted.

3. Amongst contraceptive methods, only condoms are effective in preventing HIV transmission. The contraceptive prevalence rate is also useful in tracking progress in other health, gender and poverty goals. Because the condom use rate is only measured amongst women in union, it is supplemented by an indicator on condom use in high-risk situations (indicator 19a) and an indicator on HIV/AIDS knowledge (indicator 19b).

4. This indicator is defined as the percentage of population aged 15–24 who correctly identify the two major ways of preventing the sexual transmission of HIV (using condoms and limiting sex to one faithful, uninfected partner), who reject the two most common local misconceptions about HIV transmission, and who know that a healthy-looking person can transmit HIV. However, since there are currently not a sufficient number of surveys to be able to calculate the indicator as defined above, UNICEF, in collaboration with UNAIDS and WHO, produced two proxy indicators that represent two components of the actual indicator. They are the following: a) Percentage of women and men 15–24 who know that a person can protect herself from HIV infection by "consistent use of condom". b) Percentage of women and men 15–24 who know a healthy-looking person can transmit HIV. Data for this year's report are only available on women.

5. Prevention to be measured by the percentage of children under 5 sleeping under insecticide treated bednets; treatment to be measured by percentage of children under 5 who are appropriately treated.

6. An improved measure of the target is under development by ILO for future years.

7. Under development by WHO.

Source: http://www.developmentgoals.org/About_the_goals.htm

Despite growing awareness of water issues, international economic support for water projects of all kinds is marginal and declining. Official development assistance (ODA) for water supply and sanitation projects from countries of the Organization for Economic Co-operation and Development (OECD) and the major international financial institutions has actually declined over the past few years (Table 1.1), from approximately $3.5 billion per year (average from 1996 to 1998) to $3.1 billion per year (average from 1999 to 2001). Moreover, those most in need receive the smallest amount of aid. Ten countries received around half of all water-related aid, while countries where less than 60 percent of the population has access to an improved water source received only 12 percent of the money (OECD 2002).

The 3rd World Water Forum in Japan (described in detail in the "Water Briefs" section of this volume) provided an opportunity for pledges of serious and substantial new commitments of financing and action to meet the MDGs. Such pledges fell far

TABLE 1.1   Aid to Water Supply and Sanitation by Donor (1996 to 2001)

Country or Organization	US$ (millions)	
	1996–1998 average	1999–2001 average
Australia	23	40
Austria	34	46
Belgium	12	13
Canada	23	22
Denmark	103	73
Finland	18	12
France	259	148
Germany	435	318
Ireland	6	7
Italy	35	29
Japan	1442	999
Luxembourg	2	8
Netherlands	103	75
New Zealand	1	1
Norway	16	32
Portugal	0	5
Spain	23	60
Sweden	43	35
Switzerland	25	25
United Kingdom	116	165
United States	186	252
Subtotal Countries	**2906**	**2368**
African Development Fund	56	64
Asian Development Bank	150	88
European Community		216
International Development Association	323	331
Inter-American Development Bank, Special Operations Fund	46	32
Subtotal Multilateral Organizations	**575**	**730**
TOTAL Water Supply/Sanitation Aid	**3482**	**3098**

Source: OECD 2002.

short of the level conceivably needed to actually meet the targets. Box 1.2 lists some of the more serious and substantive commitments that came out of the Water Forum. As Box 1.2 shows, together with Table 1.1, commitments to the MDG remain inadequate.

Consequences: Water-Related Diseases

Although water-related diseases—associated with the lack of access to clean and safe water and sanitation—have largely been eliminated in wealthier nations, they remain a major concern in much of the developing world. While data are incomplete and unreliable, the World Health Organization regularly publishes water- and health-related information in the *Human Development Report* series and the annual *World Health Report* (UNDP 2003, WHO 2003). In the early 1990s there were on the order of 250

Box 1.2 MDG Commitments Coming Out of the 3rd World Water Forum, Kyoto, 2003

UN-HABITAT (the United Nations Human Settlements Program aimed at providing sustainable housing for all) signed a memorandum of understanding with the Asian Development Bank (ADB) to create a program to build the capacity of Asian cities to secure and manage pro-poor investments and to help the region meet the water-related Millennium Development Goals (MDGs). The program will cover $10 million in grants from ADB and United Nations Habitat for the first two phases and $500 million in ADB loans for water and sanitation projects in cities across Asia over the next five years. The Government of Netherlands also made additional funding for Water for Asian Cities available to United Nations Habitat.

The United Nations Development Program (UNDP) commited to a Community Water Initiative, aimed at building on the power of the local community to solve water and sanitation challenges. The aim of the initiative is to provide innovative communities with small grants to expand and improve their solutions to the water and sanitation crisis. The Community Water Initiative has an estimated target budget of $50 million for 2003–2008.

The Netherlands will concentrate its international water aid on Africa and assist 10 countries in the development of their national plans. Further, it is committed to support the African Water Facility.

Australia committed about $60 million (converted to U.S. dollars) in the current financial year for water activities, primarily in countries in the Asia-Pacific region.

Source: Press Release of the World Water Council 2003.
http://www.worldwatercouncil.org/download/PR_finalday_23.03.03.pdf

million cases of water-related diseases annually, excluding common diarrheal diseases. In the *2000 World Health Report*, the United Nations reported 4.37 billion cases of diarrhea annually (Annex Table 9 in Murray et al. 2001). The true extent of these diseases is unknown and many cases of water-related illnesses are undiagnosed and unreported.

Water-related diseases can be placed in four classes: waterborne, water-washed, water-based, and water-related insect vectors. The first three are most clearly associated with lack of improved domestic water supply and adequate sanitation. Box 1.3 lists the prevalent diseases associated with each class.

Waterborne diseases include those in which transmission occurs by drinking contaminated water, particularly contamination by pathogens transmitted from excreta to water by humans. These include most of the enteric and diarrheal diseases caused by bacteria, parasites, and viruses. Waterborne diseases also include typhoid and over 30 species of parasites that infect the human intestines. Seven of these are distributed globally or cause serious illness: ameobiasis, giardiasis, *Taenia solium* taeniasis, ascariasis, hookworm, trichuriasis, and strongyloidiasis. Evidence also suggests that waterborne disease contributes to background rates of disease not detected or reported explicitly as outbreaks.

Water-washed diseases occur when there is insufficient water for washing and personal hygiene or when people come in contact with contaminated water. These include trachoma and typhus, and diarrheal diseases that can be passed from person to person.

Water-based diseases come from hosts that live in water or require water for part of their life cycle. These diseases are passed to humans when they are ingested or come into contact with skin. The most widespread examples in this category are schistosomiasis and dracunculiasis. Schistosomiasis currently infects 200 million people in 70 countries. Safe drinking water systems will eliminate exposure to these hosts.

Box 1.3 Water-Related Diseases

Waterborne diseases: caused by the ingestion of water contaminated by human or animal feces or urine containing pathogenic bacteria or viruses; include cholera, typhoid, amoebic and bacillary dysentery, and other diarrheal diseases.

Water-washed diseases: caused by poor personal hygiene and skin or eye contact with contaminated water; include scabies, trachoma, and flea-, lice- and tick-borne diseases.

Water-based diseases: caused by parasites found in intermediate organisms living in water; include dracunculiasis, schistosomiasis, and other helminths.

Water-related insect vector diseases: caused by insects that breed in water; include dengue, filariasis, malaria, onchocerciasis, trypanosomiasis, and yellow fever.

The final category—*water-related insect vectors*—includes those diseases spread by insects that breed or feed near contaminated water, such as malaria, onchocerciasis, and dengue fever. These diseases are not typically associated with lack of access to clean drinking water or sanitation services, and they are not included here in estimates of water-related deaths. It must be noted, however, that their spread may be facilitated by the construction of large-scale water systems that create conditions favorable to their hosts.

Measures of Illness from Water-Related Diseases

As tragic and unnecessary as are water-related deaths, there are other significant health consequences of failing to provide adequate water services. These consequences include morbidity, lost workdays, missed educational opportunities, official and unofficial health care costs, and draining of family resources. These are poorly understood and even more poorly measured and assessed. Nonetheless, there are a variety of ways of reporting on the prevalence and severity of illnesses that result from water-related diseases. The most common ways have been morbidity (number of reported incidents of sickness) and mortality (deaths). These measures, however, have some important limitations. There is a significant difference (in both perception and economic impact) of a death of a child, who may lose decades of productive life, and the premature death of an older person, who may lose a few months or years to a late-developing disease. Illnesses that cause blindness or other incapacitating effects should be considered—and measured—differently than illnesses that are only temporarily or mildly debilitating.

In 1993 the Harvard School of Public Health in collaboration with the World Health Organization and the World Bank began a new assessment of the "global burden of disease" (GBD). This effort introduced a new indicator—the "disability adjusted life year" (DALY)—to quantify the burden of disease. The DALY is a measure of population health that combines in a single indicator years of life lost from premature death and years of life lived with disabilities. One DALY can be thought of as one lost year of "healthy" life. International policy interest in such indicators is increasing, and the WHO *World Health Reports* now use this as a basic measure of well-being and health (Mathers et al. 2003).

Table 1.2 lists deaths and DALYs from selected water-related diseases, as reported by the World Health Organization for 2000. This table excludes diseases associated with insects (water-related insect vectors) such as malaria and dengue. As Table 1.2 shows, total estimated DALYs associated with water-related diseases exceed 75 million annually. The vast majority of these are the result of diarrheal diseases. We do not discuss the broader health consequences of these diseases except to note their importance and to urge that better data be collected on the true economic and social costs of the failure to provide adequate water of appropriate quality.

Mortality from Water-Related Diseases

Deaths from water-related diseases are inadequately monitored and reported. A wide range of estimates is available in the public literature, ranging from 2 million to 12 million deaths per year (see Table 1.3). Current best estimates appear to fall between 2

TABLE 1.2   Annual Deaths and DALYs from Selected Water-Related Diseases

	Deaths	**DALYs**
Diarrheal diseases	2,019,585	63,345,722
Childhood cluster diseases		
Poliomyelitis	1,136	188,543
Diphtheria	5,527	187,838
Tropical-cluster diseases		
Trypanosomiasis	49,129	1,570,242
Schistosomiasis	15,335	1,711,522
Trachoma	72	3,892,326
Intestinal nematode infections		
Ascariasis	4,929	1,204,384
Trichuriasis	2,393	1,661,689
Hookworm disease	3,477	1,785,539
Other Intestinal Infections	1,692	53,222
TOTAL	**2,103,274**	**75,601,028**

Notes: DALYs: The DALY is a measure of population health that combines in a single indicator years of life lost from premature death and years of life lived with disabilities. One DALY can be thought of as one lost year of "healthy" life. This table excludes mortality and DALYs associated with water-related insect vectors, such as malaria, onchocerciasis, and dengue fever. Trachoma: While few deaths from trachoma are reported, approximately 5.9 million cases or 3.9 million DALYs of blindness or severe complications occur annually.
Source: World Health Organization 2001.

and 5 million deaths per year. Of these deaths, the vast majority is of small children struck by virulent but preventable diarrheal diseases.

The current best estimate of water-related deaths from diarrheal diseases is around 2 million per year (as shown in Table 1.3), but this estimate must be qualified. First, huge numbers of cases of diarrheal diseases are not reported at all, suggesting that some—perhaps many—deaths may be misreported as well. Second, deaths are categorized by the primary diseases, but it is well known that diarrhea is a contributing cause of death in many circumstances (Box 1.4). Third, other deaths from water-related diseases are also poorly monitored in some places and for some diseases.

Scenarios of Future Deaths from Water-Related Diseases

The number of deaths anticipated from water-related diseases over the next two decades depends on many factors, including total global population, the relative rates of mortality from various diseases, the incidence of those diseases, interventions on the part of the health community, and future efforts to changes these factors.

Current projections of population growth worldwide are available from several sources, including the United Nations and the United States (U.S.) Bureau of the Census. In addition, each of these sources provides a range of estimates to account for uncertainty and changes in birth and death rates in the future.

According to the U.S. Bureau of the Census international data group and United Nations population estimates, the global population between 2000 and 2020 will grow from just over 6 billion to as much as 7.5 billion, with most of the increase in developing countries of Africa and Asia (United Nations 2000b, U.S. Census Bureau 2002). Pro-

TABLE 1.3   Estimates of Water-Related Mortality

Source	Deaths per Year (millions)
World Health Report WHO 2001	2.1 (diarrheal and other water-related deaths)
World Health Organization 2000	2.2 (diarrheal diseases only)
World Health Organization 1999	2.3
WaterDome 2002	more than 3
World Health Organization 1992	4
World Health Organization 1996	more than 5
Hunter et al. 2000	more than 5
UNDP 2002	more than 5
Johannesburg Summit 2002	more than 5
Hinrichsen et al. 1997	12

Notes:
- WHO 2001. The "version 2" update of the *World Health Report* notes 2 million diarrheal deaths and another 100,000 deaths from intestinal nematodes, childhood "cluster" diseases, and tropical diseases associated with unsafe water.
- WHO 1996. "Every year more than five million human beings die from illnesses linked to unsafe drinking water, unclean domestic environments and improper excreta disposal."
- Johannesburg Summit 2002. "More than 5 million people die each year from diseases caused by unsafe drinking water, lack of sanitation, and insufficient water for hygiene. In fact, over 2 million deaths occur each year from water-related diarrhea alone. At any given time, almost half of the people in developing countries suffer from water-related diseases."
- WHO World Health Report 1999. Statistical Annex. Totals of 2.3 million excluding several diseases.
- WaterDome 2002. "More than 3 million die from diseases caused by unsafe water… . Currently, about 20 percent of the world's population lacks access to safe drinking water, and more than 5 million people die annually from illnesses associated with unsafe drinking water or inadequate sanitation. If everyone had safe drinking water and adequate sanitation services, there would be 200 million fewer cases of diarrhea and 2.1 million fewer deaths caused by diarrheal illness each year."
- Hunter et al. 2000. "Excluding deaths from malaria and other diseases carried by water-related insect vectors, the best international estimates of total water-related disease mortality range from 2.2 to 5 million annually. We use this range in our calculations of future mortality to represent the uncertainty in projections. No "best estimate" is provided."
- UNDP 2002. "We know it is the poor who are most affected, with 800 million people undernourished and 5 million dying each year because of polluted water, lack of sanitation, and waterborne diseases alone…"
- WHO. 1992. "Lack of sanitary conditions contributes to about two billion human infections of diarrhea with about 4 million deaths per year, mostly among infants and young children."
- Hinrichsen, D., Robey, B., and Upadhyay, U.D., 1997. " Water-borne diseases are "dirty-water" diseases—those caused by water that has been contaminated by human, animal, or chemical wastes. Worldwide, the lack of sanitary waste disposal and of clean water for drinking, cooking, and washing is to blame for over 12 million deaths a year." This estimate seems high and may mistakenly include deaths from malaria and other diseases not associated with inadequate and unsafe drinking water and sanitation.

Box 1.4   A Comment on Fatalities from Diarrheal Diseases

A very large (but uncertain) number of deaths result every year from diarrheal diseases that occur in conjunction with non-water-related diseases, such as measles and AIDS. When this occurs, deaths are categorized by the primary disease. Thus, many cases of early death that may be caused by lack of access to clean water and proper sanitation are not included in the estimate of approximately 2 million diarrheal deaths each year.

jections of future water-related deaths will depend on these future population estimates as well as a wide range of other factors.

Two scenarios are presented below as simple ways to think about the consequences of these factors, including meeting the United Nations Millennium Development Goals for water. Unfortunately, there is no consensus on the best way forward toward these goals. Traditional thinking says that large financial commitments for centralized infrastructure will be required (Camdessus 2003). Yet there is a growing realization that the funds for such investments are not being made available, that past infrastructure investments have not satisfied the objectives of universal coverage, and that alternative approaches may be more effective, faster, and less expensive. The more expensive estimates assume a cost of around $500 per person—typical of the costs of centralized water systems in developed countries—while field experience suggests that safe and reliable water supply and sanitation services can be provided in urban areas for $35 to $50 per person and in rural areas for less than that when local communities build appropriate-scale technology (WSSCC 2000). The lack of agreement about how best to proceed, however, makes it increasingly unlikely that the goals will be met.

Scenario A: No Action: Proportional Deaths

A relatively simple estimate of future deaths from water-related diseases comes from assuming that the current proportion of deaths to the population without access to safe water services today will remain constant and that the population without access to water services will grow in proportion to global population growth. Currently, approximately 18 percent of the world's population lacks access to safe drinking water

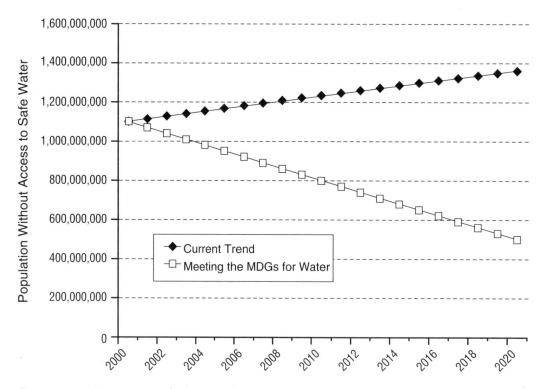

FIGURE 1.1 Future populations without access to safe water, with and without the MDGs.

(1.1 billion people out of a total of 6 billion in the year 2000). As total population grows between 2000 and 2020 to around 7.5 billion, the total population without access to safe drinking water can be expected, without action on the part of the water community, to grow to over 1.3 billion. While this assumes no actions are taken to provide new access, it is also not the worst case because it ignores the fact that the populations without access to water may be growing at a faster rate than the world average.

If the total annual death toll from water-related diseases remains proportional to the population without access to clean water (around 0.19 percent of the population without access to safe water dies annually: 2.1 million deaths per year out of 1.1 billion people), the annual death toll will reach almost 2.6 million per year in 2020, and a total of nearly 50 million people will die from these diseases between 2000 and 2020 (Figures 1.1 and 1.2).

Scenario B: Meeting the United Nations Millennium Goals

It is reasonable to hope and assume that the MDGs will cause additional actions to be taken in the next few years to accelerate the rate at which access to safe water is provided, although as noted, the actions taken so far suggest that we will fail to meet the MDGs. Nevertheless, assuming that the Millennium Goals are met, the proportion of the population without access to safe water will drop in half by 2015. Thus, the current proportion—18 percent of the world's population—will drop to 9 percent. In 2015, 9 percent of the world's population will still be 650 million people without access to safe water. Extending this trend through 2020 shows that the global population will exceed 7.5 billion and that 6.6 percent will still lack basic water services—a total of approximately 500 million people.

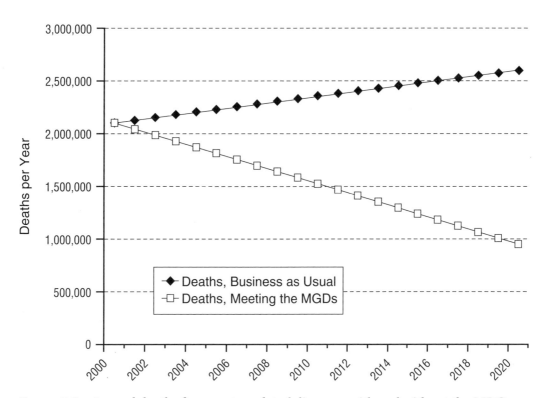

FIGURE 1.2 Annual deaths from water-related diseases, with and without the MDGs.

Figure 1.1 shows the downward curve of population without access to safe water associated with meeting the MDGs compared to the population without access to safe water associated with Scenario A. If the ratio of deaths to the population without access to safe water remains the same, at around 0.19 percent annually, the total number of deaths from water-related diseases will drop to around 1 million per year. Summing the number of deaths between 2000 and 2020 gives a total of 32 million deaths (Figure 1.2). While this scenario represents 18 million fewer cumulative deaths than Scenario A, it still represents a tragedy of an extreme order.

Conclusions

The failure to meet basic human needs for water is widely acknowledged to be a major development failure of the twentieth century. In recognition of this failure, the United Nations and the world community adopted a set of ambitious development goals to try to address unmet issues of poverty and human development, including two goals to provide safe and reliable access to water and sanitation services. Despite this laudable goal, and fine rhetorical efforts, practical actions and commitments to provide universal coverage for water and sanitation continue to be inadequate. The price for this will be paid by the poorest populations of the world in sickness, lost educational and employment opportunities, and for a staggeringly large number of people, early death. Even if the official Millennium Goals set for water are met—which is unlikely given the current level of commitments by national governments and international aid agencies—as many as 32 million people, and perhaps many more, will die by 2020 from preventable water-related diseases. This is morally unacceptable in a world that values equity and decency. At present, it appears unavoidable.

REFERENCES

Camdessus report. 2003. *Financing water for all: Report of the world panel on financing water infrastructure.* World Water Council, Global Water Partnership. (This report, written by J. Winpenny, is often called the *Camdessus report* after its chair, Michel Camdessus.) March 2004. http://www.gwpforum.org/gwp/library/FinPanRep.MainRep.pdf

Gleick, P. H. 1998. *The world's water 1998–1999: The biennial report on freshwater resources.* Washington, D.C.:Island Press. (Chinese edition published in Beijing, 2001.)

Gleick, P. H. 2000. *The world's water 2000–2001: The biennial report on freshwater resources.* Washington, D.C.:Island Press.

Gleick, P. H. with Burns,W. C. G., Chalecki, E. L., Cohen, M., Cushing, K. K., Mann, A. S., Reyes, R., Wolff, G. H., and Wong, A. K. 2002. *The world's water 2002–2003: The biennial report on freshwater resources.* Washington, D.C.:Island Press.

Gleick, P. H. 2002a. Is the skeptic all wet? *Environment*, Vol. 44, no. 6:36–40.

Hague. 2000. *Ministerial declaration of The Hague on water security in the 21st Century.* The Hague, Netherlands. January 2004. http://www.waternunc.com/gb/secwwf12.htm

Hinrichsen, D., Robey, B., and Upadhyay, U. D. 1997. Solutions for a water-short world. *Population Reports*, Series M, No. 14. Baltimore: Johns Hopkins School of Public Health, Population Information Program, December.

Hunter, P. R., Colford, J. M., LeChevallier, M. W., Binder, S., Berger, P. S. 2000. *Emerging Infectious Diseases Journal*, Vol. 7, no. 3 Supplement:544–545. Panel summary on waterborne diseases from the 2000 emerging infectious diseases conference. Atlanta. January 2004. http://www.cdc.gov/ncidod/eid/vol7no3_supp/hunter.htm

Johannesburg Summit 2002. Johannesburg summit secretary-general calls for global action on water issues. January 2004. http://www.johannesburgsummit.org/html/media_info/pressrelease_prep2/global_action_water_2103.pdf

Lomborg, B. 2001. *The skeptical environmentalist.* Cambridge, United Kingdom:Cambridge University Press.

Mathers, C. D., Bernard, C., Iburg, K. M., Inoue, M., Ma Fat, D., Shibuya, K., Stein, C., Tomijima, N., Xu, H. 2003. *Global burden of disease in 2002: Data sources, methods and results.* Global Programme on Evidence for Health Policy Discussion Paper No. 54. Geneva:World Health Organization. December.

Murray, C. J. L., Lopez, A. D., Mathers, C. D., Stein, C. 2001. *The global burden of disease 2000 project: Aims, methods and data sources.* Global Programme on Evidence for Health Policy Discussion Paper No. 36. Geneva: World Health Organization. November. (Revised).

Organisation for Economic Co-operation and Development (OECD). 2002. *Creditor reporting system, aid activities in the water sector: 1997–2002.* Paris:OECD, Development Assistance Committee.

United Nations. 2000a. *The millennium goals.* United Nations Millennium Declaration A/RES/55/2 8 September 2000. New York:United Nations.

United Nations. 2000b. *World population prospects: The 2000 revision.* Volume 3. Analytical Report ST/ESA/SER.A/200. January 2004.
http://www.un.org/esa/population/publications/wpp2000/wpp2000_volume3.htm

United Nations. 2002. *Report of the world summit on sustainable development* Johannesburg, South Africa, August 26–September 4. New York:UN Document. A/CONF.199/20. January 2004.
http://ods-ddsny.un.org/doc/UNDOC/GEN/N02/636/93/PDF/N0263693.pdf?OpenElement

United Nations Development Programme (UNDP). 2002. Statement by UNDP Administrator Mr. Mark Malloch Brown. January 2004.
http://www.undp.org/dpa/frontpagearchive/2001/june/05june01/

United Nations Development Programme (UNDP). 2003. *Human development report 2003.* New York:United Nations.

United States Census Bureau. 2002. http://www.census.gov/ipc/www/world.html

WaterDome 2002. Johannesburg Summit. January 2004. http://www.waterdome.net/

Water Supply and Sanitation Collaborative Council. 2000. *Vision 21: A shared vision for hygiene, sanitation, and water supply.*
http://www.worldwatercouncil.org/Vision/Documents/VISION21FinalDraft.PDF

World Health Organization (WHO). 1992. *Our planet, our health: Report of the WHO Commission on Health and Environment.* Geneva: World Health Organization.
http://www.ku.edu/~hazards/foodpop.pdf. Cited in R. Hopfenberg and D. Pimentel. 2001. Human population numbers as a function of food supply. *Environment, development and sustainability* 3, 2001:1–15.

World Health Organization (WHO). 1995. Community water supply and sanitation: Needs, challenges and health objectives 48th World Health Assembly, A48/INEDOC./2, April 28, Geneva.

World Health Organization (WHO). 1996. *Water and sanitation.* January 2004.
http://www.who.int/inf-fs/en/fact112.html WHO Information Fact Sheet No. 112. Geneva.

World Health Organization (WHO). 1999. *World health report 1999.* January 2004.
http://www.who.int/whr/1999/en/pdf/StatisticalAnnex.pdf

World Health Organization (WHO). 2000. *Global water supply and sanitation assessment 2000 report.* January 2004.
http://www.who.int/watersanitationhealth/Globassessment/GlobalTOC.htm

World Health Organization (WHO). 2001. *World health report 2001—mental health: New understanding, new hope.* Version 2 data tables on the global burden of disease. Geneva. January 2004. http://www.who.int/whr2001/2001/

World Health Organization (WHO). 2003. *World health report 2003—shaping the future.* January 2004. http://www.who.int/whr/en/

The Myth and Reality of Bottled Water

Peter H. Gleick

Sales and consumption of bottled water have skyrocketed in recent years. From 1988 to 2002, the sales of bottled water globally have more than quadrupled to over 131 million cubic meters annually (BMC 2003). Bottled water sales worldwide are increasing at 10 percent per year, while the volume of fruit drinks consumed is growing less than 2 percent annually and beer and soft drink sales are growing at less than 1 percent per year (Bottled Water Web 2003). More than 50 percent of Americans drink bottled water occasionally or as their major source of drinking water—an astounding fact given the high quality and low cost of U.S. tap water.

Why the great growth in bottled water sales? Bottled water typically costs a thousand times more per liter than high-quality municipal tap water. Are consumers willing to pay this price because they believe that bottled water is safer than tap water? Do they have a real taste preference for bottled water? Or is the convenience of the portable plastic bottle the major factor? Are they taken in by the images portrayed in commercials and on the bottles?

The answers are consequential. We estimate that total consumer expenditures for bottled water are approximately $100 billion per year—a vast sum that both indicates consumers are willing to pay for convenient and reliable drinking water and that society has the resources to make comparable expenditures to provide far greater quantities of water for far less money by investing in reliable domestic supplies.

Ironically, despite its cost, users should not assume that the purity of bottled water is adequately protected, regulated, or monitored. Even where regulations exist, bottled water plants typically receive far less scrutiny from inspectors than other food plants or municipal water systems. In many places, such as the United States, bottlers themselves do most sampling and testing, which opens the door to fraud, misreporting, and inadequate protection. Ultimately, the provision of clean water to all will not come from sales of bottled water but from effective actions of communities, governments, and municipal providers to provide a safe and reliable domestic water supply.

This chapter reviews the recent history of and trends in bottled water, the regulations governing bottled water production and sale, and growing concerns about the costs and implications of bottled water use. We also address growing concerns in both industrialized nations where high-quality tap water is readily available and in poorer developing countries where the high cost of bottled water raises questions about equity and access to basic water services for all.

Bottled Water Use History and Trends

The global consumption of bottled water is growing faster than 10 percent per year with substantial growth in sales volumes on every continent. The slowest growth is occurring in European countries, where bottled water has long had a commercial foothold. Even there, growth rates of five to ten percent per year are common. The highest growth rates are occurring in Asia and South America, with annual sales increases of 15 percent or more in places as diverse as Egypt, Kuwait, the United States, and Vietnam.

In 2002, the Beverage Marketing Corporation (BMC) estimated total consumption of bottled water at more than 131 billion liters, up from 72 billion liters in 1996. Table 2.1 shows annual global consumption from 1996 to 2002 (estimated), along with the annual percent increase. Figure 2.1 shows the trend in global consumption since 1996.

According to the Beverage Marketing Corporation (2003), global per-capita bottled water use has risen from 12.6 liters per-capita per year (lpcy) in 1996 to over 21 lpcy in 2002 (see Figure 2.2). Changes in per-capita consumption of bottled water are even more dramatic, however, when evaluated on a regional or national basis. Figure 2.3 shows trends in continental per-capita consumption over this period. The rate of increase is extremely high in South America, where use has doubled from 14 to 28 lpcy, and in Asia, where use is growing by 20 percent per year and has increased from under 4 to more than 8 lpcy. Total per-capita use, however, is still dominated by consumers in North America and Europe, where annual use is 85 and 64 lpcy, respectively.

The top ten bottled water-consuming countries are shown in Data Table 6 (in the Data Section) and Figure 2.4 for 1997 through 2002. By far, the greatest consumption occurs in the United States, followed by Mexico. In the past few years, however, consumption in China has grown enormously. In 1997, China was the ninth largest consumer

TABLE 2.1   Total Global Consumption of Bottled Water, 1996 to 2002

Year	Thousands of Cubic Meters	Percent Change
1996	72,675.62	–
1997	80,649.10	11.00
1998	87,838.89	8.90
1999	97,848.00	11.40
2000	107,381.48	9.70
2001	117,876.24	9.80
2002(P)	131,412.11	11.50

Note: Data for 2002 are preliminary.

Source: Courtesy of the Beverage Marketing Corporation (2003).

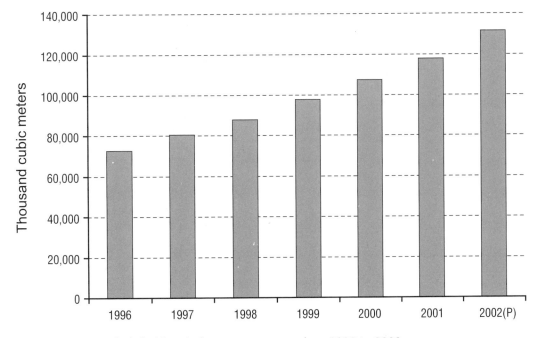

FIGURE 2.1   Total global bottled water consumption, 1996 to 2002.

Note: Data for 2002 are preliminary.

Source: Beverage Marketing Corporation, with permission.

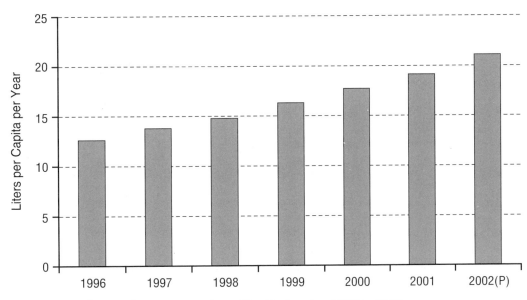

FIGURE 2.2   Per-capita consumption of bottled water, 1996 to 2002.

Note: Data for 2002 are preliminary.

Source: Beverage Marketing Corporation, with permission.

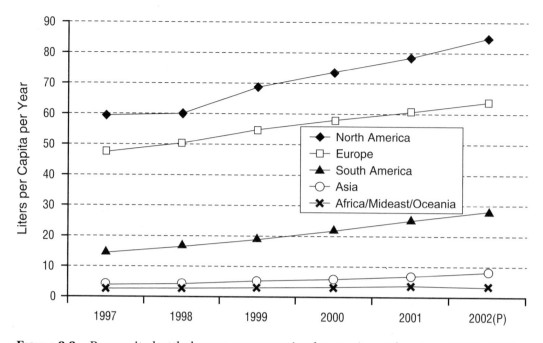

FIGURE 2.3   Per-capita bottled water consumption by continental region, 1997 to 2002.
Source: Beverage Marketing Corporation, with permission.

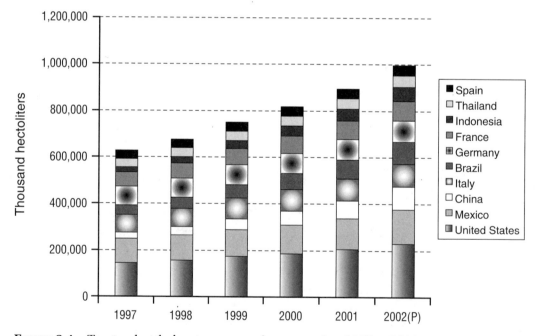

FIGURE 2.4   Top ten bottled water consuming countries, 1997 to 2002.
Source: Beverage Marketing Corporation, with permission.

TABLE 2.2 Bottled Water Sales in China, 1997 to 2002

	Sales (million liters)
1997	2,750
1998	3,540
1999	4,610
2000	5,993
2001	7,605
2002	9,887

Note: Data for 2002 are preliminary.

Source: Beverage Marketing Corporation, with permission.

of bottled water. By 2002, China had moved up to become the third largest bottled water consuming country, going from 2.7 billion liters in 1997 to 9.9 billion in 2002 (Table 2.2).

While there are thousands of bottling companies, the industry is undergoing a rapid consolidation as major bottlers tighten their holds on key markets. Nestle S.A., for example, owns dozens of brand names including Arrowhead and Poland Springs in the United States (with the third and fourth largest market shares in the United States in 2001), and the well-known brand, Perrier. In 2001, the largest selling brands in the United States were Aquafina (a Pepsi product), with revenues of $645 million, Dasani (a Coca-Cola product) with revenues of $560 million, Poland Spring, Arrowhead, and others as shown in Table 2.3. Box 2.1 shows the three leading U.S. purveyors of bottled water.

Box 2.1 Largest U.S. Sellers of Bottled Water in 2001

Nestle S.A.'s water division sells 70 bottled water brands in 160 countries. Its North American subsidiary sells nine domestic brands, including Arrowhead, Poland Spring, and Deer Park, and five imported brands, including San Pellegrino and Perrier. Nestle Waters North America, Inc. had revenues of $2.1 billion in 2001. Its market share is 32.5 percent and growing.

PepsiCo comes in second place with its Aquafina product, which currently has a 14 percent market share but is the top-selling, single-serve bottled water in the United States. In 2001, Aquafina sales grew nearly 45 percent and comprised 4 percent of all of Pepsi's beverage sales.

Coca-Cola sales are in third place with Dasani a 12 percent market share in 2001. Sales grew 90 percent or so in 2002. Coca-Cola recently entered into a production, marketing, and distribution partnership with France's Groupe Danone, owner of several brands, including Evian.

Sources: Bobala 2003, McKay 2002.

TABLE 2.3 Leading United States Bottled Water Brands, 2001

	2001 U.S. Market Share (%)	2001 Volume Growth (%)
Aquafina	13.80	45
Dasani	12.00	95.50
Poland Spring	11.20	29
Arrowhead	6.60	50
Aberfoyle	5.60	33
Crystal Geyser	5.50	15
Evian	3.80	–5

Note: Nestle S.A. owns both Arrowhead and Poland Spring (and many other brands, see Box 2.1).
Source: McKay 2002.

TABLE 2.4 Summary Results of Bottled Water Price Surveys

	Average Price per Cubic Meter (US$)
California Tap Water 2003	0.50
Nepal 2003	206
India 2003	267
France 2004	332
Spain 2003	411
Malawi 2004	825
South Africa 2003	857
Italy 1999	879
California 2003	995
Switzerland 1998	1,616

Note: Prices vary. These reflect the average of all the options available in surveyed stores. Currency conversions were done using rates at time of survey, uncorrected for inflation.
Source: Local price data was provided to the author by Moench, M., Water Nepal, M., Turton, A., Smets, H., and Lane, J. Prices for Switzerland 1998, Italy 1999 provided by http://www.bottledwaterweb.com/pricescan.html. Prices for California 2003 from three surveys by author.

The Price and Cost of Bottled Water

By any standard, bottled water is hugely expensive to consumers compared to reliable high-quality municipal water supply. In regions where no such municipal supply is available, bottled water may provide a temporary and vital source of safe drinking water. The key word, however, is "temporary"—bottled water should not be considered a permanent alternative to reliable municipal supply for many reasons, including cost, control, and equity. Failure to provide municipal supply often affects the poorest populations of peri-urban areas, leaving them to pay the inflated prices for water provided by private vendors or bottled water purveyors.

Surveys on the prices of bottled water are limited. Data from a series of surveys conducted between 1998 and 2003 in the United States, Europe, Nepal, South Africa, Malawi, and India are presented in Table 2.4 and Figure 2.5, compared to the price of high-quality municipal tap water in California. As this table shows, bottled water in most industrialized nations costs between $500 and $1000 per m^3. Even in Nepal, where per-capita annual income in 2002 was only about $230 (The World Bank 2002),

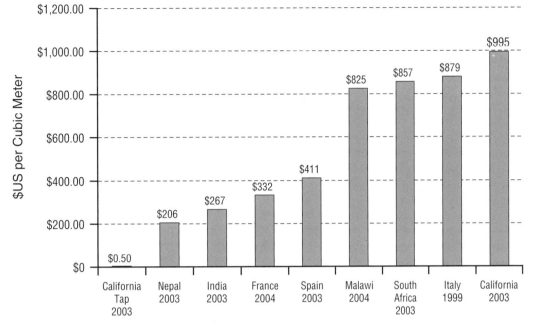

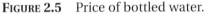

FIGURE 2.5   Price of bottled water.

Note: Prices found in a survey for Switzerland exceeded $1,500 per m³, but may not reflect average prices there.

bottled water costs over $200 per m³. This is in stark contrast to the cost of high-quality municipal water, which is usually well below $1 per m³.[1] Even water from desalination facilities, considered the most expensive source of municipal supply, is rarely more than $2 per cubic meter and the price of desalination has been dropping.

What justifies such a high price? Quite simply, consumers are willing to pay it. The cost of production for bottled water is quite low—spring owners may sell water to bottlers for only a few pennies per liter. The cost of labor, bottling, plastic bottles, transportation, and marketing greatly increases the overall costs, but bottled water remains a product with high profits—profit margins of 25 percent or more have been reported (Berberoglu, no date; Olson 1999), providing a major incentive for beverage companies to push bottled water production and sales.

Overall, we estimate that total consumer expenditures for bottled water approach $100 billion annually, assuming an average of $750 per cubic meter for the 131 billion liters sold annually. This sum greatly exceeds the money that would be required to produce a comparable amount of water from high-quality, reliable municipal systems.

The Flavor and Taste of Water

Highly subjective preferences for taste and flavor in water help drive the market for bottled water. Water has different flavors and tastes depending on its origin, type and duration of storage, treatment, and method of delivery. Other than concerns (valid or not) about water quality, the most common reason offered to explain the growing use of bottled water is dissatisfaction with the taste of locally available tap water.

---

1. The author pays around $600 per acre-foot, or around 50 cents per cubic meter³ for high-quality drinking water in California.

The taste of water reflects different concentrations of minerals and trace elements. Highly mineralized water can sometimes taste metallic. High levels of bicarbonates can taste salty. Water with hydrogen sulfide smells and tastes like rotten eggs. Certain types of plastic bottles can impart a plastic taste. For carbonated waters, customers are willing to accept higher levels of minerals (Bottled Water Web, *Water Taste*, 2003).

Professor William Bruvold of the University of California at Berkeley conducted and published findings on the taste of minerals in water in the 1960s. His study showed that certain combinations and concentrations of minerals in water were more acceptable than others (Pomento 2001). In 1998, the Metropolitan Water District of Southern California surveyed customers' perceptions of tap water quality and flavor. At that time 56 percent of the customers rated their tap water as fair and poor. Consumers believed that taste problems were increasing over time and 77 percent of consumers agreed that local water utilities "should be expected to provide tap water that looks and tastes as good as bottled water." Only 22 percent were very satisfied with how their tap water "looks and tastes" (Suffet 2000). Taste and odor concerns appear more important to consumer confidence in Southern California than meeting quality standards, a finding similar to that found in an evaluation of seven consumer surveys across the U.S. (Torobin et al. 1999). Consumers who reported fair or poor aesthetic water quality were the same ones who had less favorable perception of the safety of their local tap water.

The subjective nature of water tastes has been revealed regularly during taste testing. Several taste testings have been held to judge both tap and bottled water, with revealing results. At a blind water tasting held by the *San Francisco Chronicle* in 1980, samples of municipal waters from around the San Francisco Bay area were collected and rated by a panel of food and wine experts. Included among the tap waters was an example of a French non-carbonated bottled water. All of the judges gave it poor marks except for a French wine maker, who remarked, "This water reminds me of home." Indeed, this particular water was actually what he drank at home, demonstrating that we all develop a preference for what we are used to drinking, and that our brains have strong taste recall (Pomento 2001).

At a water tasting in Atlanta that became famous for its tongue-in-cheek rating system, ten Southern U.S. municipal waters were rated on a scale from zero (sludge), to 13 (nirvana). Memphis won with comments such as, "...On the nose, at first it was cottony...a refreshing texture." Judges rated New Orleans, "...for its neutrality, this is Swiss of the waters." Dallas was said to be, "...crude, with an edge." About Houston, judges noted, "...bring on the chlorine...It was like a chemistry lab...", and for one of the judges, "...that brought back unpleasant memories." Atlanta's water, was described "...like having a gulp of a swimming pool." Water from Charlotte, North Carolina was described as tasting "like when you have a Band-Aid on your finger and you get in the shower and you get out and suck the water out of the Band-Aid...It's like a wet Band Aid." And of Orlando, Florida's water, judges said, "...It's the reason most people don't drink water" (Bottled Water Web, *Water Taste*, 2003).

In another, unscientific but blind tasting conducted by the Pacific Institute at its December 2003 holiday party, 40 individuals expressed no clear preference among tap water, two spring waters, and highly processed bottled municipal water sold at supermarkets. While a majority of people preferred what turned out to be the local tap water (and the fewest number disliked this water), there was little agreement about taste and almost random success at trying to identify which water was which.

Bottled Water Quality

The public perception, and probably the reality, is that bottled water is regularly of high quality. This belief is encouraged by publicly reported problems with tap water and by aggressive advertising by bottled water companies and water filter sales pitches. In 2004, a company that produces filters for tap water was advertising that their product can produce tap water "the quality of bottled water," as though that guarantees an improvement (Procter and Gamble 2004). In many developing countries with unreliable and inconsistent municipal water supplies, bottled water quality may be a better source of safe water than tap water. But this cannot be universally assumed, nor should it be the goal of water providers. Standards vary from place to place, testing is irregular and inconsistent, and contaminated source water may lead to contaminated products. Over the past few years, there have been an increasing number of reports of water-quality problems with bottled water.

In a highly publicized case, a newspaper in Mumbai, India contracted with an independent water-quality laboratory to conduct tests for pesticides in bottled water samples and raw-water samples from bottled water plants (Mid-Day Mumbai 2003). The results were compared to European Economic Community standards (Directive 80/778/EEC), which provide water-quality guidelines for 62 parameters on the "quality of water intended for human consumption" and is used as a norm at the European level.

Altogether, 26 samples of 13 bottled water brands and raw water samples collected from different bottled water plants in Mumbai were tested for 12 organochlorine pesticides and 8 organophosphorus pesticides most commonly used in India. The results of the study showed pesticides in every sample, whether raw or bottled. The organochlorine Lindane was detected in all the raw water samples with a minimum concentration of 0.0007 mg/l and a maximum of 0.0042 mg/l. The organophosphorus pesticide Chlorpyrifos, a moderately persistent insecticide used against mosquito and fly larvae, aphids, and other insects was also detected in all the raw water samples analyzed (Center for Science and Environment 2003).

Similar problems were found with the bottled water samples. Lindane was present in all 26 samples. A minimum concentration of 0.0005 mg/l and maximum concentration of 0.0041 mg/l was detected. This maximum concentration is 40 times higher than the prescribed EEC limit of 0.0001mg/l. Chlorpyrifos was also detected in all 26 samples analyzed. The minimum concentration found was 0.0001mg/l, just at the limit permitted by the EEC Drinking Water Directive. The maximum concentration detected was 0.0075 mg/l, 75 times above the EEC limit. The average concentration of total pesticides analyzed in 26 bottled water samples was 0.0036 mg/l, 7.2 times higher than the total recommended pesticide limit of 0.0005 mg/l (Center for Science and Environment 2003). In a follow-up study, similar problems were then found in bottled water samples in India's capital (Mathur et al. 2003).

Pip (2000) describes water-quality testing of 40 domestic and imported brands of bottled water purchased in Manitoba, Canada. These waters were examined for total dissolved solids (TDS), chloride, sulfate, nitrate-nitrogen, cadmium, lead, copper, and radioactivity. The samples showed great variation in quality, with some violations of the Canadian Water Quality Guidelines for drinking water for TDS, chloride, and lead. A number of deficiencies were found with respect to product labeling. In the United

States, periodic problems with bottled water lead to recalls of product from stores (see Box 2.3). These examples hint at the kind of problems that already exist and that would be found with more regular and consistent monitoring.

Regulating Bottled Water

Perhaps the most charitable thing that can be said about the myriad and complex regulations around bottled water is that they are byzantine. Four levels of contradictory and complex regulation apply to the bottled water industry: international, national, local, and trade associations standards. These different levels are not consistent with each other. As a result, the practical effect of regulation varies widely from place to place. In some regions, consumers of bottled water can be assured of reliable and consistently high water quality. In others, no such assurances are possible. While many countries have national standards for bottled waters and some have national certification schemes, no universally accepted international standard certification scheme yet exists.

In the United States, the Food and Drug Administration (FDA) regulates bottled water as a food. Some state and industry standards also apply. And most U.S. bottled water is as safe and reliable as regulated tap water, albeit vastly more expensive.

In Canada, bottled water is also regulated as a food product, under the Canadian Food and Drugs Act and Regulations. These regulations set standards for microbiological quality, composition, and labeling, which are then enforced by the Canadian Food Inspection Agency (CFIA). Bottled water is also subject to the requirements of the Consumer Packaging and Labeling Act and Regulations. The CFIA also sets and enforces requirements under these acts to protect consumers against fraud in bottled water content, packaging, labeling, and advertising. As part of its enforcement role, the CFIA can inspect products, labels, and establishments involved in the sale, manufacture, and distribution of bottled water. In addition, some Canadian provincial and municipal ministries and agencies may regulate and inspect bottled water (Health Canada 2003).

For Australia and New Zealand, minimum standards for bottled water are set by the Food Standards Code of *The Food Standards Australia New Zealand* (FSANZ). The primary Standard is 2.6.2 "Non-Alcoholic Beverages & Brewed Soft Drinks." Bottlers must also comply with codes on labeling, contaminants, microbiological and processing requirements, and food safety regulations. Bottlers in Australia and New Zealand may choose to be subject to standards developed by the Australasian equivalent to the International Bottled Water Association (IBWA)—the Australasian Bottled Water Institute (ABWI). ABWI members must meet standards that are, in some cases, stricter than the FSANZ standards. The ABWI has a model code for processing, third-party audits, and labeling, but their mission also includes promoting bottled water use. ABWI is a member of the International Council of Bottled Water Associations (ICBWA). A fundamental requirement of membership is that each member association must have in place an approved, third-party audit program and certified model code and be in compliance with that code (ABWI 2003).

The intergovernmental body for the development of internationally recognized standards for food is the Codex Alimentarius Commission (CAC). The World Health Organization, one of the co-sponsors of the CAC, has advocated the use of the *Guide-*

lines for Drinking-Water Quality as the basis for all bottled water standards (WHO 2000, Fact Sheet 256). This chapter is not the place for a comprehensive survey or review of all bottled water standards, though such a survey would be valuable. The focus below is on United States standards and their application with some reference to similar standards in other countries and regions.

United States Federal Regulations

In the United States, bottled water and tap water are regulated by different federal agencies. The United States Environmental Protection Agency (EPA) regulates municipal water or public drinking water under the federal Safe Drinking Water Act.[2] EPA's Office of Ground Water and Drinking Water is responsible for issuing regulations that cover the production, distribution, and quality of drinking water, including regulations on source water protection, operation of drinking-water systems, contaminant levels, and reporting requirements. Bottled water is regulated by the FDA under the Federal Food, Drug, and Cosmetic Act (FFDCA).[3]

FDA regulations cover packaged water sold in individual containers at retail outlets and larger containers sold to residential and office markets. Under the FFDCA, manufacturers are responsible for producing safe, wholesome, and properly labeled food products, including bottled water products. It is a violation of the law to introduce into interstate commerce adulterated or misbranded products that violate the various provisions of the FFDCA (Posnick and Kim 2002).

The FDA has established specific regulations for bottled water in the Code of Federal Regulations (C.F.R.)—21 Code of Federal Regulations—including (i) standard of identity regulations that define different types of bottled water, (ii) standard of quality regulations that establish allowable levels for contaminants (chemical, physical, microbial, and radiological) in bottled water, and (iii) Current Good Manufacturing Practice (CGMP) regulations for processing, bottling, and labeling. Bottled water is one of the few "foods" for which the FDA has developed specific CGMP regulations or such a detailed standard of quality (Posnick and Kim 2002).

Standards of Identity

Because of the great variety of "types" of water, certain FDA "standards of identity" have been set for labeling and identifying bottled water. Standard of identity requirements are designed to ensure that bottled water manufacturers meet the requirements of specific definitions for labeling. The FDA's labeling rules establish standardized definitions for specific terms found on bottled water labels such as "artesian," "distilled," "deionized," "drinking," "mineral," "purified," "sparkling," "spring," and "sterile." (See Box 2.2). For example, under the standard of identity regulations, bottled water may only be labeled "mineral water" if it: (1) contains not less than 250 parts per million (ppm) total dissolved solids; (2) comes from a source tapped at one or more boreholes or springs; (3) originates from a hydrogeologically protected source; and (4) contains no added minerals.[4] If these terms appear on a label, the consumer should be able to assume that specific meanings apply.

---

2. 42 U.S.C. § 300f (United States Code, Title 42, Chapter 6A, Subchapter XII, Part A)

3. 21 U.S.C. § 301 (United States Code, Title 21, Chapter 9, Subchapter 1)

4. 21 C.F.R. § 165.110(a)(2)(iii)

Box 2.2   United States FDA Product Definitions for Bottled Water

The Food and Drug Administration (FDA) Standards of Identity guide what can be put on bottled water labels in the United States. If a bottler calls water "glacial" it has to come from a glacier. "Artesian" water has to flow above the water table and "naturally sparkling" has to come from a natural carbonated spring. Sodium declarations such as "sodium free," "very low sodium," and "low sodium" have explicit meanings: "sodium-free" must be less than 5 milligrams of sodium per serving (usually 360 ml); "very low sodium" may contain 35 milligrams or less of sodium per serving; and "low sodium" may contain 140 milligrams or less per serving. These kinds of claims trigger the inclusion of the Nutrition Facts panel as required by the Nutrition Labeling and Education Act of 1990. The FDA product definitions for bottled water are:

Artesian Water/Artesian Well Water

Bottled water from a well that taps a confined aquifer (a water-bearing underground layer of rock or sand) in which the water level stands at some height above the top of the aquifer is identified as Artesian water.

Drinking Water

Drinking water is water sold for human consumption in sanitary containers. It must have no calories or sugar. Drinking water may be sodium-free or contain very low amounts of sodium. Flavors, extracts, or essences may be added to drinking water, but they must comprise less than one percent by weight of the final product or the product will be considered a soft drink. "Carbonated water," "seltzer water," "soda water," and "tonic water," are considered soft drinks.

Infant Brands

Special labeling is required for products marketed for infants. If a product is labeled "sterile," it must be processed to meet the FDA requirement for commercial sterility. Otherwise, the labeling must indicate it is not sterile and should be used in preparation of infant formula only as directed by a physician or according to infant formula preparation instructions (Bottled Water Web Regulations 2003, at http://www.bottledwaterweb.com/regulations.html).

continues

Mineral Water

Mineral water is distinguished from other types of bottled water by "its constant level and relative proportions of mineral and trace elements" at the source. According to Title 21 (21CFR165.110) of the *Federal Register*, "mineral water" must contain at least 250 parts per million (ppm) of total dissolved solids (TDS). No minerals can be added to this product. Sources must be "tapped at one or more bore holes or springs, originating from a geologically and physically protected underground water source." If the TDS content of mineral water is below 500 ppm or greater than 1,500 ppm, the statement "low mineral content" or "high mineral content," respectively, must appear on the label.

Municipal Water

Municipal water is used as a source for approximately 25 percent of the bottled water sold in the United States. Water bottled from municipal water supplies must be clearly labeled as such. Municipal water that has received further processing and treated to the appropriate level can be labeled as "distilled" or "purified" water.

Natural

The word "natural" is allowed for bottled water derived from springs or wells where the natural chemical (mineral and trace elements) composition of the water has not been altered.

Purified Water

Water that has been produced by distillation, deionization, or reverse osmosis and that meets the definition of purified water in the United States Pharmacopoeia may be labeled as purified bottled water.[1] Bottled water treated by one of these processes may also be called "distilled water" if it is produced by distillation, "deionized water" if the water is produced by deionization, or "reverse osmosis water" if the process used is reverse osmosis.

---

1. The United States Pharmacopoeia is a nongovernmental, standards-setting organization that advances public health by ensuring the quality and consistency of medicines, promoting the safe and proper use of medications, and verifying ingredients in dietary supplements (www.usp.org).

continues

Box 2.2 *continued*

Sparkling Water

Sparkling water is water that contains the same amount of carbon dioxide that it had at the source, though it can be removed and then replaced. Soda water, seltzer water, and tonic water are not considered bottled waters. They may contain sugar and calories and are regulated separately as soft drinks. In 1990, the FDA made Perrier drop the words "Naturally Sparkling" from its label when it was revealed that Perrier artificially carbonated its water after taking it out of the ground (Mowen and Minor 2001).

Spring Water

The term spring water is restricted to water collected from a spring that originates from an underground formation from which water flows naturally to the surface of the earth, or from a borehole that connects to the formation. Spring water collected from a borehole must have all the physical properties, before treatment, and be of the same composition and quality as the water that flows naturally to the surface of the earth. Controversy has arisen when bottlers drill wells near springs in order to extract water more quickly than natural flow rates. In the United States, the spring must continue to flow even when water is pumped from the same aquifer. European bottlers using boreholes do not have to maintain a flow from the spring. This issue became the focus of debate at a meeting of Codex Alimentarius in Bern, Switzerland.[2] On one side was the National Spring Water Association, which is lobbying in both the United States and Europe for the term "spring" to only be used in bottled water products using water that flows from a natural opening. In contrast, the International Bottled Water Association (IBWA) endorses the use of boreholes as an acceptable method of spring water extraction. In either case, bottled water products cannot use the term "spring" if the water is substantially processed or from a municipal source (von Wiesenberger 2003).

Well Water

Well water is simply bottled water from a bored, drilled, or otherwise constructed hole in the ground, which taps the water of an aquifer.

---

2. The Codex Alimentarius Commission was created in 1963 by FAO and WHO to develop food standards, guidelines, and codes of practice. The main purposes of this Codex are protecting health of the consumers and ensuring fair trade practices in the food trade, and promoting coordination of all food standards work undertaken by international governmental and non-governmental organizations (see http://www.codexalimentarius.net/).

Sources: United States Code of Federal Regulations 21, Bottled Water Web (2003), Posnick and Kim 2001, United States Pharmacopoeia (2003).

Other terms, however, have no clear definitions and can be misleading to the consumer. For example, terms like pure, purest, pristine, premium, mountain water, and clean are advertising descriptors with no official meaning. These terms do not accurately describe the source or purity of the water, nor do they certify that the water is safe (Suffet 2000). Images on bottled water can also contribute to confusion and misunderstanding about contents. Aquafina (a Pepsi product), like many other bottled waters, puts images of mountains and snow on the label, despite the fact that Aquafina is bottled using processed municipal water. Such images are not adequately regulated. Seltzer, soda water, and tonic water are considered soft drinks and are excluded from these regulations.

Food misbranding provisions require that food labeling contain certain information, such as the name of the manufacturer, and not contain certain other types of prohibited information, such as false or misleading statements. For bottled water, labels must contain a statement of identity, the name and location of the manufacturer, the net weight of the contents, and ingredients if the product contains more than one ingredient. Depending on what nutrients and minerals are present or added, some bottled waters have to bear nutrition labeling. The FDA requires that all nutrient content claims, such as sodium free, and health claims comply with specific definitions. In 1924 the Supreme Court ruled that the Food and Drugs Act must address statements that may mislead or deceive. In part, this decision was the result of unsubstantiated health claims for bottled waters. Today in the United States, health claims are allowed only when there is a proven link between the nutrient and certain health conditions, such as significant amounts of calcium and protection against osteoporosis.

Bottled water is also misbranded if its labeling contains false or misleading statements. For instance, a label that falsely states that the product is free from a certain contaminant would be misbranded under this provision. A bottle of water with more than 0.3 mg/l of iron would be misbranded unless its label stated "Contains Excessive Chemical Substances" or "Contains Excessive Iron."[5]

Despite these protections, the information provided by labels in the United States is still limited. For example, the "Nutrition Facts" panel (common to all United States food products) for water tends to show that water has no fat, carbohydrates, and proteins, but the label carries no other mineral analysis. European labels carry a more informative mineral analysis that provides consumers with information on the levels of calcium, magnesium, potassium, and other nutrients. Although certain mineral waters may be useful in providing essential micronutrients, such as calcium, the World Health Organization *Guidelines for Drinking-Water Quality* do not make recommendations regarding minimum concentrations of essential compounds because of the lack of convincing evidence on the beneficial effects of consuming such mineral waters (WHO 2000).

Quality Standards

Bottled water products must comply with the FDA's Quality Standards in Section 165.110(b) of Title 21 of the *Code of Federal Regulations* (CFR). Bottled water manufacturers must ensure that their products meet some, though not all, of the federal

---

5. 21 C.F.R. § 165.110(c)

standards of quality for tap water, which establish allowable levels of substances related to microbiological quality (e.g., limits on coliform organisms), physical quality (e.g., turbidity, color, and odor), organic and inorganic chemical quality, and radiological quality. The FDA has established allowable levels in the standard of quality for approximately 75 substances.

Bottled water regulations set testing methodologies and time frames to determine compliance with FDA Quality Standards for both source water and product water. Source water—that is the water taken for processing and packaging—must be obtained from an approved source and conform to applicable state and local laws and regulations. Bacteriological analysis of source water must be done at least weekly. Chemical analysis of source water must be done at least annually. Radiological analysis must be performed at least once every four years.

Product water—that is the bottled water itself—is also subject to regulation. Representative bacteriological analysis of product samples must be done at least once per week for each type of water produced. Chemical, physical, and radiological analysis must be done at least annually on an appropriate sample from a batch or segment of a continuous production run for each type of water produced. All records as well as government approvals of source water must be available for official review.[6]

Good Manufacturing Practices

The Food and Drug Administration promulgates Good Manufacturing Practice (GMP) requirements. Bottled water is subject to GMPs for both food and specific bottled water processing and bottling practices. General food GMPs govern such areas as plant construction and ground maintenance, sanitary maintenance of buildings and fixtures, and sanitary facilities, water supply, plumbing, and sewage disposal. These also apply to bottled water for quality control of receiving, inspecting, transporting, segregating, preparing, manufacturing, packaging, and storing product.[7]

In addition to general food GMPs, there are specific bottled water GMPs that apply to facilities, sanitation, product quality and testing, and record keeping. A fundamental requirement of the bottled water GMPs is that all product water be obtained from a source that has been "…inspected and the water sampled, analyzed and found to be of a safe and sanitary quality according to applicable laws and regulations of state and local government agencies having jurisdiction…."[8] GMPs specific to bottled water processing and bottling also contain rules on use of approved test and sample methods; adequate sanitization of all product water contact surfaces; protection of single service containers, caps, and seals; and sanitary filling, capping, closing, sealing, and packaging of containers. Bottlers must also maintain and retain records of product sample analysis, source approvals, and all records that may be required for official review.[9] If bottlers do not comply with these regulations, their product can be considered adulterated and they may be subject to regulatory action.

Food Adulteration

The FDA deems a food to be adulterated if it contains an added "poisonous or deleterious substance which may render it injurious to health" [FFDCA Section 402(a)(1)].

---

6. 21 C.F.R. § 120.80(h).

7. 21 C.F.R. Part 110

8. 21 C.F.R. § 129.3(a)

9. 21 C.F.R. Part 129

Bottled water may be deemed adulterated if it contains industrial contaminants, unapproved pesticides, or other substances harmful to health. The FDA deems a food to be adulterated "if it consists in whole or in part of any filthy, putrid or decomposed substance, or if it is otherwise unfit for food" [Section 402(a)(3)]. Bottled water could be deemed adulterated under this provision if it becomes contaminated during processing by dirt or other debris.

The FDA also deems bottled water to be adulterated if it has been "prepared, packaged, or held under unsanitary conditions whereby it may have become contaminated with filth, or whereby it may have been rendered injurious to health" [Section 402(a)(4)]. Bottled water may be deemed adulterated if it was not manufactured in substantial compliance with the Good Manufacturing Practices. A food is adulterated if a bottled water container "is composed, in whole or in part, of any poisonous or deleterious substance which may render the contents injurious to health" [Section 402(a)(6)].

Bottled water is also considered adulterated if any valuable constituent has been removed from the product, any substance has been wholly or partially substituted, damage or inferiority has been concealed in any manner, or if any substance has been added to the product or its packaging to increase the product's weight or value or reduce its quality or strength.[10] The FDA has tools available to it, in theory, to enforce these provisions, but the provisions for sampling and testing have some important and worrisome loopholes.

Sampling and Testing, and FDA Inspections of Bottled Water Plants

Bottled water plants are subject to FDA monitoring and inspection. Specific items inspected in bottled water establishments include: (1) verifying that the plant's product water and operational water supply are obtained from an approved source; (2) checking whether source claims on the label comply with FDA definitions; (3) inspecting washing and sanitizing procedures; (4) inspecting the filling, capping, and sealing operations; and (5) determining whether the firms analyze source water and product water according to the required schedules. Despite these inspection requirements, bottled water plants generally are given low priority for inspection because of their relatively good safety record compared to other food plants, and because of financial constraints on the ability of the FDA to monitor at all (Posnick and Kim 2002).

Even more problematic, most sampling and testing of bottled water are done by the bottlers themselves. Samples may be collected by the FDA during inspections if the inspector's observations indicate it is necessary or if the facility has a previous history of contamination. Samples are also collected in response to trade or consumer complaints. Samples of foreign bottled water products offered for entry into the United States may also be collected and tested to determine if they are in compliance with applicable United States laws and regulations.

FDA laboratories may test the water for microbiological, radiological, or chemical contamination. Individual samples are not tested for all possible contaminants cited in the quality standard, but for selected contaminants, depending on why a sample may have been taken. The FDA also may review the labeling on bottled water samples.

While bottled water companies are required to comply with specific quality standards, as described above, there is a major loophole to the testing requirements: bottlers are not required to use FDA-listed methods in their own facilities, nor are they

---

10. FFDCA §§ 402(a),(b)

required to have independent water laboratories run the tests. In theory, bottlers are responsible for ensuring that their bottled water could pass the tests used by the FDA in its own laboratories, but such independent testing rarely occurs.

FDA Enforcement/Regulatory Action

The FDA has a selection of enforcement tools available to it for violation of food standards. These have occasionally, though rarely, been applied to the bottled water industry. If a product is adulterated or misbranded and a company declines to comply with applicable requirements or declines to take action to correct the violation, the FDA may take civil and criminal action. Typically, however, the FDA first sends "warning letters" and requests for voluntary recalls. If these mechanisms do not work, the FDA may ask the Department of Justice to bring either a civil seizure or an injunction against the products and/or company involved. If the conduct warrants criminal prosecution, the FDA may seek such action from the Department of Justice as well. These criminal cases may be either misdemeanors or felonies, depending on the circumstances involved, and may result in monetary penalties and jail sentences. In all cases, the FDA may issue warnings to the public as a means to protect public health.

State Regulations

In addition to FDA regulatory requirements, the bottled water industry is subject to state regulatory requirements. A significant responsibility of the states is inspecting, sampling, analyzing, and approving sources of water. States also have the responsibility to certify testing laboratories and the authority to perform unannounced spot inspections. Some states perform annual inspections as well and some states (e.g., California, Pennsylvania, and Florida) have adopted regulations that are stricter than federal requirements (Fine Waters 2003; Bottled Water Web 2003).

Bottled Water Industry Associations: Standards and Rules

There are a number of international bottled water industry groups that also maintain their own memberships, standards, and rules. The International Bottled Water Association (IBWA), for example, is active in monitoring and reviewing bottled water standards and in supporting commercial bottled water activities. IBWA has established a quality assurance program comprising a set of standards called the model code. In some cases, the model code establishes tougher requirements than United States federal and state authorities and may also provide a model for countries where regulatory authority over bottled water is weak or non-existent.

For example, as a condition of membership in the IBWA, bottlers are subject to an annual, unannounced plant inspection administered by an independent, internationally recognized, third-party inspection organization. This inspection audits quality and testing records, reviews all areas of plant operation from source through finished product, and checks compliance with FDA Quality Standards, Good Manufacturing Practices, and any state regulations (IBWA 2003).

The International Council of Bottled Water Associations (ICBWA) is a group of groups, including the IBWA, the Canadian Bottled Water Association, the Latin American

Bottled Water Association, and several others (see www.icbwa.com). The ICBWA requires its members to meet "Codex Alimentarius Commission, national, regional, and industry standards for bottled water." Each member association is required to have a Model Code outlining good manufacturing practices and quality control standards. Each bottling production plant is required to undergo an annual, unannounced plant inspection to determine compliance with standards set forth in the Model Code. These inspections are conducted by International Council-approved, third-party organizations that audit quality and testing records, and review plant operation from source to finished product. Two other conditions of bottler membership within each member association are regular microbial testing using qualified personnel and an annual water analysis administered by an independent laboratory covering more than 150 possible compounds (http://www.icbwa.org/standards.htm).

International Standards—The Codex Alimentarius

The closest thing to a universally accepted international certification scheme is the intergovernmental body for the development of internationally recognized standards: the Codex Alimentarius Commission (CAC). The WHO, one of the co-sponsors of the CAC, has advocated use of the Guidelines for Drinking-Water Quality as the basis for standards for all bottled waters. Neither the CAC nor the WHO offer certification of any bottled or mineral water products.

The Codex Alimentarius, or the food code, was initiated by the WHO and the Food and Agriculture Organization (FAO) in 1961 as the principal tool for drawing attention to food safety and quality at the international level. It serves as a reference point for consumers, food producers and processors, and national food regulatory agencies. The Codex Alimentarius system presents an opportunity for countries without their own ability to generate detailed regulations to formulate and harmonize food standards and the codes and regulations governing food safety.

The significance of the Codex for consumer protection was highlighted in 1985 by United Nations Resolution 39/248, which advised that "Governments should take into account the need of all consumers for food security and should support and, as far as possible, adopt standards from the…Codex Alimentarius."

The Codex Alimentarius is relevant to the international production and trade in bottled water. The advantage of having uniform standards for the protection of consumers is evident, as long as those standards are strict enough to provide real and consistent protection. It is not surprising, therefore, that various international agreements (such as the Agreement on the Application of Sanitary and Phytosanitary Measures [SPS] and the Agreement on Technical Barriers to Trade [TBT]) encourage the international harmonization of food standards. The SPS Agreement, a product of the Uruguay Round of multinational trade negotiations, describes Codex standards, guidelines, and recommendations as the preferred international approaches for aiding international trade in food.

While the growing world interest in Codex activities indicates growing acceptance of the idea of harmonization, consumer protection, and facilitation of international trade, it is difficult in practice for many countries to accept Codex standards as law. Differing legal formats and administrative systems, varying political conditions and national attitudes, and concepts of sovereign rights slow harmonization and hinder the acceptance of Codex standards. Despite these difficulties, a number of countries are modifying national food standards, or parts of them, based on the Codex Alimentarius (FAO 1999, Codex Alimentarius 2001).

The CAC has developed the Codex Standard for Natural Mineral Waters and an associated code of practice. The Codex Standard describes the product and its labeling, composition and quality, hygiene, and packaging. The CAC health and safety recommendations are recognized by the World Trade Organization as representing the international consensus for consumer protection, but they are not mandatory.

The CAC also has a Codex Standard for Bottled/Packaged Waters to cover drinking water other than natural mineral waters. Under the existing Codex Standard and Code of Practice, natural mineral waters must conform to strict requirements concerning, for example, their collection and bottling without further treatment from a natural source, such as a spring or well. In comparison, the Codex Standard for Bottled/Packaged Waters includes waters from sources other than springs and wells, and covers treatment to improve their safety and quality. The distinctions between these standards are especially relevant given the growing tendency to sell bottled water that is little different in quality from municipal supplies. Ultimately, however, these standards give enormous leeway to national standards and are not in themselves likely to form the basis for specific legislation and wording (Codex Alimentarius 2001).[11]

Comparison of United States Standards for Bottled Water and Tap Water

Many people purchase bottled water because of concern over the quality of their tap water. Is this concern valid? Unfortunately, it is not possible to draw broad or definitive conclusions because of the wide range of products available and the great differences in tap waters. It is, however, possible to offers some comments on the relative standards themselves. While bottled water and tap water standards are roughly comparable, there are differences in scope, enforcement, and monitoring of those standards that make it impossible to say that bottled water offers any guaranteed improvement over tap water.

As noted above, U.S. bottled water manufacturers must ensure that their products meet FDA standards of quality. These standards establish allowable levels for substances including microorganisms such as coliform, physical parameters such as turbidity, color and odor, and radiological quality for such substances as radium 226. There are also limits specified for individual chemicals, including metals, inorganics, volatile organics, pesticides, trihalomethanes, and fluoride. The FDA standards of quality also establish testing time frames for the various categories of contaminants and specify both the methodologies required for conducting the analysis and the records bottlers must maintain.

Historically, the FFDCA required that, whenever the EPA prescribed a drinking water regulation, the FDA was to consult with the EPA to determine whether that standard should be applied to bottled water. Within 180 days, the FDA was to either adopt a comparable regulation or publish a reason for not doing so. In 1996, the FFDCA

---

11. See the Codex for detailed information on European standards and practices: ftp://ftp.fao.org/codex/standard/en/cxs_227e.pdf

requirement was strengthened such that if the FDA fails to act within the time provided, the drinking water regulation will automatically apply to bottled water (FDCA § 410, 21 U.S.C. 349).

A report from the Natural Resources Defense Council in 1999 compared FDA and EPA allowable limits for chemical, radiological, and microbiological compounds in drinking water (Olson 1999). At the time of this study, there were 13 substances where FDA bottled water standards were considered weaker than EPA standards for tap water; three bottled water standards that were stricter; and a range of other discrepancies. For example, many unregulated substances must still be monitored in tap water, but not in bottled water.

There are important differences related to water testing as well. Drinking water providers are required to use laboratories certified by a state in accordance with EPA criteria.[12] This system has flaws, since many states permit water systems to collect and submit samples for testing—a potential way to circumvent water quality problems. But the system for bottled water is even less strict. The FDA permits water bottlers to conduct their own tests and to select their own, uncertified laboratories. In 1991, the General Accounting Office criticized this practice, saying "FDA lacks assurance that such [bottled water] tests are done correctly or that the results are reliable. FDA regulations specify that either 'qualified bottling plant personnel' or 'competent commercial laboratories' use approved water-quality test methods…[but] has not defined qualified personnel or competent laboratories, and it does not require that such personnel or laboratories be certified or otherwise establish their qualifications to do the required tests. In contrast, for public drinking water, EPA requires certified laboratories…." (GAO 1991).

Other Concerns Associated with Bottled Water

Recalls

Even with the limited independent testing done for bottled water, problems are periodically discovered. These include the inappropriate use of poor-quality source water, accidents with bottling equipment or water-handling machinery, or even intentional acts. If contamination is detected, actions can be taken to prevent the consumption of the unsafe water, including recalls and impoundments of shipments. Rules and regulations for dealing with these situations vary around the world with little consistency or reliability.

The United States EPA regulations on tap water require providers to report violations of water-quality standards. In contrast, the FDA relies almost exclusively on voluntary action by bottlers to initiate recalls. Some consumer groups argue that the lack of a stronger requirement for reporting is a major flaw in bottled water protection and regulation. The Natural Resources Defense Council, for example, states, "FDA rules include no provision obligating a bottler to notify FDA or a state of test results, contamination problems, or violations, even in the case of contamination that could pose a serious health threat" (Olson 1999).

---

12. 40 C.F.R. § 141.28

Box 2.3   Examples of Bottled Water Recalls or Contamination

1990   One of the most famous cases of a food recall in history occurred in 1990 when a few bottles of Perrier in North Carolina were discovered to contain traces of benzene, a carcinogen. The initial response of Perrier was that the source of the benzene was cleaning fluids used inappropriately on bottling equipment in the United States. After a delay, the company recalled 70 million bottles of water. Shortly thereafter, officials in Denmark and the Netherlands announced the discovery of benzene contamination in Perrier sold in their countries as well, leading to a worldwide recall of 160 million bottles of Perrier. The source turned out to be the failure to replace filters that eliminate "naturally occurring" benzene from carbon dioxide in the water. This incident has become a classic case study in the field of food protection, recalls, and public relations (Browning et al. 1990, Mowen and Minor 2001, University of York, 2003).

1990   Recall of Newton Valley Distilled Drinking Water and Newton Valley Artesian Water in one-gallon plastic bottles because of contamination with an unidentified chemical odor and taste. The recall was initiated by the firm in Manitowoc, Wisconsin, and affected about 2,500 cases, or around 15,000 gallons, of product.
http://www.fda.gov/bbs/topics/ENFORCE/ENF00024.html

1990   Recall of bottled water of the Island Waters Division, Aqua Vie Beverage Corporation. The water was contaminated with algae and *Pseudomonas aeruginosa*. *Pseudomonas aeruginosa* infection may lead to gastrointestinal illness or other more serious consequences. More than 13,600 cases were recalled.
http://www.fda.gov/bbs/topics/ENFORCE/ENF00008.html

1996   Natural Springs bottled water, distributed in Arkansas, Illinois, Kentucky, and Tennessee by Marion Pepsi-Cola Bottling Co., was recalled due to bacterial contamination at levels exceeding federal and state water-quality standards. The presence of coliform bacteria was considered an indicator of problems with the Marion Pepsi-Cola bottled water production lines.
http://www.idph.state.il.us/public/press96/water.htm

JULY 1999   Coca-Cola recalled mineral water in Poland due to bacterial contamination after mold was discovered in bottled water. Sampling at a plant in Sroda Slaska discovered E. coli. An estimated 180,000 plastic half-liter containers of Bonaqa Plus mineral water were pulled off store shelves for failing to meet quality standards.
http://www.morningsun.net/stories/070399/usw_0703990018.shtml

continues

2000 Canadian Food Inspection Agency issued a warning about Mount Pelion brand bottled water contaminated with *Pseudomonas aeruginosa.*
http://www.inspection.gc.ca/english/corpaffr/recarapp/2000/20001020e.shtml

2000 The Pennsylvania Department of Environmental Protection (DEP) expanded its recall on water bottled by Roaring Springs/Global Beverage Systems Inc. of Latrobe, Westmoreland County, after it was discovered the company distributed bottles of coliform-contaminated water to several retailers.
http://www.waterindustry.org/New%20Projects/bottled-water.htm

2000 More than 30,500 cases (with 6-gallon bottles per case) of Safeway brand drinking water, manufactured by the Safeway Bottled Water Division, Tempe, Arizona, were recalled due to particulate matter contamination.
http://www.fda.gov/bbs/topics/ENFORCE/2001/ENF00679.html

2000 20,000 gallons of Bareman Dairy Crystal Clear Drinking Water sold in Michigan and northwestern Indiana were recalled after customer complaint led to testing that discovered contamination from an equipment sanitizer. The sanitizer was made of a mix of peroxiacetic acid and hydrogen peroxide.
http://www.wndu.com/news/contact16/contact16_1498.php,
http://www.fda.gov/ora/about/enf_story/archive/2001/ch4/default.htm

2002 The New Hampshire Department of Health & Human Services' Bureau of Food Protection ordered a recall of Granite State Artesian bottled water due to the presence of coliforms. The bottled water, produced by Turner's Dairy in Salem, New Hampshire, was distributed in Massachusetts and in New Hampshire stores located in Nashua, Manchester, Pelham, Salem, and East Derry.
http://www.fda.gov/oc/po/firmrecalls/granite07_02.html

2002 Recall of bottled drinking water, packaged in 12 packs of 20-ounce plastic PET bottles due to contamination with excessive chlorine. The manufacturer was the Southern Bottled Water Company, Anniston, Alabama, with distribution in Florida, Texas, South Carolina, Georgia, Arkansas, and Louisiana. Around 14,000 cases were recalled.
http://www.fda.gov/bbs/topics/enforce/2002/ENF00755.html

2002 Recall of bottled Spring Water labeled American Fare Premium Water Quality contaminated with particulate matter. The water was produced by the Glacier Clear LP Company of Orange Springs, Florida, and covered around 4,000 cases, with 12 bottles per case.
http://www.fda.gov/bbs/topics/enforce/2002/ENF00755.html

Under the FDA recall program, a recall can be initiated when a firm believes a product may be in violation of FDA standards. In such a case, the firm should (but is not required to) notify the FDA and provide information on: (1) the identity of the product; (2) the reason for the recall and when the problem was discovered; (3) evaluation of the risk involved; (4) time and amount of production; (5) total amount in and nature of distribution; (6) a copy of any recall communication already done or proposed; (7) proposed strategy for the recall; and, (8) the name and coordinates of a company contact.[13] Upon review of this information, the FDA assigns a recall classification for the ensuing action.

The regulations identify three classes of recall, with different requirements for the extent of the recall, the degree of public warning, and verification and monitoring of the recall. The most serious level is Class I, which is applied when there is a reasonable probability that the product may cause death or illness. A Class I recall leads to individual consumer notification through the media and aims at 100 percent effectiveness. A Class II recall would be used when the health consequences expected are less serious and the supporting activities would be dictated by the situation. Class III is the least serious recall, and may only extend to the wholesale level.

If the FDA determines that a product poses a serious health risk and a company declines to recall that product, the FDA can take further legal action to have that product removed from the market. FDA guidelines recommend, though they do not require, that all food producers, including bottled water companies, have a written recall plan in the event products must be withdrawn from the market because they are adulterated or misbranded.

Some examples of recalls in several countries show the types of water-quality problems that may be found in bottled water and the actions that were taken in response (see Box 2.3).

Selling Bottled Water to the Poor

One of the most controversial aspects of the growing trend toward bottled water is the fact that the populations most in need of improvements in municipal water systems, and hence most likely to need to use bottled water for legitimate health purposes, are those least able to afford the high costs of bottled water. Some worry that increased bottled water sales will reduce pressures on governments to provide necessary improvements in basic water infrastructure (Beck 1999). This controversy has grown in recent years as marketing efforts to sell bottled water have expanded in developing countries.

Perrier, for example, one of the world's leading name brands for water, is now owned by the world's largest food company, Nestle. In 1998, Nestle launched a new initiative named Pure Life to expand bottled water sales to the poorest consumers. Pure Life was launched in Pakistan and soon appeared in Brazil, Argentina, Thailand, the Philippines, China, and Mexico in 2000. In 2001, India, Jordan, and Lebanon followed, and in 2002, Egypt, Uzbekistan, and the United States (Nestle 2003). Nestle's effort to expand in India failed, and sales there ceased in 2003 (Datta 2003).

Nevertheless, sales of bottled water in developing countries are skyrocketing. As noted earlier, sales in China are growing faster than anywhere else, and companies see huge markets developing in Latin America, India, and elsewhere. As Figure 2.3 shows,

---

13. 21 C.F.R. § 7.46

more bottled water is consumed in Asia today than was consumed in North America in the late 1990s, and while North America and Europe still dominate overall sales, rates of growth are faster in Asia and South America.

Environmental Issues

More than 1.5 million tons of plastic are used to bottle water. Plastics are made from oil and natural gas, both of which are non-renewable resources. The processes used to make plastics can cause serious pollution affecting both the environment and human health if left unregulated. PET (polyethylene terephthalate, a plastic resin and the substance that most water bottles are made of) requires less energy to recycle than glass or aluminum, and releases fewer emissions into the atmosphere, but most plastic bottles are not being recycled, leading to serious landfill and garbage disposal problems. In 2002, only 20 percent of the total PET production available for recycling was actually recycled (NAPCOR 2002). Since plastic degrades at a very slow rate, plastic bottles will remain a waste problem for a long period.

Another growing concern is the local impact of water-bottling plants on the sustainability of groundwater aquifers and their effect on local streams (Glennon 2002). Some localities have mobilized against bottling plants in recent years when the pumping of groundwater for export from a basin was thought to threaten other local waters. In Wisconsin, local opposition to a Nestle's Perrier bottled water plant near Mecan Springs led to the relocation of the plant to Michigan, thought to be more complacent about the risks of this industrial activity. Yet a judge there ordered the Perrier pumping operation shut down when it threatened to affect local surface waters (WMEAC 2003). Michigan water law prohibits the reduction in flow of a stream by water users that export the water from the local watershed. (See also chapter 4 in this book.)

Conclusions

The recent dramatic upward trend in the consumption of high-priced bottled water is likely to continue, barring any dramatic improvement in global access to safe, reliable, and inexpensive domestic water, or any highly publicized, widespread incident of contamination that diminishes the reputation of bottled water. Moreover, the high profit margin for bottled water, the extensive advertising campaigns bottlers can afford to produce, and legitimate advantages of convenience all suggest that bottled water use is here to stay. This is fine. But is bottled water an acceptable alternative to reliable municipal supply? The quality of bottled water is inadequately monitored and regulated in many regions. Its cost is far higher than good quality tap water—often a thousand times more expensive. There are real environmental impacts of extracting large volumes of water from local aquifers and of producing and disposing of plastic containers. We estimate that $100 billion annually are, conservatively, spent to purchase bottled water worldwide. Our failure to meet basic human needs for water should not open the door to replacing a public good with a private commodity, but rather should motivate us to spend the same resources to produce a more widely available, and far less costly, public product.

REFERENCES

Australasian Bottled Water Institute, Inc. (ABWI). 2003. Bottled water regulations.
    http://www.bottledwater.org.au/Regulations/regulations.html.

Beck, E. 1999. Popular Perrier—Nestle Pitches Bottled Water to World's Poor. *Asian Wall Street Journal.* Indian Express Newspapers (Bombay) Ltd.
    http://www.financialexpress.com/fe/daily/19990623/fec23069.html

Berberoglu, H., no date. Bottled Water: A profit center for restaurants. Food reference web site.
    http://www.foodreference.com/html/artbottledwater.html

Beverage Marketing Corporation. 2003. Personal communication of data to Peter H. Gleick.

Bobala, B. 2003. Water wars: The Motley Fool commentary. January 6, 2004.
    http://www.fool.com/news/commentary/2003/commentary030310bb.htm

Bottled Water Web. December 2003. http://www.bottledwaterweb.com. See Bottled water regulations, http://www.bottledwaterweb.com/regulations.html. See, bottled water "Water Taste," http://www.bottledwaterweb.com/watertaste.htm. See Bottled water FAQs, http://www.bottledwaterweb.com/qna.html.

Browning, E. S., Freedman, A. M. and King, T. R. 1990 Perrier expands North American recall to rest of globe, *Wall Street Journal,* February 15, B1, B4.

Bruvold, W.H. 1992. Sensory research for the establishment of secondary maximum contaminant levels. AWRA Paper Number 91085, *American Water Resources Association Journal,* 28:No. 4 (August):775–782.

Center for Science and Environment. 2003. CSE report on pesticide residues in bottled water.
    http://www.chalomumbai.com/articleimages/images17/mumbai.pdf

Codex Alimentarius. 2001. General standard for bottled/packaged drinking waters (Other than natural mineral waters). Codex Stan 227–2001.
    http://www.codexalimentarius.net/standard_list.asp#.

Datta, P. T. Jyothi. 2003. Distribution problems hit "Pure Life—Nestle India to exit water business. *Hindu Business Line.* August 26.
    http://www.thehindubusinessline.com/2003/08/27/stories/2003082702690300.htm

Fine Waters. 2003. Fine waters FAQ.
    http://www.finewaters.com/FAQ/Federal,_State_Industry_Regulations.asp

Food and Agricultural Organization. 1999. Understanding the Codex Alimentarius. Rome.
    http://www.fao.org/docrep/w9114e/w9114e00.htm.

General Accounting Office. 1991. *Food safety and quality: Stronger FDA standards and oversight needed for bottled water.* GAO/RCED-91-67, Washington, D.C. March.

Glennon, R. J. 2002. *Water follies: Groundwater pumping and the fate of America's fresh waters.* Washington, D.C.:Island Press.

Health Canada. 2003. Bottled water questions and answers.
    http://www.hc-sc.gc.ca/food-aliment/mh-dm/mhe-dme/e_faqs_bottle_water_eng.html

International Bottled Water Association (IBWA). 2003. Industry regulation: The IBWA model code.
    http://www.bottledwater.org/public/model_main.htm

Mathur, H. B., Johnson, S., Mishra, R., Kumar, A., Singh, B. 2003. Analysis of pesticide residues in bottled water [Delhi region]. CSE/PML-6/2002. New Delhi:Centre for Science and Environment. Report available in full at
    http://www.cseindia.org/html/lab/Delhi_uploadfinal_sn.pdf

McKay, B. 2002. Pepsi, Coke take opposite tacks in bottled water marketing battle. *Wall Street Journal,* April 18. http://www.refrigeration-magazine.com/2002june1.htm

Mid-Day Mumbai. 2003. Is bottled drinking water safe? February 4.
    http://www.chalomumbai.com/news/city/2003/february/43698.htm

Mowen, J., and Minor, M. 2001. Chapter 4: Rebuilding the Perrier brand. In *Consumer Behavior: A Framework.* http://www.consumerbehavior.net/Cases/Case%204%20Perrier.doc

National Association for PET Container Resources. 2002. Final Report on post consumer PET container recycling activities. http://www.napcor.com/2002_report.pdf.

Nestle. 2003. Bottled water. http://www.nestle.com/html/brands/bottled.asp

Olson, E. 1999. Chapter 2. Bottled water: Pure drink or pure hype? Natural Resources Defense Council. http://www.nrdc.org/water/drinking/bw/chap2.asp

Pepsi/Aquafina. 2003. FAQs. http://www.aquafinaindia.com/faqs.aspx#w4

Pip, E. 2000. Survey of bottled drinking water available in Manitoba, Canada. *Environ Health Perspectives,* 108:863–866.
    http://ehpnet1.niehs.nih.gov/docs/2000/108p863-866pip/abstract.html

Pomento, J. 2001. The taste of water. *Aqueduct Magazine Cubed,* 1, issue 3, December.
http://www.mwdh2o.com/Aqueduct/dec2001/taste.htm

Posnick, L. M., and Kim, H., 2002. Bottled water regulation and the FDA. *Food Safety Magazine.*
(August/September). The Target Group. http://www.cfsan.fda.gov/~acrobat/botwatr.pdf

Procter and Gamble. 2004. Pûr water filtration system. January 13, 2004. www.purwater.com.

Suffet, I. H. 2000. Bottled water. University of California Los Angeles.
http://www.ioe.ucla.edu/publications/report01/BottledWater.htm

Torobin, M. G., Johnson, T. D., and Suffet, I. H. 1999. The importance of water quality aesthetics in
consumer confidence in the safety of drinking water supplies. Proceedings of AWWA Water
Conference. June, Section Tu–21, Paper 3. Chicago.

United States Pharmacopoeia. 2003. General information at www.usp.org

University of York. 2003. Green issues:benzene. November 6.
http://www.uyseg.org/greener_industry/pages/benzene/BenzeneGreen.htm

von Wiesenberger. 2003. Reading between the lines of bottled water labels.
http://www.bottledwaterweb.com/articles/avw-0002.htm.

West Michigan Environmental Action Council (WMEAC). 2003. Judge's ruling on Perrier water rights
case, press release. http://www.wmeac.org/haps/press/current/perrier.asp

World Bank. 2002. Gross national income estimates.
http://www.worldbank.org/data/databytopic/GNIPC.pdf

World Health Organization. 2000. Fact sheet no. 256, bottled drinking water. Geneva.
http://www.who.int/inf-fs/en/fact256.html

Water Privatization Principles and Practices

Meena Palaniappan, Peter H. Gleick, Catherine Hunt, Veena Srinivasan[1]

As the twenty-first century unfolds, billions of people are still struggling without access to the most basic water services—safe drinking water and adequate sanitation services. At the same time, national governments and local municipalities are struggling to find ways of providing these services, efficiently, effectively, and quickly. New ideas—good and bad—are being considered and implemented.

Among the most controversial of these new ideas is that water should be considered an "economic good," increasingly subject to the rules and power of markets, prices, multinational corporations, and international trading regimes. As described in the Water Brief on the March 2003 Kyoto World Water Forum, how to address the economic aspects of water while protecting the public interest is one of the most contentious challenges now facing the water community. In the last decade, various economic approaches for water have been put into practice in many different ways in many different places. Prices are being charged for water previously provided for free. Private water companies are taking over the management, operation, and sometimes even the ownership of previously public systems. The water industry is consolidating and evolving rapidly; one source indicates that the global water service market is a $260 billion market growing at 6 percent per year (Pictet 2001). In the 1990s, international development agencies that previously worked with governments to improve water services began pushing privatization efforts as part of financial restructuring activities.

In 2002, the Pacific Institute released a comprehensive assessment of the risks and benefits of water privatization (Gleick et al. 2002a), and we discussed these issues in two chapters of the last volume of *The World's Water*, also published in 2002. That work

---

1. This work builds on the earlier work of the Institute on the risks and benefits of water privatization: *The New Economy of Water* by Peter H. Gleick, Gary Wolff, Elizabeth Chalecki, and Rachel Reyes, published in February 2002. This study is available at www.pacinst.org/reports/new_economy.htm). We also thank Panha Chheng, Andrew Hill, Daniel Seamans, and Tracy Wates for early research on several of these case studies.

included a set of principles and rules to guide any proposed privatization efforts. These principles have received considerable attention: the *Financial Times Global Water Report* has called them "The Pacific Institute Principles." The World Bank requested briefings on the applicability of the principles in Latin America and as a tool to re-evaluate their own efforts. Discussions over the principles have been held with groups as diverse as local anti-privatization activists and the multinational water company Suez. The principles were also presented at the March 2003 Third World Water Forum in Tokyo, Japan.

This chapter presents short vignettes from around the world where public water systems, private water companies, or partnerships between the two have successfully met one or more of the Pacific Institute's principles. The good news is that we believe these principles are achievable for every water system around the world. The bad news is that we have not found any single case where all of these principles have yet been applied. A more comprehensive summary of these cases will be available from the Institute at www.pacinst.org.

Update on Privatization

Public water companies still provide around 95 percent of the water and wastewater services worldwide. But in the last 12 years, private management of water resources has grown tremendously. In 1990, private companies provided water to 51 million people. By 2002, the number of people whose water was provided under some sort of privatization agreement went up nearly sixfold to 300 million. The International Consortium of Investigative Journalists tracked six of the most active water companies and found that their scope had expanded from 12 countries in 1990 to over 56 countries by 2002 (CPI 2003a). According to the World Bank, the top five private sponsors of water and sewerage projects in developing countries from 1990 to 2001 accounted for 45 percent of the total projects in that category (World Bank and PPIAF 2002).

Privatization has been proposed as one solution to the numerous woes facing water utilities, including corruption, inefficiencies, and the lack of capital for needed service improvements and infrastructure maintenance. But privatization has not yet delivered on all of its promises. By the beginning of 2003, two highly publicized and lauded municipal privatization contracts had ended prematurely. In December 2002, Maynilad Water, owned by a subsidiary of the French water company Suez Lyonnaise and the Lopez family, announced it was pulling out of the long-term water concession contract it signed in 1997 to provide water to the western zone of the capital city of Manila in the Philippines (Waterwatch 2002). Within five years from when Maynilad and the other private operator in the eastern zone, Manila Water Company, took over the waterworks in 1997, the companies had connected about 2 million more customers (CPI 2003b). But when Maynilad Water was unable to recover costs as anticipated, the company decided to return the concession to the government. Arbiters now need to decide whether the government or Maynilad will pay penalties to the other party for the termination of the concession.

In January 2003, the city of Atlanta terminated its 20-year water services contract with United Water Services Atlanta, a subsidiary of Suez Lyonnaise (Cox 2003a). When the contract was signed in late 1998, it was the largest privatization of its kind in the United States. According to some analysts, the contract was cancelled for a number of reasons, including the company's failure to return the promised cost savings, evidence of poor service, improper billing, and water-quality problems (CPI 2003c).

At the same time, researchers were discovering (or rediscovering) that revitalization of water systems, through either public or private efforts, could bring better water quality and services and improve health. Galiani et al. (2002) found in their study of water privatization in Argentina that child mortality fell 5 to 7 percent in areas that privatized all of their water services. The improvement in access to water impacted the poor the most, with child mortality falling by 24 percent in the poorest areas. The same kinds of improvements happened when water systems were converted from private to public ownership in the southwestern United States in the late 1800s. In this case, public takeovers led to dramatic reductions (in some cases as great as 90 percent reductions) in water-related diseases, specifically typhoid among African Americans, who were twice as likely to get typhoid as white Americans. Newly formed public water companies at the turn of the century extended water mains to the most disenfranchised members of society, in this case African Americans, and greatly reduced water-borne disease rates (Troesken 2001).

The Buenos Aires water concession of Aguas Argentinas, a consortium that included Vivendi and Suez, is often cited as an example of successful privatization, with significant improvements including greater coverage, better service, more efficient operations, and lower prices for consumers. From 1992 to 2000, some 1.46 million people were provided with piped drinking water, and 583,000 more were provided access to sewerage (Alcazar et al. 2000). While the privatization effort was successful on many indicators, the government failed to ensure transparency and public access, which has led to dwindling public confidence over time. At the low point of the economic crisis of the late 1980s, 59 percent of Buenos Aires' residents favored privatization, and just 16 percent opposed it. Four years after the concession was put in place, those numbers had essentially reversed. According to a case study on the Buenos Aires Water Concession by The World Bank: "…public confidence in the process has eroded. The Buenos Aires concession shows how important transparent, rule-based decision making is to maintaining public trust in regulated infrastructure" (Alcazar et al. 2000).

Principles and Standards for Water

The importance of improving water supply and sanitation services is indisputable. But as these mixed examples show, no secret or universal formula for success has been identified. Ultimately, we believe the answer is not to focus on "public" or "private" water systems, but on developing water systems that meet clear standards and principles for equitable, efficient, and reliable operation and management.

Despite the growing opposition to water privatization, proposals for public-private partnerships in water supply and management seem likely to become more numerous in the future. There are many forms of water privatization, or public-private partnerships, making unilateral support for, or opposition to, privatization illogical. Figure 3.1 shows some of the range of systems involving private participation.

We believe that the primary responsibility for providing water and water services should rest with local communities and governments, and that efforts should be made to strengthen the ability of governments to meet water needs. As described in *The New Economy of Water* (Gleick et al. 2002a), the potential advantages of privatization are often

Option	Asset Ownership	Management	Capital Investment	Contract Duration
Management Contract	Public	Private	Public	2–5 years
Lease or Affermage	Public	Private	Public	8–20 years
Concession	Public	Private	Private	20–30 years
Divestiture	Private	Private	Private	Fixed or open term

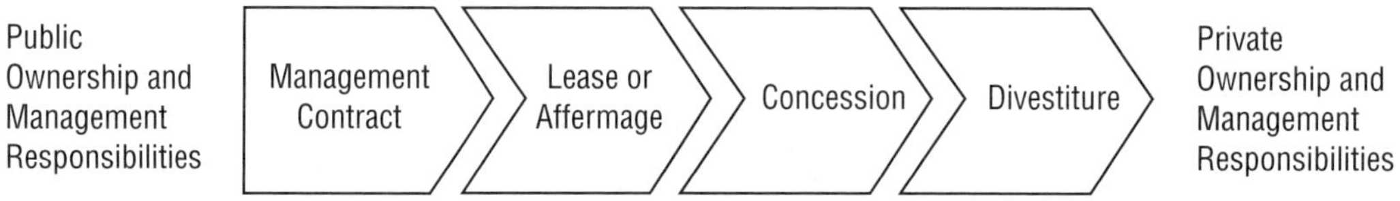

Public Ownership and Management Responsibilities → Management Contract ⟩ Lease or Affermage ⟩ Concession ⟩ Divestiture ⟩ → Private Ownership and Management Responsibilities

FIGURE 3.1   Unbundling the types of private participation.

greatest where governments have been weakest and failed to meet basic water needs. Where strong governments are able to provide water services effectively and equitably, the comparative advantages of privatization often decrease. Unfortunately, the greatest risks of privatization are also present when governments are weak—when they are unable to provide the oversight and management functions necessary to protect public interests. This contradiction poses the greatest challenge for those who hope to make privatization work successfully. Hence, the need for "principles."

Can the Principles Be Met?

Is it possible to design a private water system, or a public-private partnership, that can adequately meet the Pacific Institute Principles and protect the public good? On this, the jury is still out. Yet, demands for unmet water and wastewater services must be satisfied. Closing the sanitation and drinking water gap requires that more than 1 billion additional people be given access to safe and affordable drinking water, and an additional 2.4 billion or even more gain access to adequate sanitation services. Aging infrastructure in the developed and developing world will potentially require hundreds of billions of dollars to maintain and expand. These needs, coupled with the increasing pressures on limited water resources and increasing urban populations, create urgent challenges that demand resolution.

A wide range of different approaches has been taken to manage water resources around the world, from completely public systems to completely private ownership and management. No single management approach works all the time in all circumstances. The experience described here suggests, however, that there are important lessons to be learned from every approach, and these experiences can provide insights to water managers, local communities, and policymakers to improve future water management. This chapter describes the principles and give examples where some of them have been applied successfully.

Principle 1. Manage Water as a Social Good

Water is both a social and an economic good. While the application of economic tools is essential to the efficient and successful operation of water systems, social goods must continue to be protected, especially the goal of meeting basic human and ecological needs for water.

1.1. Meet basic human needs for water. All residents in a service area should be guaranteed a basic water quantity under any privatization agreement.

Basic water requirements for humans must be met as a top priority, whether water provision is under public or private control or management.

A diverse array of individuals, professional groups, private corporations, and public governmental and non-governmental interests have recently stepped up arguments that access to a basic water requirement is a fundamental goal, responsibility, and even human right (McCaffrey 1992, United Nations Development Programme 1998, Gleick 1996, 1999, 2000, see also Water Brief 4 in this book, The Human Right to Water). Both the 1977 Mar del Plata statement and the 1986 UN Right to Development set a goal of meeting "basic" needs.

Given resource limitations, ecological constraints, and economic and political factors, how much water is necessary to satisfy a basic human need? And how are institutions to provide it? Answers to these questions can only come from on-the-ground experience that informs international discussions. One such experience is the effort underway in Durban, South Africa, where basic human needs for water are being defined and provided, and where explicit attention is being given to ensuring that economic subsidies are carefully evaluated and applied only where necessary for reasons of extreme poverty.

Durban, South Africa

South Africa is a study of contrasts, with both a sophisticated and well-developed water system and millions of people who lack clean water and sanitation. This situation is largely the result of apartheid-era water policies that failed to provide basic water services to black Africans, with devastating consequences for health and well-being.

After apartheid, the new South African government set as a top priority a fundamental reallocation of water rights and water access for humans and the environment. Water provisions in the new Constitution and the new National Water Law have led to a dramatic shift in approaches and new attention to rural and marginal urban communities. One example of the new efforts to meet community needs is the water program of the Durban metropolitan Council and Durban Metro Water Services.

Durban is the busiest port in Africa and the economic center of the KwaZulu-Natal province in South Africa. Currently, the Durban metropolitan area has a population of around 3 million and is growing at a rate of approximately 2 percent per year. In recent years the area of the city has grown to include marginal and poor peri-urban neighborhoods and households, many with fewer urban services than the Durban average (Durban Metro 1999, Dahane and Vargas 2000).

Traditionally in South Africa, the municipality served as the service provider as well as the water authority. As a result of the new 1998 Water Act, some regional water authorities have sought to enter into contracts with private companies or public utilities to outsource the water provision or water-services management responsibilities. In response to the problems with managing water service in rural areas, a limited public-private partnership (PPP) for the Durban area was set up in March 1999 with the South African Water Research Commission, Umgeni Water, Mvula Trust (an NGO serving as a community liaison), and Vivendi Universal, with the Durban Metropolitan Council providing overall oversight.

Major objectives of the PPP were to improve the level of water service for customers in newly incorporated low-income areas not yet receiving coverage and to improve water-system management. As part of the partnership, Vivendi implemented electronic metering systems to manage the exploitation of the free basic water system and to improve efficiency. In turn, these systems allow the water supplier to better understand consumer water needs. Actual water provision remains a public responsibility (The World Bank 2001).

Even before the implementation of the partnership, problems materialized with the cost of administering the monthly billing and payments for the poorer populations connected to ground tanks and water manifolds, a system to control delivery of water. In 1997, non-payment for basic services—a holdover from the days of apartheid— remained rampant, and ethical questions were being raised about cutting off services for non-payment.

A decision was taken by the Metropolitan Council to reduce these problems by providing basic water free to consumers using ground tanks, to a limit of 6,000 liters per household per month. A year later, this free water allocation was extended to all domestic consumers in Durban by the introduction of a stepped tariff. The 6,000 liters a month is calculated to enable the poorest of the poor to have a free water supply of 25 liters per person per day in a household of eight. This subsidy (around Rand 30 million per year) is to be covered by consumers in a new tariff system (Ingegneria Senza Frontiere 2002). By providing a free allotment, the water agency gained a greater authority to implement unpopular disconnections, since disconnections only applied when payments were withheld for quantities of water above the guaranteed free allotment (Ingegneria Senza Frontiere 2002).

While concerns remain whether 25 liters per person per day is adequate to support a large family or home gardening, the PPP has been effective at improving service. Areas that previously lacked service are now being reached; practically 100 percent of poor households in the area now receive free basic water service, and the principle has been extended nationwide. According to Guy Preston, a water advisor to the Minister for Water Affairs and Forestry, the free basic water program is working well. Preston said people who did not pay water bills in Durban were still given the 6,000 liters a month free, but were put on a system that allows them no more than the basic amount. Better tracking of water use and losses has permitted the city to keep water consumption to the level it was in 1996 despite extension of service to many more people. This in turn has permitted them to delay construction of a dam that was planned on the Mooi River at a cost of Rand 300 million (Gosling 2001).

Based on the success of the basic water supply policy in Durban and other South African cities, in February 2001, South Africa's Minister of Water Affairs and Forestry announced that all households would be guaranteed a basic supply of water free of charge. The South African Cabinet approved this policy based on the principle of meeting basic human needs (SADWAF 2001). Soon afterwards, national guidelines were developed that help the local authorities to implement the free basic water service and national funding was provided to help ensure that local communities could cover the subsidies required. In the 2003/2004 fiscal year, as many as 27 million South Africans—nearly 60 percent of the population—were estimated to be receiving the free water allocation (SADWAF 2003).

1.2. Meet basic ecosystem needs for water. Natural ecosystems should be guaranteed a basic water requirement under any privatization agreement.

Basic water-supply protections for natural ecosystems must be put in place in every region of the world. This includes protection of watersheds that provide water for urban needs.

Along with the desire to meet basic human needs for water comes the need to provide natural ecosystems with their basic requirements. All over the world natural ecosystems are at risk of degradation and destruction because of the way water resources are managed and used to meet human needs. Excessive water withdrawals, land mismanagement, and water pollution have led to the loss of entire species of fish, plants, and animals, and even ecosystem collapse. A fundamental principle for sustainable long-term water management, whether public or private, is that natural ecosystems should be guaranteed sufficient water to protect basic health and functions.

This principle has received increased attention in the past few years as studies have begun to define and quantify minimum ecosystem water needs. On an institutional level, more regions are beginning to consider these water requirements. For example, the constitution of South Africa acknowledges both human and ecosystem water needs and explicitly calls for the provision of minimum water commitments. More western United States watersheds are evaluating and restoring basic flows to protect commercial fisheries and threatened and endangered species. Any water management agreement, whether public or private, should include explicit protections of environmental water needs, enforced by government oversight. Recent experience in New York City provides an example of fundamental protection of the watershed to protect drinking water.

New York, USA

New York City has implemented a watershed program that permits comprehensive land management designed to protect both the source of the city's water and overall water quality. The city has one of the finest public water systems in the world, serving 9 million residents with freshwater supply and wastewater services. New York City's water, derived from the Catskill, Delaware, and Croton watersheds, is under the public jurisdiction of the city's Department of Environmental Protection (DEP). The watersheds' resources provide an estimated 7.6 billion liters of water per day to city residents.

The combined system spans 5,100 square kilometers from the city into upstate New York (NYDEP 2002b, National Research Council 1999).

Over the past few years, New York has led a highly successful effort to manage watershed lands in order to protect ecosystems and water quality as an alternative to building large and expensive new water-treatment facilities. In the mid-1990s, new concerns about contamination from *Giardia* and *Cryptosporidium*, chlorine-resistant contaminants, and other pollutants in drinking water around the United States prompted the federal Environmental Protection Administration (EPA) to release the 1989 Surface Water Treatment Rule (SWTR) mandating mechanical filtration of all surface water supplies (NRC 2000). Mechanical filtration greatly reduces the risk that these contaminants will enter drinking water supplies, but requires very expensive capital equipment at a time when municipalities are often strapped for finances.

The SWTR permitted exceptions for municipalities that could employ other methods that would produce comparable water quality. In lieu of constructing a filtration plant that would have cost city residents $6 to $8 billion, New York elected to design and implement an innovative Watershed Protection Plan to protect water quality. In 1997, the EPA granted New York City a waiver, which allowed the city to undertake a broad-based initiative involving governmental and non-governmental groups in both the city and upstream watershed communities.

Relationships between the city and watershed communities have not always been friendly (NRC 2000). In order to address concerns about economic prosperity, commercial and industrial development, and local property values, the New York City DEP launched programs to "change the landscape from hostility and destruction to one of trust, respect, and credibility" (DEP Commissioner M. Gelber quoted in Calhoun 1997). The culmination of months of negotiations between upstate property owners and city officials resulted in a formal Memorandum of Agreement (USEPA 1997b) to keep contamination out of the water supply in the first place by working within the watershed to establish best farming practices, land-use changes, and new construction polices. These efforts were designed to reduce or eliminate the introduction of pollutants (i.e., pesticides, animal fecal matter, etc.) into downstream water. The program also included efforts to improve wastewater treatment plants in the upper watershed, leading to better overall regional water quality.

This approach saved the city $4 to $6 billion and brought many of the watershed's stakeholders into a collaborative process that both respected watershed landowners' economic well-being and protected the city's water quality. The city's watershed protection strategy has permitted the city to continue to provide high-quality potable water to 9 million residents without the need for massive new mechanical infrastructure. The success of the land management and wastewater treatment upgrade programs resulted in a renewal of the EPA waiver in June 2002, contingent on continued improvements in wastewater treatment plants, repair of septic systems, new land acquisition around reservoirs, and construction by 2009 of an ultraviolet treatment plant for city water (*US Water News 2002*, WAC 2001, United Nations Habitat 2001).

The city's system stands out as a model case of publicly controlled water and sewage systems. The city has succeeded in combining the interests of the region's stakeholders into an environmentally sound and economically advantageous system. New York City's watershed protection plan acts as the keystone in the city's success. Albert Appleton, former DEP Commissioner, touts the city's multi-faceted success:

"What you have accomplished is not just a great example of how to create a win-win partnership between agriculture and the environment, or between upstate rural and downstate urban interests. It is a marvelous example of why democracy is a good idea, of why citizens and communities are things to be cherished, and of why family farm communities are rightfully recognized as particularly embodying those hopes and values" (Appleton 2001).

1.3 The basic water requirement for users should be provided at subsidized rates when necessary for reasons of poverty.

Subsidies should not be encouraged blindly, but some subsidies for specific groups of people or industries are occasionally justified. One example is subsidies for meeting basic water requirements when a minimum amount of water is unaffordable due to poverty.

Details of this principle and its application can be found in the summary of Principle 1.1 and those related to economics and pricing.

Principle 2 Use Sound Economics in Water Management

A second set of principles for privatization involves ensuring that the economic aspects of water are balanced with the social goods aspects. Water needs to be affordable for people while at the same time making certain that water agencies are financially healthy and able to expand service coverage and improve quality. Decisions around water supply and management also need to identify the most cost-effective method of providing water, which is no longer limited to traditional expansion of supply options. The first three principles in this section address the need to provide water at fair and reasonable rates and to develop appropriate subsidies to provide water to the poor. The last principle identifies the need for water companies to consider the role of conservation, efficiency, and innovative supply in providing needed water.

2.1 Water and water services should be provided at fair and reasonable rates.

Provision of water and water services should not be free. Appropriate subsidies should be evaluated and discussed in public forums. Rates should be designed to encourage efficient and effective use of water.

2.2 Whenever possible, link proposed rate increases with agreed-upon improvements in service.

Experience has shown that water users are often willing to pay for improvements in service when such improvements are designed with their participation and when improvements are actually delivered. Even when rate increases are primarily motivated by cost increases, linking the rate increase to improvements in service creates a performance incentive for the water supplier and increases the value of water and water services to users. Ideally, consumers and residents should be involved in the water service and rate decisions that affect them. Consumers are in the best position to weigh service improvements against rate increases to determine the best strategy that meets their needs and fits their budgets.

The issue of rate increases is often one of the most controversial issues around privatization. As the world learned during the upheaval over rate increases after water privatization in Cochabamba, Bolivia, rapid and large increases in water rates can cause strong social and political reactions. Public protests and political demonstrations over price increases have also taken place in Tucuman, Argentina; Puerto Rico, United States; Johannesburg, South Africa; and elsewhere.

The actual record of rate increases under privatization is mixed—both rate increases and rate decreases have occurred. In cases where rate changes need to be made, improved services should be clearly described and rate changes should be tied to comprehensive consumer education and information programs describing the changes and the reasons. Consumers are in the best position to weigh service improvements against rate increases to determine the best strategy that meets their needs and fits their budgets. Such a demand-driven planning approach has been termed a "Neighborhood Deal" by some researchers (Whittington et al. 2002). This "Neighborhood Deal" would work with communities and consumers to design services that fulfill their needs and are financially attractive, technically feasible, and affordable to local governments or other financiers. This improves public health, along with increasing the revenues available to finance more connections.

Tegucigalpa, Honduras

Tegucigalpa, the capital city of Honduras, has experienced rapid growth over the last 20 years, and more than half of the population lives in settlements on the outskirts of the city. Provision of water and sanitary services has been especially difficult because of the local geography: the settlements tend to be located on steep hillsides, very little surface water is available, and the groundwater is for the most part either polluted or too deep. The fact that these settlements are at much higher elevations than the central city makes it prohibitively expensive to extend the city water system to them. Another problem is that the soil does not percolate well enough to allow for the use of septic tanks.

A United Nations sponsored program offered peri-urban populations flexibility and choice in determining whether to apply for loans to implement water and wastewater projects in their homes and communities and which technologies best suited their needs and budgets. Before the interventions described here, the primary source of water for most *barrio marginales* (peri-urban settlement) residents was private vendors, whose prices were 10 times higher than the official prices paid by those in wealthier neighborhoods with connections to the public water system. Eighty percent of families spent between 11 percent and 20 percent of their monthly income on water (Torres 1996, United Nations Habitat 2002). Another source of water was collected rainwater, often kept in drums previously used to store toxic materials such as pesticides.

In 1987 UNICEF, together with SANAA, the National Autonomous Water and Sewage Authority, began a program to supply water and sewerage to the approximately 225 peri-urban communities that surround Tegucigalpa proper (called Executive Unit for Settlements in Development (EUBD). The Cooperative Housing Foundation (CHF), together with UNICEF and local non-government organizations (NGOs), also initiated a program of loans for sanitary services in 1991 (the Urban Family Sanitation Program). Both of these programs offer community-driven solutions to the water and sanitation problems of the *barrios marginales.*

To join the program, a community needs to formally apply for assistance from EUBD, outlining its contributions of labor and materials, a plan for loan repayment through collection of water tariffs, and proof of land tenure. The community retains ownership of the distribution system that it constructs with the assistance of EUBD.

The community must establish a Water Board (*Juntas de Agua*) to collect tariffs, administer the water system, and provide operation and simple maintenance. The water boards usually hire one plumber and one administrator, whose salaries are paid from the water tariffs, while the elected board members volunteer their time. The community members involved in the organizing efforts are often women, and women fill about one-third of the positions on the water boards, often as committee presidents or financial controllers (Metell 1998).

The financial framework for both water supply and sewerage projects is the same, with the communities repaying (without interest) the revolving funds for investment in both materials and technical assistance. All communities need to repay the loans within seven years, and some have done so in two or three years. After the project has been completed, the EUBD offers periodic follow-up and support, usually in accounting and funds management. This follow-up has proven to be essential to the long-term success of the programs (Torres 1996).

The water-supply alternatives available to each community vary according to the local geography, including drilling wells connected to elevated tanks using SANAA tankers to bring water to the neighborhood and distributing water through a central cistern. When possible, the communities make connections to the SANAA water network. In the case of sewerage systems, EUBD also is flexible in choosing which type of system best accommodates each community (Mooijman 1998).

The complementary CHF program in Tegucigalpa is a finance and education program that works with individual families to improve sanitation. The program employs promoters, generally from the community, who raise awareness of sanitation issues in general and of the availability of loans from the program. When a family expresses interest in the program, they complete a loan application and receive an on-site consultation from a technician. At this point, the family is made aware of a menu of possible improvements. The CHF program will finance whatever project the family decides to undertake, so long as it is technically and fiscally feasible. After the project is completed, the promoter makes several follow-up visits in order to make sure that the improvements are properly used and maintained, and to make any necessary clarifications about the loan repayment.

The water component of the EUBD, in place since 1987, has been a proven success. By the end of 1997, more than 150,000 persons in 95 communities had benefited from some aspects of the water program, leaving only 20 urban communities without a water system (Mooijman 1998). Clients with in-house connections from EUBD save a significant amount on water services: a UNICEF evaluation of the program showed that for each dollar invested by individual families, the families saved $16.60 in water expenses (Aasen 1994).

The Tegucigalpa model was successful for a number of reasons. People are more likely to use and maintain a service that they choose and help design and plan. People are also more likely to pay for a service when they have a chance to agree to its costs. The ability of families to choose from a menu of possible improvements makes it more likely that the alternative chosen will be best suited to their needs and budgets.

2.3 Subsidies, if necessary, should be economically and socially sound.

Subsidies are not all equal from an economic point of view. For example, subsidies to low-income users that do not reduce the price of water are more appropriate than those that do because lower water prices encourage inefficient water use. Similarly, mechanisms should be instituted to regularly review and eliminate subsidies that no longer serve an appropriate social purpose.

Water and wastewater utilities need to be financially sustainable in order to meet the needs of growing populations in urban centers throughout the world. To fill the water and sanitation gap in the developing world, these utilities need to expand service coverage to the poor and underserved, maintain existing and build new infrastructure, provide reliable service, and ensure water quality. Setting appropriate water rates is an important step toward assuring the financial stability and sustainability of the water and wastewater utility so that it can meet current and long-term needs of consumers.

The need for utilities to be financially sustainable should be balanced with social and cultural requirements related to water. Water rates must play a variety of roles. Tariffs charged for water should reflect the cost of service, allow water and sanitation coverage for all, promote conservation, and be transparent. Several of the cases studied include a discussion of how to design rate structures for water service. In England and Wales, for example, serious opposition to privatization arose after 1989 when rates began to rise substantially. Eventually, strong government regulatory oversight helped stabilize rates and link rate increases with improvements in service. Similarly, approaches to rate designs in Bolivia, South Africa, and Chile demonstrate different ways of addressing issues of equity and fairness.

Increasing Block Tariffs: La Paz/El Alto, Bolivia

One approach to rate equity is the use of increasing block rates, where the charge per unit volume goes up in increments as the volume of water used increases (Mitchell et al. 1994). Increasing block rates can promote customer conservation, because customers who can consume less are rewarded by a lower unit (and total) cost. In La Paz/El Alto, Bolivia, a progressive rate structure was developed that subsidized low-volume residential users and imposed an increasing four-block rate—the more water used, the higher the tariff. Industrial customers pay a single rate, equal to the long-run marginal cost. Two tiers were set for commercial users (see Table 3.1). The potential problem with this model is that most households in El Alto use much less than 30 cubic meters per month (the lowest block rate), and the sale of water to these households, because it is subsidized, is an economic cost to the service provider. Thus, the service provider has an incentive to serve industrial, commercial, and high-volume residential customers before low-volume residential customers (Komives 1999, 2001).

In other cases, increasing block tariffs may have unintended effects on the poor. For example, when low-income families band together to purchase water to save on connection costs, they may fall in a higher tariff bracket and end up with an unaffordable water bill (Wegelin-Shuringa 1998).

Critics have also pointed out that cross-subsidies inherent in the rising block tariff rely on "subsidizing" customers to remain in the system to provide revenue to support the "subsidized" customers. If the tariff for the subsidizing customers is too high, they may choose to exit the system altogether, which can further worsen the financial

TABLE 3.1   Tariff Structure for Aguas del Illimani, Bolivia (US\$/m^3)

Tariff	Residential	Commercial	Industrial
0.2214	1 to 30 m^3		
0.4428	31 to 150 m^3		
0.6642	151 to 300 m^3	1 to 20 m^3	
1.1862	Above 300 m^3	Above 20 m^3	All water

Notes: 99 percent of all residential customers use less than 150 m^3 per month. The long-run marginal cost is estimated at \$1.18 per m^3.
Source: Komives 2001.

condition of the utility (Yepes 1999), though this is difficult with the typical monopoly situation of water provision.

Water Stamps: Santiago, Chile

In Santiago, Chile, an innovative "water stamps" scheme provides subsidies that are effective at targeting the poor, without reducing the price of water or leading to inefficient use of water. Introduced in the early 1990s, water stamps in Santiago cover part of the cost of water purchases for the poorest residents. Until the late 1980s, Chile had used a cross-subsidy program to address the needs of the poor, but the water utility was not recovering the costs of providing water service, and was having difficulties extending service to rapidly growing peri-urban populations.

In Empresa Metropolitana de Obras Sanitarias (EMOS), the public water utility serving the capital city of Santiago, water rates covered less than half of the cost of providing water service. Although EMOS records reported water coverage of 98 percent and sewage coverage of 88 percent in 1985, this figure did not include informal settlements in the outskirts of the city. At the same time, the utility's finances were deteriorating and it was not in a position to extend coverage (EMOS 1995, Gomez-Lobo 2001).

In 1988, Chile embarked on reform of its water sector, including partial privatization and the introduction of a new tariff scheme (Serra 2001). The country was divided into pricing zones that grouped together areas with relatively homogeneous costs. The tariffs were changed to more accurately reflect the cost of water supply in each zone. When Santiago privatized its water system, tariffs went up by 90 percent in four years. At the same time, a subsidy scheme was introduced targeted to the poorest section of the population. The objective of the subsidy scheme was to meet the WHO standard that no family should spend more than 5 percent of its income on water and sanitation services.

Eligible customers are issued "water stamps" for anywhere between 25 to 80 percent of their water bill, depending on the tariff in that area and their need. The maximum amount of water eligible for subsidy was set at 15 m^3, per month per household. The subsidy is allocated from the central government budget to state governments, which in turn distribute the funds among municipalities. The municipal government reimburses the utility directly for the value of the water stamps once the utility submits proof that the customer has paid his or her part of the bill. By the end of 1998, nearly 100 percent of the population had running water and 17.4 percent received subsidies. The number of customers with service increased from about 700,000 in 1985 to over a million in 1997 to include the previously unserved peri-urban settlements. The rate of return on assets increased considerably in the same period from 3.6 percent in 1988 to 11 percent in 1996 (Gomez-Lobo 2001, 2003).

After including subsidies, the average water bill of the poorest tenth of the population was about 8 percent of their average income, in excess of the WHO standard but still an improvement over unserved customers, who paid as much as 15 percent of their income to private vendors. EMOS hopes to reduce this percentage further. By the late 1990s, around 450,000 customers representing 95 percent of the target population were using water stamps. About 77 percent of the subsidy went to the poorest section of the population while about 23 percent "leaked" to moderate and higher-income customers (EMOS 1995, Serra 2001, Gomez-Lobo 2003). Moreover, 99 percent of the urban population was estimated to have access to water services.

Water stamps in Santiago provide an innovative subsidy that goes directly to the poor, does not reduce the incentive of water agencies to expand services to poorer populations, and does not inordinately place the burden of expanding the system on existing or new customers.

2.4 Private companies should be required to demonstrate that new water-supply projects are less expensive than projects to improve water conservation and water-use efficiency before they are permitted to invest and raise water rates to repay the investment.

Privatization agreements should not permit new water-supply projects unless such projects can be proven to be less costly than improving the efficiency of existing water distribution and use. When considered seriously, water-efficiency investments can earn an equal or higher rate of return than that earned by new water-supply investments. Rate structures should permit companies to earn a return on efficiency and conservation investments.

There are vast gains to be made in efficiency and conservation improvements in the water sector. Many of these gains require a shift from a traditional engineering-driven approach to a social-driven approach that focuses on educating and empowering water users with tools to be more effective consumers of water. Demand management takes a flexible approach to meet the water-related needs of populations—for example, for personal hygiene, food, and drinking water—rather than merely supplying water. When a water utility forecasts demand for water-supply needs, these projections should include recognition of the various approaches to meet demand, including the ways in which efficiency can meet that demand. One example of demand management is reduction of water losses in the water system, as is well demonstrated in Singapore.

Singapore

Efficiency has proven to be a cost-effective way to meet the water needs of the island nation of Singapore. The city has no natural rivers or lakes, and depends upon rainwater stored in reservoirs and water imported from Malaysia for its water supply. The Singapore Public Utilities Board (PUB) serves the 2.8 million residents in the island nation of Singapore. PUB has often been described as a model public utility with high accountability, well-trained staff (staff ratio is low at 2.2 per 1,000 connections) and extremely high-quality service.

With no rivers or lakes to tap for fresh water, Singapore's only indigenous source of water is rainfall collected in its 14 reservoirs. The country relies on imports from Malaysia and in the future intends to increase reliance on desalination and reclaimed water. Because of its dependence on outside sources, the utility has made an effort in recent years to improve its efficiency and to reduce waste, especially in the area of unaccounted for water (UfW).

Unaccounted for water (UfW) is the difference between the water delivered to the distribution system and the water sold. UfW has two basic components: physical losses, such as water lost from pipes and overflows from tanks; and commercial losses, which include water used but not paid for. Commercial losses can include illegal and unmetered connections, faulty meters, and poor billing and collection systems. In both cases, the result is a loss of revenues. UfW percentages vary considerably from as high as 60 percent (e.g., Manila in 1990) or as low as 6 percent (e.g., Singapore, Netherlands). Data Table 18 in Gleick et al. (2002b) lists unaccounted for water for a wide range of cities and countries.

Developing country cities can have high rates of unaccounted for water, sometimes as high as 40 to 60 percent of the water produced. This represents a huge economic loss for utilities in these countries. The World Bank estimates that avoidable water losses in developing countries are one-quarter of the total water supply. Some economists estimate that the total annual cost of loss of water is $4 billion to developing country governments (De Moor 1997). Some cities control water losses by keeping water in the system for very short periods of time. In Madras, India, water is provided in the system for 2 hours per day at low pressure. If this were to be increased to 12 hours per day at higher pressure, water losses would be higher than the total current daily supply in Madras (Briscoe, no date).

Through a consistent monitoring program, Singapore has achieved an UfW of only six percent—an impressively low rate. The main features of Singapore's UfW reduction program include:

Metering and Monitoring

- Achieve universal metering with high meter accuracy on all meters.
- Replace domestic meters every seven years and industrial meters every four years.
- Measure and bill for water used for firefighting.
- Identify low- or high-consumption patterns and notify customers of excessive consumption.
- Identify inconsistent meter readings and promptly replace faulty meters.

Leak Detection

- Test the entire system annually and check over-the-surface pipes three times a year. Troublesome spots are hydraulically isolated and checked more frequently.
- Line pipes in the distribution system with concrete to prevent corrosion.
- Replace pipes if they report more than 3 breaks a year. As a result of this program, the number of pipe breaks has decreased from 12 per 100 km/ year in 1985 to fewer than 4 breaks per 100 km/year in the mid-1990s.

Applying these approaches, Singapore was able to reduce unaccounted for water from 10 percent in 1989 to 6 percent in 1995. The annualized cost of these investments over a 30-year period works out to S$ (Singapore Dollars) 5.8 million and the program has proven to be cost-effective. Overall, the efficiency programs and efforts to reduce lost water have permitted Singapore to reduce investments in desalination plants and have also given them flexibility in their negotiations with Malaysia over water supply.

Principle 3. Maintain Strong Government Regulation and Public Oversight

Local communities and governments should have the final responsibility and authority for providing water and water services. Governments must have the power and the resources to effectively regulate water-service providers. Ultimately, the success of public-private partnerships depends upon the strong role of governments in protecting public health, water users, and the water resource itself.

3.1. Governments should retain or establish public ownership or control of water sources.

The "social good" dimensions of water cannot be fully protected if ownership of water sources is entirely private. Permanent and unequivocal public ownership of water sources gives the public the strongest single point of leverage in ensuring that an acceptable balance between social and economic concerns is achieved.

Some of the private water companies insist that their goal is not to own water or water infrastructure, but to provide services. For example, the Gerard Mestrallet (CEO of Suez/Ondeo) is on record as saying that

> "I think we should be able to agree that water is a common good, one of the basic public goods. At Suez, we are opposed to the private ownership of water resources precisely because, in our eyes, water is not a commodity. We do not trade in water. We do not sell a product. We provide a service" (Mestrallet 2002).

Such a philosophy should be encouraged. Nevertheless, most private water companies do own water rights themselves. This remains a serious concern. While some privatization contracts and proposals do not lead to any formal change in water rights, a growing number either intentionally or unintentionally change the status quo. Some even explicitly transfer ownership of water resources from public to private entities. For example, the Edwards Aquifer Authority in the central United States has considered selling water rights for either a limited period of time (e.g., one year) or in perpetuity (EAA 2001) (Box 3.1). Granting perpetual withdrawal rights would reduce the public's ability to ensure that the aquifer is managed as a social good.

Despite numerous legal challenges, these and other actions to establish public ownership of underground water in Texas have been upheld. Most strikingly, the

Box 3.1 Establishing Public Property Rights for *In-situ* Water

The Edwards Aquifer of South Central Texas is the sole source of drinking water for 1.5 million people in parts of eight counties, including all of San Antonio, the ninth largest city in the nation (according to the 2000 United States Census—www.census.gov). The aquifer provides 300 million cubic meters of irrigation water annually for about 34,000 hectares of agricultural land. It also supports an extremely diverse wildlife population in surface springs and underground. At least nine endangered species rely on spring-flows for their survival; baseflow in the Guadalupe and San Antonio rivers depends in part on the aquifer; and its subterranean aquatic ecosystem is believed to be the most diverse in the world.

Historically, Texas law granted complete ownership of groundwater to the landowner above it, unless the groundwater is flowing in an underground stream or river, in which case the laws governing surface water apply. This common law rule was replaced long ago in most other U.S. states. Several serious droughts (1984 and 1996), legal decisions to enforce the Endangered Species Act (between 1990 and 1996), and citizen action that raised public understanding of the importance of the aquifer, led the Texas legislature to gradually impose public control over (and hence partial public ownership of) water in this and other aquifers in Texas. In 1993, the Texas Legislature created an Edwards Aquifer Authority to limit water pumping, penalize violators, issue permits, control the transfer of water rights, and institute water-quality programs.

Supreme Court of Texas rejected a claim that action creating the Edwards Aquifer Authority deprived landowners of a property right vested to them by the Texas Constitution. Establishment of the Edwards Aquifer Authority is an excellent example of the type of changes in property rights and rules that are necessary if water is to be managed effectively as both a social and an economic good (EAA 2001). However, the existence of such public bodies does not ensure sound water management. The Edwards Aquifer Authority itself has allowed some water rights holders to sell those rights in perpetuity (www.edwardswater.com), thereby reducing the public's ability to ensure that future water from the aquifer is managed as a social good. Full implementation of public ownership of water at the source requires that ownership cannot be permanently transferred to private hands.

Changes in access and water rights may also occur without explicit agreement. One of the causes of tensions in Bolivia over the proposal to privatize the water systems in Cochabamba was an effort to restrict unmonitored groundwater pumping by rural water users and to bring them into the private system. While this may make sense from a purely economic and efficiency perspective, it imposed a fundamental change in the historical use rights in the region.

Another challenge associated with privatization is the degree to which the process of privatization leads to the transfer of government or public assets into the hands of those who are friends of government, or already wealthy. When privatization results in a redistribution of wealth in an inequitable way, there will be strong pressure to oppose or cancel reforms. Confidence in the fairness of the process, in turn, depends on both the design and the transparency of the rules and legal system (Yergin and Stanislaw 1999).

3.2 Public agencies and water-service providers should monitor water quality. Governments should define and enforce water-quality laws.

Water suppliers cannot effectively regulate water quality. Although this point has been recognized in many privatization decisions, government water-quality regulators are often under-informed and under-funded, leaving public decisions about water quality in private hands. Governments should define and enforce laws and regulations. Government agencies or independent watchdogs should monitor, and publish information on, water quality. Where governments are weak, formal and explicit mechanisms to protect water quality must be even stronger.

Private suppliers of water have few economic incentives to address long-term (chronic) health problems associated with low levels of some pollutants. In addition, private water suppliers have an incentive to understate or misrepresent to customers the size and potential impacts of problems that do occur. As a result, there is widespread agreement that maintaining strong regulatory oversight is a necessary component of protecting water quality. Concerns about the ability of private water providers to protect water quality led the National Council of Women of Canada, a non-partisan federation of organizations, to adopt a policy in 1997 of opposition to the privatization of water purification and distribution systems (NCWC 1997). The Water Environment Federation in the United States supports "national policy to encourage public/private partnerships (privatization)" but with appropriate public oversight (WEF 2000).

When strong regulatory oversight exists, privatization can lead to improvements in water quality. For example, Standard and Poor's notes that water and wastewater quality have improved in the United Kingdom after water privatization (S&P 2000). Indeed, prior to privatization, there was a distinct reluctance of government agencies to monitor and fine other government water providers who were violating water-quality standards—a classic conflict of interest. Governments that own, operate, and finance water and wastewater utilities have shown that they cannot also always properly regulate them. Privatization has the potential to reduce those conflicts and permit governments to regulate. In the United Kingdom, government regulators have greatly increased their successful prosecutions for violations (Orwin 1999). But as we see in the following case study, these opportunities for regulating water suppliers can only be captured if there is strong, well-funded, and well-defined regulatory oversight.

United Kingdom

The principle of the need and value of strong regulatory oversight has been put into practice in the United Kingdom after the initial efforts to privatize the water system in 1989 ran into problems. In particular, the lack of clear oversight protecting the public interest became apparent with rapid rate increases, inadequate water-quality monitoring, and concerns about customer service and environmental protection. These problems led to significant improvements in oversight by three government agencies. Now, while problems and concerns still remain, the value of strong government oversight and regulation in the water sectors in England and Wales has been reinforced and re-emphasized.

The water systems of England and Wales were privatized as part of a wave of government-promoted privatization efforts in the late 1980s. The acts gave the privatized water companies 25-year licenses for water and sanitation services. The form of competition was initially set forth via "yardstick competition," in which the economic regulator, The Office of Water Services (OFWAT), was given the mandate to determine consumer tariffs based on company performance in, for example, water quality and leakage rates. Under the initial licenses, the operators had exclusive rights to their area of service. Since then, two forms of limited competition have been permitted: 1) inset appointments (whereby an external water and/or sewerage supplier can provide services within another operator's service area), and 2) common-carriage permits (whereby two operators share the same infrastructure assets, such as water pipes). There was no formal public consultation at the time the industry was privatized despite (or because of) polls that suggested that 75 percent of the public did not support privatization (Saunders and Harris 1990). Since 1997, the new Labor government has made an effort to widen public participation around the review of the licenses, operator performance, and customer services.

As a result of the initial structure and form of privatization, a variety of problems materialized early that led to changes, modifications, and revisions in the government agencies responsible for oversight, customer protection, and regulation. This section is not the place for a comprehensive review of the British privatization experience. Several such reviews are already available (see, for example, Lobina and Hall 2001, Green 2003 and all of the annual OFWAT reports). Nevertheless, we offer a summary of the most relevant issues that arose and the responses by public agencies. Among the problems:

- Tariffs rose sharply following privatization, necessitated by investments in water-system improvements, with little public input. The different regulatory authorities with different mandates sent conflicting signals to the water companies.

- Public opinion was divided on how much should be spent on environment protection.

- The rise in tariffs led to an increase in costs and an inability on the part of poorer customers to pay, which in turn led to disconnections, drawing widespread public criticism.

- There was public anger over the fact that water companies were continuing to earn substantial profits even in drought years, when drought measures, such as consumption restrictions, had been imposed on the public.

A 1996 study by The Save the Children Fund showed that 70 percent of low-income customers were taking health-endangering measures to reduce consumption, such as flushing less frequently, sharing baths, and washing clothes less often. It concluded that vulnerable groups could not make any further reductions in household water consumption without eliminating essential uses of water. Another study by the British Medical

Association correlated the rise in dysentery rates with water disconnections (BMA 1994). These studies served to consolidate the negative public image of water companies.

As initially designed, water privatization gave inadequate attention to the public and social good aspects of water, leading to rapid and vociferous opposition from citizens and calls for redesign and restructuring of the water systems. Reforms began to be put in place in the early 1990s, with a focus on improving public participation and input and on monitoring and regulating certain aspects of the business. In 1997, a new Labor government was voted into power. Because it had little incentive to defend existing privatization arrangements, further regulatory oversight was implemented.

After considerable reworking in the 1990s, the British developed a more effective system of regulation that minimized jurisdictional overlap by separating environmental, economic and social, and drinking-water quality into three independent bodies. The National Rivers Authority had the responsibility for environmental quality and protection until 1996, after which it became the responsibility of the Environment Agency (EA). The Office of Water Services (OFWAT) was created to ensure that asset investment is appropriate, that costs to the customer are appropriate and fair, and that certain customer service standards are maintained. The Drinking Water Inspectorate (DWI) safeguards drinking water quality. Each of the responsible agencies has worked to minimize overlapping duties and close significant gaps in regulatory roles.

These agencies could be considered to be reasonably free from political interference and are independent of government funding. The water resource management activities of EA are funded from abstraction license charges. OFWAT is funded from a charge on each customer's bill.

OFWAT is the major agency responsible for monitoring water system costs and customer service across all of the water companies. OFWAT sets prices using a price-cap mechanism and "yardstick competition." The price-cap mechanism allows the companies to keep profits from efficiency gains for five years until the next tariff cycle. Comparing the performance of the different companies for indicators of efficiency and customer service allows OFWAT to reward superior performers. Profit levels in the United Kingdom were higher than the public was comfortable with during the first five years of privatization, but came down to more reasonable levels as OFWAT limited the maximum rate companies could charge consumers.

Another new development has been the rise of a category of agents who act as representatives of the public but participate in setting policy. The OFWAT Customer Service Committee (CSC) is one such example. Consumers in each of the 10 water and sewerage service areas are represented in committees that report to OFWAT. The CSCs are expected to represent the interests of all current and potential consumers and investigate complaints by individual customers.

All water companies have a responsibility to promote the efficient use of water and annually report indicators of water and sanitation services to OFWAT and EA. The results are compiled and published by OFWAT every year. In the case of poor performers, OFWAT can demand interim reports. The efficiency improvements contribute to the overall OFWAT "score," which is used to determine the maximum consumer tariffs, thereby indirectly influencing a company's financial return.

OFWAT's regulatory measures of industry performance—indicators mentioned in addition to unplanned and planned interruptions in service and responses to written consumer complaints—have led to significant progress in the sector. As evidenced in

OFWAT reports, the number of written complaints unanswered in a set period of time fell from 30 percent in the early 1990s to its current level of under 1 percent.

The DWI and EA have stringent reporting requirements to keep tabs on the environmental performance of water companies. These agencies have powers to serve notices, file cases, and impose fines on offenders. The government also provides information on the availability and allocation of water resources so that the needs of abstractors and the environment can be balanced in consultation with locally interested parties.

The public has also unarguably been a good watchdog in contributing to initiatives and requiring rigorous and robust monitoring of performance. The improvements since 1989 in customer service and system reliability, reductions in leakages, and new efforts on water conservation all reinforce the importance of government oversight and regulation.

3.3 Contracts that specify the responsibilities of each partner are a prerequisite for the success of any privatization.

Contracts must protect the public interest; this requires provisions ensuring the quality of service and a regulatory regime that is transparent, accessible, and accountable to the public. Good contracts will include explicit performance criteria and standards, with oversight by government regulatory agencies and non-governmental organizations.

Contracts for private participation in water systems must include provisions to protect the public interest and to ensure that the responsibilities of each party are clearly defined and constrained. Experiences from the long history of public-private partnerships in Côte d'Ivoire show the importance of explicitly defining the separate responsibilities of each partner.

Côte d'Ivoire

One of the difficulties in operating a sustainable and efficient water system is to clearly set the role that each of the many stakeholders will, or can, play in management and oversight. Traditional approaches have more clearly defined sets of responsibilities, but have typically excluded important constituents, including both public and community representatives. In recent years, this had led to inefficient operation, or, more often, serious public opposition and unrest over water provision.

Côte d'Ivoire in Sub-Saharan Africa became independent in 1960, but continues to maintain close ties with France. The country has a population of around 16 million, with around 3 million living in and around Abidjan, the principal port and administrative center. A private operator has been providing water services to the residents of Côte d'Ivoire for many decades through a series of contracts that have evolved into a well-defined and differentiated set of responsibilities for both the private company and the related public agencies.

Between 1960 and 1995, the water distribution network grew from 176 kilometers (km) to over 10,000 km nationwide. Over 285,000 new connections were added during this period as well. Côte d'Ivoire has one of the highest rates of in-home connections in Africa at 76 percent; while 2 percent get their water from standpipes and the remaining 22 percent utilize independent providers or traditional sources. The percentage of unaccounted for water in the urban sector is between 15 and 20 percent, better than average for most developing countries. Water quality is required by contract to meet

World Health Organization standards; in 1997 it met those standards 99 percent of the time (Biemi 1996, Ménard and Clarke 2000).

Over the past several decades, various rearrangements in the responsibilities of the public and private entities have been made to fix inefficiencies, restructure financial and planning responsibilities, and address disparities in access to water services. This history highlights the importance of designing clear and consistent roles for each of the parties.

Prior to independence, an international tender was initiated to find a private company to provide municipal water services to Abidjan, a city of 300,000 at the time. A French company, Société pour l'aménagement urbain et rural (SAUR), won the bid and a local company, Société de distribution d'eau de la Côte d'Ivoire (SODECI), was formed with SAUR as the main shareholder. SODECI was given responsibility for operation, maintenance, and water investment. The national agency Direction de L'eau (DdL) retained ownership of water infrastructure. Under the initial arrangement, DdL had no responsibility for financing, which created an incentive to propose unnecessary and expensive projects funded through loans from international banks and international funding organizations. In 1967, DdL re-established jurisdiction over planning financial investment.

In 1974, SODECI was awarded a new 15-year lease contract to provide water for all urban systems in Côte d'Ivoire to help address disparities in access to and costs of water. In 1964, more than 70 percent of villages lacked clean water. By 1985, more than 80 percent of the population, both urban and rural, was supplied with potable water. SODECI has developed a comprehensive set of employee benefits, including special funds for employee loans, an extensive training program, and social events.

A number of problems began to appear in the 1980s, partly due to national economic problems and partly due to inadequacies in the institutional arrangements of the water agreement. Concerns about financial stability and the disparate responsibilities in the water sector led the water partners to discuss reform and ultimately develop a new arrangement in the late 1980s. The main reform centered on investment responsibility. Initial discussions focused on creating a full concession contract where SODECI would bear responsibility for investment financing and debt servicing. Doubts were raised, however, whether sector revenues would be sufficient to cover debt servicing (Kerf 2000).

In December 1987, a revised agreement and contract was negotiated between the government and SODECI, which went into effect the following July. This new contract changed water tariffs, the responsibility for investment and planning, and the structure of government oversight of SODECI. The new contract gave planning authority to SODECI and set up a system of financing from a special fund financed by a surtax on water fees. The complex situation involving responsibilities for financial arrangements, planning, and regulation revealed the importance of having clear rules for decision making and for resolving disputes (Ménard and Clarke 2000).

Côte d'Ivoire is often cited as a privatization success story. Rather than a comprehensive success, however, its value as a case study derives from how the responsibilities for oversight, financing, planning, and management have evolved over the past several decades. Equally important has been the success at periodically revisiting those responsibilities to finetune operations. Over several decades, modifications of the agreement between the public and private sector have helped build a water system that is reliable, extensive, and relatively efficient. The water provided by SODECI is clean, prices are reasonable and equitable, and access and monitoring are comprehensive. One of the primary characteristics of Côte d'Ivoire that has helped greatly with the success of the public-private system has been the country's political stability—a characteristic that helped attract investment and international support to the country. It remains to be seen if the turmoil of the past two years will erode the gains made over the past decades.

3.4 Clear dispute-resolution procedures should be developed prior to privatization.

Dispute resolution procedures should be specified clearly in contracts. It is necessary to develop practical procedures that build upon local institutions and practices, are free of corruption, and difficult to circumvent.

As described above, initial privatization efforts in England failed to adequately address how complaints from customers and disputes over management and service would be handled. Indeed, for the first few years after privatization, vast numbers of complaints arose and were inadequately addressed. In recent years, however, regulatory agencies have clarified expectations and procedures in this area and pushed for fast and responsive handling of complaints.

3.5 Independent technical assistance and contract review should be standard.

Weaker governments are most vulnerable to the risk of being forced into accepting weak contracts. Many of the problems associated with privatization have resulted from inadequate contract review or ambiguous contract language. In principle, many of these problems can be avoided by requiring advance independent technical and contract review.

When a government is considering privatization of its water and wastewater service, the affected community needs to have access to independent technical assistance to ensure that the privatization or re-engineering process and contract is in their best interests, is well regulated, and will improve their service and protect their rights.

Oversight and monitoring of public-private agreements are key public responsibilities. Far more effort has been spent trying to ease financial constraints and government oversight, and to promote private-sector involvement, than to define broad guidelines for public access and oversight, monitor the public interest, and ensure public participation and transparency. Ultimately, weaknesses in monitoring progress can lead to ineffective service provision, discriminatory behavior, or violations of water-quality protections.

Weaker governments are most vulnerable to the risk of being forced into accepting weak contracts. Many of the problems associated with recent privatization efforts in the United States and many other parts of the world have resulted from inadequate contract review or ambiguous contract language. In principle, many of these problems can be avoided by requiring advance independent technical and contract review.

New Orleans, United States

When New Orleans considered privatization of its water and wastewater system, a local independent research group, the Bureau of Government Research (BGR), played a critical role in providing independent technical assistance to the community in assessing the impact of a proposed privatization, and ensuring a procurement process that would maximize community benefits.

New Orleans, Louisiana, is a mid-size city of just under half a million people (2000 census), which sits on a land area of 11,000 square kilometers, 140 kilometers from the mouth of the Mississippi River. The current New Orleans Sewerage and Water Board (S&WB) is made up of 13 members: the mayor (serving as president), 3 members of the city council, 7 citizens appointed by the mayor with city council approval, and two members from the New Orleans Board of Liquidation and City Debt. While, by law, the S&WB must set rates for services, these are subject to ratification by the city council, which makes it difficult to increase rates. Even when rate increases are recommended by the S&WB, in some cases the City Council fails to act on the rate increases in a timely manner: in one instance, no action was taken for 18 months after the S&WB voted to increase rates (BGR 2002).

The S&WB provides services to approximately 145,000 retail customers in a total population of 440,000. The board sold approximately 90 million cubic meters (23 billion gallons) of water in fiscal year 2000. Certain parts of the S&WB operations have been privatized since 1992. A private company, US Filter, operates both wastewater treatment plants. Contractors also manage computer systems , security, and janitorial services (SWB 2002).

The city has had difficulty meeting federal environmental standards covering the operation of some of its facilities. In 1998, a federal consent decree settled a complaint brought by the EPA for violations of the Clean Water Act and the Clean Air Act at the East Bank wastewater treatment plant. The S&WB was required to pay a $1.5 million civil penalty, comply with a 13-year process to fix the aging sewer system, and comply with federal mandates (United States District Court 1998). The S&WB estimated initially that the cost of the capital requirements to fulfill the consent decree would be $250 million, but initial work on the improvements revised that estimate to $455 million (BGR 2000).

The consent decree caused a financial strain at the S&WB, which was also under pressure to update the aging infrastructure of the water and sewer system, meet new drinking water standards, and implement capital improvements to improve drainage. In order to cover some of the costs of the mandated improvements, the S&WB increased sewer rates by 30 percent in March 2000 (BGR 2000).

In October 1999, a consortium of legal and consulting firms recommended that the S&WB pursue privatization of the operations, management, and maintenance of the water and wastewater systems. While S&WB had previously privatized both sewage treatment plants; privatizing the entire collection and distribution network would be a significant change for the S&WB. The proposed procurement was the largest ever in United States, comprising both the wastewater and water systems over a 10- to 20-year period, with an estimated value of $1 billion (BGR 2001).

After extended public comment, in August of 2001, the S&WB released a final Request for Qualifications (RFQ) and a draft Request for Proposals (RFP) soliciting proposals for either management only or the management, operation, and maintenance of the water and wastewater systems. The procurement was designed as a managed competition and permitted a team of current S&WB employees to submit a proposal to take over the system.

In response to the RFP, the S&WB received proposals from US Filter, United Water, and the Managed Competition Employee Committee (MCEC). At the end of February 2002, a committee of the S&WB heard presentations and considered proposals (BGR 2002b). The Evaluation Committee evaluated the three proposals and submitted a report to the S&WB. On October 16, 2002, after spending more than three years and several million dollars to explore privatizing the city's water and sewer systems, New Orleans Sewerage and Water Board voted 6 to 5 to reject all three bids to operate the city's water and wastewater system. (http://www. rppi.org/neworleanswater.html).

Throughout the process a research group, the Bureau of Government Research played a critical role in providing independent technical assistance to the community in assessing the impact of a proposed privatization, and ensuring a procurement process that would maximize community benefits. The BGR is a private, non-profit, independent research organization dedicated to informed public policy making and the effective use of governmental resources in the New Orleans metropolitan region. Over the course of the discussion on privatization, the BGR contracted with engineering and financial consultants to research and publish three comprehensive studies at various stages in the process, which made recommendations on correcting flaws in the procurement strategy. These studies were referred to by groups at diverse ends of the spectrum, including Public Citizen (opposed to privatization) and the Reason Public Policy Institute (pro-privatization). Even within the city of New Orleans, the BGR studies were cited in letters to the editor by both the City Council President Eddie Sapir, considered a driving force behind privatization efforts, as well as the Managed Competition Employee Committee, which was opposed to privatization efforts.

Among the work of the BGR were reviews of the financial health of the New Orleans S&WB and the potential role of privatization in improving the financial outlook for the agency. BGR identified issues to be addressed during a potential privatization or re-engineering process. In reviewing the two reports produced by the financial advisor team, employed by the S&WB, BGR noted numerous flaws and inconsistencies in the analysis. BGR documented that neither report explained where savings from privatization would occur or how much the savings would be. "The tenor of the financial advisor's reports produced so far has been more a promotion of privatization than a dispassionate analysis of how it applies to the S&WB. This is a concern. Key members of the financial advisor team either participate in joint ventures holding privatization contracts in other parts of the country, or have clients who do so" (BGR 2001).

After the release of the draft RFP/RFQ for privatization in February 2001, the S&WB held public comment periods. A request by BGR to extend the comment period was carried by numerous newspapers in their editorial pages, including the *Times Picayune* and the *New Orleans City Business* paper. On March 20, 2001, the *Times Picayune* editorial board wrote in "Don't Rush to Privatize" that "BGR is asking the board to delay the process—not just Wednesday's vote, but the entire privatization process—at least by 60 or 90 days. The board should grant the group's request" (New Orleans 2001). Due to mounting public pressure, the S&WB agreed to extend the comment period by four months to June 15, 2001.

On June 1, 2001 the editor of *New Orleans City Business*, Kathy Finn, wrote:

> "Those who are tuned in to the current delicate position of the... S&WB are waiting anxiously for a pronouncement of some kind from the Bureau of Governmental Research. BGR is expected to release soon results of a study it commissioned recently of the privatization potential of the city's sewerage and water systems. BGR's voice is important in this discussion for a couple of reasons. One is that the idea of privatizing management of these massive systems is complex and multifaceted...BGR has gone to great lengths to engage knowledgeable, independent analysts to examine

the issues and help identify potential problem areas in a privatization effort...In addition, BGR itself is a much-trusted organization that has earned its reputation for integrity and independence over a period of many years. The organization may not be 100 percent "right" on every issue it examines, but the public can feel confident that the agency has given the issue its best shot, in terms of a fair and honest analysis."

On June 15, 2001, the BGR released a detailed analysis of the proposed privatization and the draft procurement documents. The study noted numerous problems with the process and procurement materials including ambiguous selection criteria in determining the appropriate contractor, constraining and unnecessarily complex rules for submitting a bid, inadequate time for due diligence, and contract rules that perpetuate the role of political influence in the subcontracting process. The BGR recommended that the scope of the procurement be properly defined, the process be better designed to promote more competition, the contract be redesigned, and more effective oversight of the contractor be established (BGR 2001).

In June 2002, the BGR released a follow-up study to the June 2001 report on the S&WB proposed privatization, concluding that the procurement documents and process had serious flaws that could limit the benefit of privatization to the S&WB and its ratepayers. On October 15, the S&WB voted to halt the privatization process.

As the city of New Orleans explored the privatization of its water and wastewater systems, the BGR organization provided the affected community with access to independent technical assistance to ensure that the privatization process and contract would be in their best interests, would be well regulated, and would improve their service, reduce their rates, and protect their rights. Numerous public interest organizations, editorial boards, and politicians at diverse ends of the spectrum referred to these studies to support their positions on privatization and, ultimately, this form of independent analysis proved critical to the final decision to halt an ill-planned move toward privatization of New Orleans' water and wastewater systems.

3.6 Decision-making in the water sector should be open, transparent, and include all affected stakeholders. Negotiations over privatization contracts should be open, transparent, and include all affected stakeholders.

Officials at multilateral lending agencies have found that lack of transparency in decisions has played a key role in the failure of many urban infrastructure projects (The World Bank 2003). When decisions in the water sector are not adequately disclosed or publicly vetted, controversy can develop around the resulting projects. Social, economic, and political factors have turned out to be just as important as technological ones and must be considered at the beginning of any potential project. When the public does not have access to documents, information, or decisions being made about the water resources upon which they depend, they may perceive that these decisions are not in their best interests, that government or the private sector is hiding potential problems or flaws in a project, or that these decisions are the result of corruption or bribery (Kaufmann 2002).

Broad participation by affected parties ensures that diverse values and varying viewpoints are articulated and incorporated into the process. It also provides a sense of ownership and stewardship over the process and resulting decisions. We recommend the creation of public advisory committees with broad community representation to advise governments proposing privatization; formal public review of contracts in advance of signing agreements; and public education efforts in advance of any transfer of public responsibilities to private companies. International agency or charitable foundation funding of technical support to these committees should be provided. There are at least two major strategies to ensure effective, transparent, and inclusive decision making in the water sector: democratizing decision making, and ensuring public access and transparency.

Democratizing Decision making: Orangi Pilot Project, Pakistan

One of the best examples of a community-driven approach to addressing water needs is the Orangi Pilot Project (OPP) in Karachi, Pakistan. The Orangi Pilot Project has been internationally recognized as a successful model of a self-help project for slum community improvement involving resident participation. Dr. Akhtar Hameed Khan, a renowned social scientist, initiated the Orangi Pilot Project (OPP) in 1980. At that time, the squatter settlement of Orangi had little or no access to sanitation.

The Orangi *katchi abadi*, or slum settlement, lies in the north of Karachi, covers 2,400 hectares, and has a population of over a million people living in about 100,000 houses in more than 6,300 lanes (Hasan 1997). The populist Bhutto government initiated various policies to regularize these *katchi abadis*. The Katchi Abadi Improvement and Regularization Program (KAIRP) was initiated in 1978 with the aim of improving the settlements by providing water, sanitation, electricity, roads, and other facilities. The development was to be financed by charging the residents "lease charges."

Progress, however, was slow. A 1989 survey indicated that only 50 percent of the squatter settlements in Karachi had access to piped water and 12 percent to sanitation services. After 1984, Orangi had access to piped water from the Karachi Metropolitan Corporation (KMC), but still no sanitation. In the absence of an underground sewer system, open sewers crisscrossed the lanes and the infant mortality rate was high at 137 deaths per 1,000 live births (Hasan 1997).

In 1979, the Bank of Commerce and Credit International Foundation (BCCI), a non-profit institution, approached Dr. Akhtar Hameed Khan, a renowned social scientist to approach the problem of the slum dwellers in Orangi. Dr. Khan agreed on condition that he be allowed to work on his own terms, which were clearly spelled out in a concept paper in 1980. His two principles were that the project would avoid any political or sectarian bias and seek out the preferences of the people.

In the first year, the focus of the OPP research team was to learn how to lower the cost of sanitation options, particularly sanitary latrines and sewage lines. Research by the OPP team revealed that the connection fee charged by KWSB was 4 to 7 times the cost of labor and material and that the cost could be reduced by simplifying the design, eliminating kickbacks and excessive profiteering by contractors, and providing technical assistance to the residents to enable them to work without hiring contractors.

The team estimated that a homeowner with a house on an 80-square-meter plot could have a sanitary latrine by investing only 1,000 Rupees (US$25), including the latrine, house connection, and share of sewage line and sewage drain. Since most homeowners had constructed their own houses, investing an average of 25,000 Rupees

(US$625), the magnitude of investment was not beyond their means (Khan 1992). OPP also concluded that with the right technical input, the lane residents could be organized and trained to finance and construct their own sewage system.

OPP employees held meetings in the lane and, with the help of slides and pamphlets, explained the concept to the people. They explained that KMC did not lay sewer costs free of charge and the charges were not affordable by the lane residents, unlike the costs for the OPP-facilitated scheme. "Lane organizations," consisting of 20 to 30 houses, were created with the assistance of OPP. The residents would select a lane manager, who on their behalf would formally ask OPP for assistance. OPP conducted technical surveys, established benchmarks, and prepared plans and labor and cost estimates. The lane manager would collect money and hold meetings. OPP supervised the process, but at no time handled money.

By 2001, residents of Orangi had built over 400 collector sewers and invested some 82 million Rupees (US$1.4 million) in their sewage system.[2] By the end of the project, nearly all of Orangi homes had in-house latrines and infant mortality rates had dropped from 137 to 37 deaths per 1,000 live births. The entire project was financed, supervised, and constructed by the local population without external subsidies. There remain reservations regarding the replicability of the OPP model, but there can be no doubt of its success in Orangi.

The OPP experience shows that the reason sanitation is unaffordable to many populations is that the costs involved in conventional donor-funded, government-run projects include high overheads, profiteering by contractors, kickbacks to government officials, and fees to foreign consultants. If these are eliminated, resulting in only the cost of labor and materials, the costs for community sanitation can be affordable. In summary, community residents and organizations in Orangi were involved in the key decisions about the design, management, financing, and on-going maintenance of a sanitation system based on their needs and on local cultural, social, demographic, and economic considerations. Key to the success of the Orangi project has been that community members felt ownership over the infrastructure through the process of constructing it, and remained invested in its ongoing maintenance.

Ensuring Public Access and Transparency: The Netherlands

Better decisions and better outcomes result from the free flow of information in the water sector. Ensuring open access to documents, information, and contracts instills public trust and inspires public confidence. As mentioned earlier, transparency in contractual negotiations also ensures that decisions are sustained from one political regime to another, and prevents corruption and collusion in contract awards. Public access to information ensures that government and potential private-sector partners are accountable for agreed-upon outcomes. Publicly available information can compel better performance. For example, in São Paulo, Brazil, the introduction of pollution tests and public reporting has led 95 percent of polluting industries to install waste-treatment units to avoid paying fines and seeing their names published (Gleick et al. 2002a).

An excellent example of public access and transparency in the water sector is the Public Limited Company Model in the Netherlands. The public has full access to the annual accounts of the Public Limited Companies (PLC), which contains the list of

---

2. From *Return Of The Drain Gang—Pakistan.* http://www.tve.org/ho/doc.cfm?aid=854

company assets, liabilities, performance, and liquidity. The board of directors of the PLC is comprised of elected mayors of municipalities, so the interests of the consumer are indirectly incorporated into the water company's management. There is also a focus on public education through a pro-active dissemination of materials and a customer information center. In PLCs with more than 100 employees, a workers' council must be formed with extensive information rights, such as the right to inspect accounts, budgets, strategic information, and contracts. It also has the right to recommend organizational changes and advise on outsourcing contracts, tenders, and major investments.

The majority of Dutch water companies (32 out of 40) are organized as PLCs with their shares owned by the municipal or provincial governments of the areas they serve. The performance of Dutch water companies is excellent by any standards. The Dutch water sector has historically been characterized by a culture of equity, commitment to full-cost recovery, and lack of political interference. Service is highly reliable; demand is well managed [an average of 130 liters per capita per day (lpcd)]; pricing is volumetric; tariffs are affordable, and over 90 percent of homes are metered.

The PLC structure ensures considerable indirect involvement through democratically elected municipal and provincial bodies. Although the consumer has become increasingly involved in recent times, the price of water is, by and large, highly affordable and of reasonable quality so that no public outcry has occurred. There is a high degree of awareness that publicly supplied water is much cheaper than bottled water; as a result bottled water consumption is low in the Netherlands compared to many other parts of Europe (see Data Table 6 in the Data Section of this book). Other segments of civil society are also involved in water issues, including labor unions, environmental and other interest groups, trade associations, the press, and financial institutions. Environmental organizations have managed to influence policy by filing cases, such as against excessive groundwater extraction in Eastern Netherlands. The government now discourages non-renewable use of groundwater.

As the Dutch PLC model demonstrates, public participation and open access to information can instill public trust and inspire public confidence in water-resource decisions, leading ultimately to better decisions and better outcomes for both consumers and water providers.

Conclusions

The Institute's Principles on Water Privatization provide standards to guide effective water and wastewater management. The numerous approaches to water management highlighted in this paper, from completely public to completely private systems, provide insight into the range of options that have had some success in providing water and wastewater services around the world. While no single approach can work in all cases, important lessons can be learned from every effort.

How can we better involve the community in decisions about water resources? How can contracts be designed that effectively define the responsibilities of all parties? How can economic and social interests be balanced? We must draw lessons on appropriate water policies from current experience in order to meet the rapidly growing needs for effective investment and efficient operation of current and planned water supply and wastewater systems. Efforts to privatize water systems will ultimately fail unless the approach changes to acknowledge the important public goods that must be protected and unless fundamental principles to protect these goods are adopted. We hope that some of the information and experience available from both public and private efforts around the world can play an important role in ensuring that basic human and environmental needs for water are met universally, efficiently, and quickly.

REFERENCES

Aasen, B., and Macrae, A. 1994. *Tegucigalpa: Protecting people and the environment through community organization.* UNICEF, Waterfront special issue, March.

Alcazar, L., Abdala, M., and Shirley, M. 2000. *The Buenos Aires water concession.* The World Bank Policy Research Working Paper 2311. Washington, D.C. April.

Appleton, A. F. 2001. Senior fellow, regional plan association, and former commissioner, New York City Department of Environmental Protection, June 6. http://www.nycwatershed.org/index_wachistory.html

Bayliss, K. 2001. *Water privatization in Africa: Lessons from three case studies.* Public Service International Research Unit. London. May.

Biemi, J. 1996. Water crises and constraints in West and Central Africa: the case of Côte d'Ivoire. University of Abidjan, Côte d'Ivoire, Chapter 6, in Rached, E., Rathgeber, E., and Brooks, D. B., editors, *Water Management in Africa and the Middle East.* International Development Research Centre. Canada. http://www.idrc.ca/books/focus/804/chap6.html

Briscoe, J. no date. *The challenge of providing water in developing countries.* Washington, D.C.: The World Bank. http://wbln0018.worldbank.org/ESSD/essdext.nsf/18DocByUnid/2B64A68CBB1F30D885256B5000691FD6/$FILE/TheChalllengeofProvidingWaterinDe,Countries.pdf

British Medical Association (BMA). 1994. Water—A vital resource. London: British Medical Association.

Bureau of Governmental Research (BGR). 2000. *Privatization of sewerage and water board operations.* April. www.bgr.org.

Bureau of Governmental Research (BGR). 2001. *Privatization of water and wastewater systems in New Orleans.* June. www.bgr.org.

Bureau of Governmental Research (BGR). 2002. Sewerage & Water Board privatization at a critical stage, news release, June 4. www.bgr.org

Bureau of Governmental Research (BGR). 2002b. *Report of Proposal Evaluation.* October 14. www.bgr.org

Calhoun, C. 1997. *A town called Olive: A perspective on New York City's water supply.* New York: Westchester Land Trust. http://www.westchesterlandtrust.org/watershed/olive.htm

Center for Public Integrity (CPI). 2003a. *The water barons.* The Center for Public Integrity, International Consortium of Investigative Journalists. http://www.icij.org/dtaweb/water/

Center for Public Integrity (CPI). 2003b. Loaves, fishes, and dirty Dishes: Manila's privatized water can't handle the pressure, In *The water barons.* The Center for Public Integrity, International Consortium of Investigative Journalists. http://www.icij.org/dtaweb/water/

Center for Public Integrity (CPI). 2003c Water privatization becomes signature issue in New Orleans. In *The water barons.* The Center for Public Integrity, International Consortium of Investigative Journalists. http://www.icij.org/dtaweb/water

Collignon, B., Taisne, R., and Kouadio, J. S. 2000. *Water and sanitation for the urban poor in Côte d'Ivoire.* Hydroconseil and the Water and Sanitation Program, The World Bank. June. http://www.wsp.org/pdfs/af_ci_urbanpoor.pdf

Côte d'Ivoire water distribution company (SODECI). http://www.globenet.org/preceup/pages/ang/chapitre/capitali/experien/coteivo.htm

Cox, C. 2003. Atlanta ends failed privatization deal. *Utne Reader.* January.

Dahane, S., and Vargas, M. 2000. *Durban shallow sewerage pilot project.* Suez Lyonnaise des Eaux. Durban, South Africa.

De Moor, A., and Calamai, P. 1997. *Subsidizing unsustainable development: Undermining the earth with public funds.* Canada:The Earth Council. http://www.ecouncil.ac.cr/econ/sud/subsidizing_unsd.pdf

Durban Metro. 1999. *Durban metropolitan area profile*, Report No. 1, November. Durban, South Africa.

EAA. 2001. Website of the Edwards Aquifer Authority. www.e-aquifer.com

Empresa Metropolitana de Obras Santarias S.A. 1995. Memoria 95, Santiago, Chile.

Galiani, S., Gertler, P., and Schargrodsky, E. 2002. *Water for life: The impact of the privatization of water services on child mortality.* Center for Research on Economic Development and Policy Reform, Working Paper No. 154. August.

Gleick, P. H. 1996. Basic water requirements for human activities: Meeting basic needs. *Water international,* 21:83–92.

Gleick, P. H. 1999. The human right to water. In *Water policy,* 1, no. 5:487–503.

Gleick, P. H. 2000. The human right to water. In *The World's Water 2000–2001,* ed: P. H. Gleick, 1–18. Washington, D.C.:Island Press.

Gleick, P. H., Wolff, G., Chalecki, E. L., and Reyes, R., 2002a. *The new economy of water.* Oakland, California:Pacific Institute for Studies in Development, Environment, and Security. IBSN No. 1-893790-07-X. www.pacinst.org/reports/new_economy.htm.

Gleick, P. H., Burns,W. C. G., Chalecki, E. L., Cohen, M., Cushing, K. K., Mann, A. S., Reyes, R., Wolf, R. H., and Wong, A. K. 2002b. *The world's water 2002–2003.* Washington. D.C.:Island Press.

Gomez-Lobo, A. 2001. *Incentive-based subsidies: Designing output-based subsidies for water consumption.* Viewpoint 232. Washington, D.C.:The World Bank.
http://rru.worldbank.org/viewpoint/HTMLNotes/232/232Gomez-531.pdf

Gomez-Lobo, A. 2003. Making water affordable: Output-based consumption subsidies in Chile. In *Contracting for public services: Output-based aid and its applications,* ed. Brook, P. J., and Smith, S. M., Washington, D.C.:The World Bank.
http://rru.worldbank.org/Documents/07ch2.pdf

Gosling, M. 2001. Free water for thousands for May 1. *Cape Times,* April 29.
http://qsilver.queensu.ca/~mspadmin/pages/In_The_News/2001/April/Free.htm

Green, Joanne. 2003. Regulation and the balancing of competing interests in England and Wales.: London:WaterAid and Tearfund.

Gutierrez, Calaguas, E. B., Green, J., and Roaf, V. 2003. New rules, new roles: Does PSP benefit the poor?London:WaterAid and Tearfund.

Hasan, A. 1997. *Working with the Government.* Karachi, Pakistan:City Press.

Hasan, A. 1999 *Understanding Karachi: Planning and reform for the future.* Karachi, Pakistan:City Press.

Idelovitch, E., and Ringskog, K. 1995. *Private sector participation in water supply and sanitation in Latin America.* Washington, D.C.:The International Bank for Reconstruction and Development/The World Bank Washington, D.C. May.
http://www.worldbank.org/html/lat/english/papers/ewsu/ps_water.txt

Ingegneria Senza Frontiere. 2002. Water pricing as a key part of a sustainability strategy: The Durban case study.
http://www.diam.unige.it/isf/archivioprogetti/Durban/Durban_price_and_level.pdf

Kaufmann, D. 2002. Transparency, incentives and prevention (TIP) for corruption control and good governance. *Empirical findings, practical lessons, and strategies for action based on international experience.* The World Bank.
http://www.worldbank.org/wbi/governance/pdf/quinghua_presentation.pdf

Kerf, M. 2000. Do State holding companies facilitate private participation in the water sector? Evidence from Côte d'Ivoire, the Gambia, Guinea, and Senegal. Policy Research Working Paper 2513. Washington D.C.:World Bank, December.

Khan, A. H. 1992. *Orangi pilot project programs.* Karachi, Pakistan:OPP-Research and Training Institute.

Khan, A. H. 1996. *Orangi pilot project: Reminiscences and reflections.* Karachi, Pakistan:Oxford University Press.

Komives, K. 1999, 2001. Designing pro-poor water and sewer concessions: Early lessons from Bolivia. World Bank. Also published in *Water policy,* 3, no. 1:61–80.
http://econ.worldbank.org/docs/977.pdf.

Lobina E., and Hall, D. 2001. UK water privatisation—a briefing. Public Services International Research Unit (PSIRU), School of Computing and Mathematical Sciences, University of Greenwich, London.

McCaffrey, S. C. 1992. A human right to water: Domestic and international implications, *Georgetown International Environmental Law Review,* V, issue 1:1–24.

Ménard, C., and Clarke, G. 2000. *Reforming water supply in Abidjan, Côte d'Ivoire: Mild reform in a turbulent environment.* Policy Research Working Paper 2377. Washington, D.C.:The World Bank. June.

Mestrallet, G. 2002. The water truce—an open letter published in *Bridging the Water Divide*. Paris:Suez/Ondeo.

Metell, K., and Mooijman, A. M. 1998. Water—and more—for the barrios of Tegucigalpa. *Waterlines*,17, no. 1:19–21, 32. UNICEF.

Mitchell, D., Hanemann, M., and Du, M. 1994. *Setting urban water rates for efficiency and conservation*. Sacramento, California:The California Urban Water Conservation Council.

Mooijman, A., and Torres, X. 1998. *The construction of low-cost sewerage systems in Tegucigalpa. A feasible solution for the urban poor?* UNICEF. http://www.efm.leeds.ac.uk/CIVE/Sewerage/articles/honduras.pdf

National Council of Women of Canada (NCWC). 1997. *Privatization Of water purification and distribution systems*. Policy 97.14EM. http://www.ncwc.ca/policies/water_waterquality.html

National Research Council. 1999. *New strategies for America's watersheds*. Committee to Review the New York City Watershed Management Strategy, Water Science and Technology Board, Commission on Geosciences, Environment, and Resources, Washington, D.C.:National Academy Press.

National Research Council. 2000. *Watershed Management for Potable Water Supply*. Committee to Review the New York City Watershed Management Strategy, Water Science and Technology Board, Commission on Geosciences, Environment, and Resources. Washington, D.C.:National Academy Press.

New Orleans City Business. 2001. Publisher's notes. March 26.

New Orleans City Business. 2001. Op-ed/From the Editor. June 1.

New York City Department of Environmental Protection (NYDEP). 2002a. Croton water filtration plant project. July 10. http://www.nyc.gov/html/dep/html/news/wsprot.html.

New York City Department of Environmental Protection (NYDEP). 2002b. Bureaus and offices web site. http://www.nyc.gov/html/dep/html/bureaus.html.

Orwin, A. 1999. *The privatization of water and wastewater utilities: An international survey*. Environment Probe, Canada. http://www.environmentprobe.org/enviroprobe/pubs/ev542.html.

Pictet. 2001. Pictet global water fund: Investment case and outlook. Pictet electronic publications. http://www.us.pictetfunds.com/us/research.Par.0031.File0.pdf/Case%20for%20Water.pdf

Saunders, P., and Harris, C. 1990. Privatization and the consumer. *Sociology*, 24: 57–75.

Serra, P. 2001. Subsidies in Chilean public utilities. World Bank electronic publications. http://econ.worldbank.org/docs/1200.pdf

Sewerage and Water Board (SWB) of New Orleans. 2002. Request for qualifications/request for proposals for the management, operations and maintenance of water and wastewater systems. January. http://www.swbno.org/rfq-rfp/

South Africa Department of Water Affairs and Forestry (SADWAF). 2001. *Free basic water: Implementation strategy document* Version 1. Department of Water Affairs and Forestry, Pretoria, South Africa. May. http://www.dwaf.gov.za/FreeWater

South Africa Department of Water Affairs and Forestry (SADWAF). 2003. Water is a human right: Fulfilling the requirements of the Constitution. http://www.dwaf.gov.za/Communications/Articles/Minister/2003/WATER%20IS%20A%20HUMAN%20RIGHT.doc

Standard and Poor's (S&P). 2000. European water: Slow progress to increase private sector involvement. September 12. http://www.sandp.com/Forum/RatingsCommentaries/CorporateFinance/Articles/eurowater.html

Times Picayune. 2001. Don't rush to privatize. March 20.

Times Picayune. 2001. SW&B urged to rethink its privatization process. June 15.

Times Picayune. 2001. No rush on privatization. June 27.

Torres, X., and Mooijman, A. M. 1996. *An update on the Tegucigalpa model*. Waterfront 9, UNICEF, December. www.un.org/documents/ecosoc/cn17/1998/background/ecn171998-right.htm

Troesken, W. 2001. Race, disease, and the provision of water in American cities, 1889–1921. *Journal of Economic History*, September.

United Nations Development Programme (UNDP). 1998. *Integrating human rights with sustainable development*. United Nations Development Programme. New York:United Nations. January.

United Nations Economic and Social Council (UNECSC). 2002. Substantive issues arising in the implementation of the International Covenant on Economic, Social and Cultural Rights. Committee on Economic, Social and Cultural Rights, E/C.12/2002/11 Twenty-ninth session, Geneva. November 26.

United Nations Habitat. 2002. Emancipation of poor communities in Tegucigalpa—Management of the water supply (Honduras). April 14. http://habitat.aq.upm.es/bpn/bp161.html

United States District Court. 1998. Consent Decree. The United States District Court for the Eastern District of Louisiana, New Orleans Division. Civil Action No. 93-3212. April.
http://www.bgr.org/Outlooks/Orleans/Privatization/consent_decree_text.html

United States Environmental Protection Agency (EPA). 1997b. January 21. New York City Watershed memorandum of agreement. New York and Washington, D.C.

United States Environmental Protection Agency (EPA). 2001. *A landscape assessment of the Catskill/ Delaware watersheds 1975–1998:New York City's water supply watersheds* EPA/600/ R-01/075, September.
http://www.epa.gov/nerlesd1/land-sci/pdf/catskill-report-1.pdf

United States Environmental Protection Agency (EPA). 2002a. Draft:New York City filtration avoidance determination. Washington, D.C. May.
http://www.epa.gov/Region2/water/nycshed/2002fad.pdf.

United States Environmental Protection Agency (EPA). 2002b. EPA proposes to allow NYC to continue to avoid filtering Catskill/Delaware water. May 23.
http://www.epa.gov/region02/news/2002/02046.htm.

University of New Hampshire (UNH). 2000. *UNH leads international effort in lighting the way to safer drinking water.* Environmental Research Group, November 1.
http://www.ceps.unh.edu/news/releases00/uvlight1100.html.

U.S. Water News. 2000. EPA says NYC not required to filter reservoir drinking water. 19, no.7, July.

Warne, D. 2002. New York City Department of Environmental Protection. Phone conversation, July 23.

Water Environment Federation (WEF). 2000. Financing water quality improvements.
http://www.wef.org/govtaffairs/policy/finance.jhtml.

Watershed Agricultural Council. 2001. Whole farm planning.
http://www.nycwatershed.org/clw_wholefarmplanning.html.
See also the *WAC 2001 Report.* Walton, New York:Watershed Agricultural Council.

Waterwatch. 2002. http://lists.cupe.ca/pipermail/waterwatch/2002-December/000098.html

Wegelin-Schuringa, M. 1998. *Water demand management and the urban poor.* Delft, The Netherlands:IRC International Water and Sanitation Centre.
http://www.irc.nl/themes/urban/demand.html

Whittington, D. 2002. *Municipal water pricing and tariff design: A reform agenda for cities in developing countries.* Washington D.C.:Resources for the Future. August.

World Bank, The. 2001. *Durban metro water: Private sector partnerships to help the poor, World Bank Water and Sanitation Program Africa,* 5. http://www.wsp.org/pdfs/af_durban.pdf.

World Bank, The. 2003. World Bank rapid response website. Canceled infrastructure projects: causes and consequences moderated by Clive Harris. www.rru.worldbank.org.

World Bank, The. and Public-Private Infrastructure Advisory Facility (PPIAF). 2002. *Private participation in infrastructure: trends in developing countries in 1990–2001.* Washington, D.C:The World Bank.

Yepes, G. 1999. Do cross-subsidies help the poor to benefit from water and wastewater services? Lessons from Guayaquil. UNDP-World Bank Water and Sanitation Program. February.

Yergin, D., and Stanislaw, J. 1999. *The commanding heights: The battle between government and the marketplace that is remaking the modern world.* New York:Simon and Schuster.

CHAPTER 4

Groundwater: The Challenge of Monitoring and Management

Marcus Moench

Groundwater pumping is among the largest human-induced changes in the hydrological cycle, leading to changes in water levels, the residence times of water in aquifers, and water quality. However, groundwater is an "invisible" resource, making its use difficult to monitor and manage. The environmental contributions it makes to stream base flows, wetlands, and surface vegetation are masked and, unlike surface water flows, impossible to observe directly. Furthermore, although pumping wells are not intentionally hidden, their dispersed, small-scale, and generally private nature disguises the true level of groundwater extraction. Over the past five decades, however, well numbers have increased exponentially in many parts of the world. As a result, in many areas, over-abstraction is severe and groundwater water levels are declining at rates that range from 1 to 3 meters per year, a pattern documented by many researchers. As researchers at IWMI comment:

> "Many of the most populous countries of the world—China, India, Pakistan, Mexico, and nearly all of the countries of the Middle East and North Africa—have literally been having a free ride over the past two or three decades by depleting their groundwater resources. The penalty of mismanagement of this valuable resource is now coming due, and it is no exaggeration to say that the results could be catastrophic for these countries, and given their importance, for the world as a whole."

Drawing on examples from India, China, and other parts of the world, this chapter explores the role groundwater plays, the nature of emerging problems, and the challenges facing any aggregate assessment of their extent or likely impacts on key human and environmental services. The development of effective responses to emerging groundwater problems is constrained by reliance on the concept of sustained yield. While this concept is important, inherent scientific and monitoring difficulties limit its utility both at local levels and in global assessments. In most cases, the data required to quantitatively apply the concept simply are not available. The chapter is not intended to be a comprehensive assessment of all groundwater data and problems. Instead, by focusing on issues in developing countries with high levels of groundwater use, my goal is to highlight the complex challenges that we must address if we are to understand and eventually manage groundwater resources on a sustainable basis.

Conceptual Foundations

Before addressing the main themes of the chapter, it is essential to explain the conceptual foundations of groundwater assessments. Most groundwater assessments are designed to quantify the water balance within a given aquifer or region (i.e., the mass balance of water flowing into and out of a given unit) and to understand the dynamics of key water-quality parameters such as salinity, arsenic, and fluoride. While detailed assessments in industrial countries generally move beyond simple mass-balance estimates and attempt to quantify flow and pressure regimes within aquifers, this is rare in the broad regional assessments done in less industrialized regions. Instead, most regional assessments attempt to quantify the components of the water-balance equation in order to evaluate the balance between available recharge (the amount of water flowing into a given aquifer that is available for pumping) and extraction (the amount of water pumped out of that aquifer). They also attempt to quantify the stock of water available in storage within an aquifer that is available for future use. Aquifer "sustained yield" is defined as the amount of water that can be pumped from a given hydrological unit without depleting the stock of water in storage. When extraction exceeds recharge aquifers are generally described as suffering from overdraft or overextraction, the primary warning sign that management may be required.

Although these concepts may seem straightforward, in practice they are extremely complex. Quantifying the water balance within aquifers, for example, requires quantitative estimates of deep groundwater inflow from other aquifers, groundwater discharge to streams, evapotranspiration by plants, and a wide variety of other factors. These factors often vary from year to year and require extensive recording periods (and assumptions of stable climatic conditions) to develop stable quantitative estimates. In addition, evaluation is complicated where boundaries of aquifers cannot be clearly identified—a common situation. Without clear boundaries, it is often difficult to accurately evaluate either recharge or extraction. Finally, the concepts are often value-laden—the quantity of recharge "available" for extraction, for example, depends on whether or not one wants to maintain contributions to streams, wetlands and other uses. In such cases, water levels (as opposed to the physical amount of water available) must also be factored into the evaluation. Despite such limitations, mass-balance concepts remain the foundation for most regional and national groundwater assessments.

Challenges in Assessment

The scientific foundation on which current understanding of groundwater resource availability rests is hydrogeological. It is, in other words, focused on the occurrence of groundwater resources and the distribution of major aquifer systems as determined by regional geologic characteristics. The only systematic regional survey of groundwater extent was put together in the 1980s by the United Nations under the Water Series publications (United Nations 1983–1990). This comprehensive work has not been updated and remains the only systematic portrait of groundwater occurrences on a country-by-country basis. Within some countries, or regions, detailed groundwater studies are occasionally, albeit rarely, prepared. For example, California recently released a comprehensive set of new aquifer assessments that cover much of the state (CDWR 2003).

These hydrogeological descriptions of aquifer systems are intended to provide regional information to support agriculture and economic development. They summarize available information on the geology and distribution of major water-bearing formations. They also contain a limited amount of information on the organizations dealing with groundwater in many countries along with occasional notes on use patterns and the characteristics of individual aquifers, including recharge zones and patterns of groundwater flow.

Despite numerous subsequent scientific studies on the dynamics of specific aquifers, understanding of groundwater resources at a global level remains limited, focused on a few major aquifer systems. Relatively little information is available on groundwater dynamics even in some of the world's best known and largest aquifer systems. In the Gangetic basin, for example, although the location of major recharge zones, such as the Bhabar zone at the base of the Himalaya, are well known, scientifically based quantitative estimates of the actual amounts of recharge have proved unavailable. Equally importantly, relatively little is known about groundwater availability and dynamics in areas falling outside the major aquifer systems.

Of equal concern is the absence of substantive information on groundwater extraction and use patterns. In locations such as India and much of Africa, however, groundwater is a major source for irrigation, domestic, and other uses. This combination of significant use and limited knowledge is of great concern because of emerging challenges of groundwater overdraft, deteriorating aquifer water quality, and the complexities of managing surface and groundwater together.

Extraction and Use

At both global and local levels, data on groundwater extraction and use are generally both fragmentary and of questionable scientific validity. The situation with Aquastat, the main global database on water use maintained by FAO, is illustrative. Globally, by volume, groundwater extraction for agricultural purposes dominates extraction for other purposes (see Table 4.1). However, that is not clear from data available in Aquastat, which records water use for irrigation, but often does not differentiate between areas irrigated from ground or from surface water sources. This is because many country governments do not collect statistical information on the source of irrigation water (FAO, personal communication, 2003). Of the 202 countries listed in the Aquastat main data table, 116 (or 57 percent) do not provide data on the source of irrigation water.[1] The situation is further exacerbated because data are absent on key countries where groundwater use for agriculture is significant, including the United States, China, and most of Western Europe. As a result, it is not possible to derive an accurate picture of overall groundwater extraction for agriculture on a country-by-country basis utilizing

---

1. http://www.fao.org/ag/agl/aglw/aquastat/dbase/index3.jsp?radio5=y&radio8=y&cont=%25&country= %25&search=Display. Accessed July 7, 2003.

the Aquastat database. Consequently, robust estimates regarding the relative contribution of surface and groundwater sources to irrigated area or the amount of groundwater extracted for irrigation are unavailable.

Groundwater Withdrawals

Despite inherent limitations, the World Resources Institute has compiled available estimates of groundwater recharge and extraction for different uses at a global level. The full table is available on their web site and a condensed version focusing on use and containing only those entries that indicate total withdrawals is shown in Table 4.1. The full table also contains recharge estimates. It is heavily footnoted with references to the original source of the estimates and interested readers are urged to consult these. Many of these footnotes highlight uncertainties and other limitation in the data from which they were derived. Of the 125 countries in the table, no data on withdrawals is given for 74. While many of these are small countries, this gap highlights the fragmentary nature of comprehensive groundwater data.

Table 4.1 highlights some of the main patterns of groundwater use. By volume, four countries (India, China, Pakistan, and the United States) together account for over 60 percent of reported global extraction. Adding in another seven countries (Bangladesh, Japan, Italy, Russia, Iran, Saudi Arabia, and Mexico) brings the total to nearly 80 percent. Groundwater clearly plays an important and, in many cases, locally dominant role as a source of domestic water supply in many regions. Within countries, domestic uses account on average for 65 percent of extraction. This percentage can, however, be misleading since domestic uses heavily dominate in areas where total extraction is small, while agricultural uses heavily dominate where extraction is large. In addition, large-scale extraction for domestic use generally occurs through municipal utilities that, for a variety of reasons, tend to monitor extraction relatively accurately. Agricultural extraction, in contrast, generally occurs through private wells owned by individual farmers. In this situation, monitoring is often far less accurate and, particularly in locations such as India where well locations are poorly documented, the volume of extraction is likely to be heavily underestimated. Thus, in the aggregate, agricultural uses heavily dominate the total volume of use.

Based on the percentages and total volumes extracted reported in Table 4.1, extraction for agriculture is equivalent to over 400 km^3 while that for domestic use is less than 100 km^3 and industry less than 50 km^3. Together, India, China, Pakistan, and the United States account for over 75 percent of total reported groundwater extraction for agriculture—approximately 320 km^3. Beyond the broad dimensions of groundwater use, one can also see from the table the regions where overdraft is emerging as a concern. All countries where reported extraction exceeds annual recharge are, for example, with the exception of Pakistan and Mauritania, in the Middle East.

Global data such as those presented in the table sometimes obscure as much as they reveal. Estimates reported in the table are often over two decades old and, as WRI emphasizes, of variable reliability. In addition, global data provide little insight regarding in-country variations. Countries such as India, China, and Mexico, where aggregate national groundwater extraction is reportedly well below total recharge, actually face some of the most challenging regional overdraft problems. In Mexico, for example, total groundwater extraction as reported in the WRI data set is only 18 percent of annual recharge. As early as the late 1980s, however, 80 of 459 identified aquifers were overexploited, with extraction exceeding long-term recharge by more than 20 percent. By 1999, the number of overexploited aquifers increased to 99 (Burke and Moench 2000). Most of Mexico's groundwater overdraft is occurring in arid portions of the country where economic activity is concentrated and demand for agriculture, people, and industry is highest. As a result, while overall national groundwater extraction is a small fraction of total recharge, regional overdraft represents a fundamental threat to livelihoods and basic needs for much of Mexico's population. The situations in India and China are similar. Although aggregate data indicate that India extracts less than 50 percent of available recharge, for example, overdraft is a major concern in arid or hard-rock zones and high-productivity regions such as Punjab, Haryana, Gujarat, Rajasthan, Tamil Nadu, and western Uttar Pradesh.

For this reason, groundwater extraction and recharge estimates are fundamentally inappropriate measures for evaluating the sustainability of groundwater vital for agricultural, domestic, industrial, and environmental uses. As argued elsewhere, most groundwater-dependent activities depend more on water level and water-quality conditions than they do on the physical availability of groundwater in any given aquifer (Burke and Moench 2000). Access to groundwater for farmers and other users depends, for example, on the economic cost of drilling and pumping, which depends more directly on groundwater levels than on the extraction-recharge balance. Furthermore, even small changes in water levels are often sufficient to affect streamflows, wetlands, and surface vegetation patterns, even when the amount of groundwater remaining in storage is huge.

In some locations of the Gangetic Basin, for example, saturated sediments extend to depths of several thousand meters. Yet, declines of even a few meters in the water table would be sufficient to eliminate dry-season flows in many rivers, while also affecting access to basic water supplies for tens of millions of domestic and agricultural users. Such declines could occur even while extraction in the basin remained well below the recharge rates for the basin as a whole.

Groundwater Quality

Groundwater quality concerns are finally beginning to receive the attention they deserve, but vast unknowns and uncertainties remain. Where regional development is especially intense, groundwater quality is just as much an issue as overall availability and reliability. Groundwater quality often declines where extraction levels are high, but it may also decline where groundwater levels are rising—another example of extraction-recharge balances giving an imperfect view of actual conditions. This is illustrated particularly well by the case of India.

In India, groundwater salinity shows a negative correlation with water-level trends—water levels are often declining where TDS (total dissolved solids—a standard measure of salinity) levels are low, and levels are rising or stable where quality is poor (The World Bank 1998). In Haryana State, official figures indicate that 11,438 square kilometers are underlain by saline groundwater out of a total area of 44,212 square kilometers for the state (MOWR 1996). More importantly, roughly 65 percent of the agricultural area of underlain by saline groundwater (Gangwar and Panghal 1989). Groundwater extraction, naturally enough, is concentrated in areas where TDS levels are low. This causes water to flow laterally in from adjacent lower-quality areas. As a result, mobilization of saline groundwater and the contamination of existing high-quality groundwater are growing concerns where pumping is concentrated.

TABLE 4.1 Annual Groundwater Withdrawals

	Year	Total Withdrawal (km³)	Percentage of Annual Recharge	Per-capita (m³ per year)	Sectoral share (percentage)		
					Domestic	Industrial	Agricultural
	1995	600–700	x	106–124			
WORLD							
ASIA (Exclude Middle East)							
Bangladesh	1990	10.7	50.9	97.6	13	1	88
China	1988	52.9	6.4	47.1	x	x	54
Georgia	1990	3	17.4	549.5	x	x	x
India	1990	190	45.4	223.2	9	2	89
Japan	1995	13.6	50.3	108.2	29	41	30
Kazakhstan	1993	2.4	6.7	143.9	21	71	8
Korea Rep	1995	2.5	18.6	55.1	x	x	17
Kyrgyzstan	1994	0.6	4.4	132	50	25	25
Malaysia	1995	0.4	0.6	19	62	33	5
Mongolia	1993	0.4	5.8	149.1	x	x	x
Pakistan	1991	60	109.1	489.5	x	x	90
Philippines	1980	4	2.2	82.8	50	50	x
Tajikistan	1994	2.3	37.7	398.7	x	x	x
Thailand	1980	0.7	1.7	15	60	26	14
Turkmenistan	1994	0.4	11.9	100.3	53	9	38
Uzbekistan	1994	7.4	37.6	334.3	33	11	57
Vietnam	1990	0.8	1.7	11.9	x	x	x
EUROPE							
Albania	1989	0.6	9	193.6	48	x	52
Austria	1995	1.4	6.2	172.5	52	43	5
Belarus	1989	1.2	6.6	115.7	52	13	28
Belgium	1980	0.8	86.4	79	55	22	4
Bulgaria	1988	5	37.3	566.1	x	x	x
Czech Rep	1995	0.5	x	48	x	x	x
Denmark	1995	0.9	3	169.8	40	22	38
Finland	1995	0.2	12.8	47.8	65	11	24
France	1994	6	6	103.8	56	27	17
Germany	1990	7.1	15.5	89.4	48	47	4
Greece	1990	2	19.4	195.7	37	5	58

Hungary	1995	1	14.5	96.5	35	48	18
Iceland	1995	0.2	0.6	558.9	x	x	x
Ireland	1995	0.2	6.5	62.3	35	38	29
Italy	1992	13.9	32.3	243.2	39	4	58
Lithuania	1995	0.2	17.1	55.1	x	x	x
Netherlands	1990	1	23.3	70.2	32	45	23
Norway	1985	0.4	0.4	97.5	27	73	x
Poland	1995	2	5.5	51.5	70	30	x
Portugal	1995	3.1	60.1	311	39	23	39
Romania	1993	3.6	43.7	158	61	38	1
Russian Federation	1988	12.6	1.6	85.5	x	x	x
Slovakia	1995	0.6	x	113	x	x	x
Slovenia	1994	0.2	x	88.9	x	x	x
Spain	1995	5.4	18.8	137.2	18	2	80
Sweden	1995	0.6	3.2	72.8	92	8	x
Switzerland	1995	0.9	33.4	126.3	72	40	x
Ukraine	1989	4	20.1	77.5	30	18	52
United Kingdom	1995	2.5	25.2	42.4	51	47	2
MIDDLE EAST & NORTH AFRICA							
Algeria	1989	2.9	167.6	117.1	46	5	49
Egypt	1995	5.3	407.7	85.1	58	0	42
Iran, Islamic Rep	1980	29	69	738.8	x	x	x
Iraq	1985	0.2	1.5	13.1	50	40	x
Isreal	1996	1.2	234	204.5	18	2	80
Jordan	1993	0.5	91.4	100.7	30	4	66
Kuwait	1994	0.3	x	142.7	0	0	100
Lebanon	1991	0.4	8.3	153.2	13	9	78
Libyan Arab Jamahiriya	1995	3.7	561.5	734.9	9	4	87
Morocco	1998	2.7	29.8	97.9	16	x	84
Oman	1985	0.4	41.9	280.7	x	x	x
Saudi Arabia	1990	14.4	1518.9	899.3	10	x	90
Syrian Arab Rep	1993	1.8	27.3	133.5	13	4	83
Tunisia	1995	1.6	39.2	181.8	10	4	86
Turkey	1995	7.6	38	124	31	9	60
United Arab Emirates	1995	1.6	1333.3	724.1	x	19	81
Yemen	1985	1.4	88.5	139.2	x	x	x

continues

TABLE 4.1 Continued

	Year	Total Withdrawal (km³)	Percentage of Annual Recharge	Per-capita (m³ per year)	Sectoral share (percentage)		
					Domestic	Industrial	Agricultural
SUB-SAHARAN AFRICA							
Chad	1990	0.1	0.8	15.7	29	x	71
Gabon	1989	0	0	0.6	100	0	0
Madagascar	1984	4.8	8.7	482.9	0	x	x
Mali	1989	0.1	0.5	11.6	x	x	x
Mauritania	1985	0.9	293.3	498.3	x	x	x
Niger	1988	0.1	5.2	17.9	58	4	39
Senegal	1985	0.3	3.3	39.2	24	x	72
Somalia	1985	0.3	9.1	45.8	x	x	x
South Africa	1980	1.8	37.3	64.9	11	6	84
Sudan	1985	0.3	4	13	x	x	x
NORTH AMERICA							
Canada	1990	1	0.3	37.3	34	11	34
United States	1990	109.8	7.3	432.3	20	5	62
CENTRAL AMERICA & CARIBBEAN							
Cuba	1975	3.8	47.5	408.3	x	x	x
Mexico	1995	25.1	18.1	275.4	13	23	64
SOUTH AMERICA							
Argentina	1975	4.7	3.7	180.4	11	19	70
Brazil	1987	8	0.4	57	38	25	38
Peru	1973	2	0.7	139.4	25	15	60
OCEANIA							
Australia	1985	2.2	3.1	143.2	x	20	67

Source: http://earthtrends.wri.org/datatables/index.cfm?theme=2&CFID=320174&CFTOKEN=77371942. World Resources Institute, Data Table FW.2 Groundwater and desalinization. Accessed July 7, 2003. Used with permission. Refer to the original for details on specific data sources.

86

Other problems associated with irrigated agriculture are water logging, salinity, or alkalinity of soils when irrigation causes increases in groundwater levels. Ironically, such conditions can exist in close proximity to areas experiencing groundwater overdraft. In the Mehsana district of Gujarat, for example, surface irrigation has waterlogged some areas although the district is well known as an extreme case of groundwater overdraft. Similar problems occur in other Indian states. Available estimates vary greatly, but clearly indicate the severity of the problem. Ministry of Agriculture estimates in 1990, for example, placed the total area affected by water logging in India due to both groundwater-level increases and over-irrigation at 8.5 million hectares. In contrast, Central Water Commission estimates for 1990, which considered just those areas affected by groundwater rises, only totaled 1.6 million hectares (Vaidyanathan 1994, The World Bank 1998). Despite the variability of the estimates (another indication of basic data problems), the widespread nature of such problems and their linkage with groundwater extraction levels is clear.

In addition to salinity, other quality problems are closely related to the level of groundwater extraction. At least in some cases, arsenic and fluoride concentrations appear to be increasing as groundwater extraction grows. In recent years, attention has focused on arsenic, which is among the more serious environmental health problems connected to the consumption of contaminated groundwater. Arsenic from geological sources has caused poisoning outbreaks in Mexico, Argentina, Chile, Taiwan, Inner Mongolia, China, Japan, India, and Bangladesh (Nordstrom 2000). The most extensively documented and well-known case of widespread arsenic poisoning is in the Gangetic basin of Nepal, India, and Bangladesh (see, for example, the review of arsenic poisoning in Gleick 2000). In Bangladesh, perhaps 21 million people are currently estimated to be at risk and some 200,000 cases of arsenic poisoning have been documented (British Geological Survey and Mott Macdonald Ltd. 1999). In West Bengal, high levels of arsenic are found in water supplies underlying nearly 39 percent of the state and millions of people may be affected (Bhattacharya et al. 1996).

Arsenic is also emerging as a major concern in districts such as Nawalparasi in the Terai region of Nepal. In this region, approximately 550,000 people may be at risk of arsenic poisoning (Arsenic Steering Committee, Nepal, personal communication, 2003). Recent data from Nepal indicate that concentrations in some village areas average over 500 parts per billion (ppb), 50 times the WHO recommended level. In addition to arsenic, fluoride is a major problem in parts of India where it has caused widespread health problems including skeletal fluorosis, a disease that causes joints to 'freeze' and cripples victims (Moench and Matzger 1994). Increases in fluoride above acceptable levels in drinking water have been directly correlated with pumping rates, water-level declines, and groundwater level fluctuations in projects in Gujarat, India (KON 1992, Wijdemans 1995). Such geochemical threats to health are pervasive (Appleton, Fuge et al. 1996) and are particularly acute in the case of naturally tainted groundwaters where abstraction can accelerate leaching, concentration, or mobilization of toxic materials (Edmunds and Smedley 1996).

Because groundwater extraction for agricultural uses heavily dominates extraction for other purposes and both water-level and water-quality changes are often related to overall extraction, understanding the dynamics of agricultural demand for groundwater is central to understanding emerging groundwater problems at a global level. For this reason, the next section focuses on groundwater in agriculture and the socio-economic dynamics underlying increasing extraction levels.

TABLE 4.2   Structure of Groundwater Use in Different Regions

Country	Annual groundwater use (km^3)	Number of groundwater structures (millions)	Average extraction per structure (m^3/yr)	Percent of population dependent on groundwater
India	150–215	19–26	7,900	55–60
Pakistan (Punjab)	45–54.5	0.5	90,000	60–65
China	75–106	3.5	21,500	22–25
Iran	29	0.5	58,000	12–18
Mexico	29	0.07	414,285	5–6
USA	100	0.2	500,000	<1–2

Source: Shah et al. 2003. The higher annual use numbers in the first data column are more recent estimates (T. Shah, personal communication, 2003).

Groundwater in Agriculture

Groundwater plays a uniquely important role as a reliable source of water supply for agriculture, particularly in Asia. Recent estimates by the International Water Management Institute (IWMI) indicate that groundwater extraction for agriculture may exceed 300 to 400 km^3 in India, Pakistan, Bangladesh, and the North China Plain—approximately half of estimated annual use at a global level. Other estimates suggest that groundwater contributes US$ 25 to 30 billion each year to Asian agricultural economies (Shah et al. 2003). In India alone, despite massive government investment in surface water supply systems over the last 50 years, over 50 percent of the irrigated area has been supplied from groundwater sources, and the economic contribution of groundwater irrigation exceeds that of surface irrigation (Roy and Shah 2003). The scale of the groundwater extraction for agriculture in Asia relative to other parts of the world is shown in Table 4.2.

In addition to the overall magnitude of groundwater use, the table highlights important structural differences between regions. In India, and to a lesser extent Pakistan, China, and Iran, groundwater extraction occurs through numerous small structures and a large proportion of the population depends directly on those structures as part of their agricultural livelihoods. For Mexico and the United States, in contrast, most groundwater extraction occurs through large wells owned and controlled by a small fraction of the population. As discussed later, this has significant implications both for the degree to which emerging problems can be quantified and, ultimately, for decisions about management. Before exploring these implications, however, it is important to understand the underlying role groundwater plays in Asian livelihoods.

Asian Livelihoods

In Asia, the so-called "green revolution" of the 1960s and 1970s was aimed at alleviating poverty and increasing food production through agricultural intensification. While the "package" of green revolution options included improved crop varieties, fertilizers, and pesticides, reliable water is considered the lead input for most intensive agricultural production. Without assured irrigation, these other options are far less effective (Ahmed 2001). This is also the case with labor and other inputs.

Groundwater is a particularly reliable source of irrigation. When farmers have access to wells, irrigation is available "on demand"—at the time and in the amounts required to meet plant growth requirements. The importance of this has been well documented. Many crops are vulnerable to moisture stress at critical periods and yields can be substantially reduced even if adequate water supplies are available following periods of shortage (Perry and Narayanamurthy 1998). Water stress at the flowering stage of maize, for example, can reduce yields by 60 percent, even if water is adequate during all the rest of the crop season (Seckler and Amarasinghe 1999). Similar impacts on onions, tomatoes, and rice have also been documented (Meinzen-Dick 1996). In addition to the direct impact of water availability on crop growth, assured supplies are a major factor in decisions to invest in other inputs to production such as labor, fertilizers, improved seeds, and pesticides (Kahnert and Levine 1989, Meinzen-Dick 1997, Seckler and Amarasinghe 1999). "On-demand" groundwater allows farmers to take the risk of investing in fertilizer and other factors to increases crop productivity.

The reliability of groundwater as a source of irrigation is also a major factor contributing to the now well-established observation that yields in areas irrigated with groundwater are substantially higher than in areas irrigated from other sources. This was documented in the 1970s and 1980s in Pakistan and India (Lowdermilk et al. 1978, Renfro 1982, Dhawan 1989b, Shah 1993, Meinzen-Dick 1997). Subsequent studies in Asia and other parts of the world suggest that this continues to be the case. An ongoing study by IWMI in the Pakistani Punjab, for example, "indicates that farmers with wells obtain 50 to 100 percent higher yields per acre and 80 percent higher value of output per acre compared to canal irrigators" (Shah et al. 2003).

Such results apply outside of Asia. In Andalusia, Spain, groundwater-irrigated agriculture is economically over 5 times more productive (in terms of pesetas/m^3) and generates more than 3 times the employment compared to surface-irrigated agriculture (Hernandez-Mora et al. 2001). According to Barraqué: "irrigation uses 80 percent of all water in Spain and 20 percent of that water comes from underground...the 20 percent, however, produces more than 40 percent of the cumulated economic value of Spanish crops" (Barraqué 1997).

Beyond productivity *per se*, groundwater plays a particularly important role during periods of drought. During the 1960s (before extensive groundwater development occurred in India), drought caused agricultural production to decline by as much as 20 percent in some affected states. Similar droughts during the 1980s had minimal impact on overall production. Furthermore, as the level of groundwater development increased over the 1960s through the 1980s, the stability of agricultural production in India increased (The World Bank 1998). Such findings are common. In an analysis of wheat cropping in the Negev desert, Tsur estimated the "stabilization value"—that is the value associated with the *reliability* of the water supply as opposed to just the value of the volume available of groundwater development as "more than twice the benefit due to the increase in water supply" (Tsur 1990). In southern California where surface water supplies are less variable than the Negev, the stabilization value in agriculture is, in some cases, as much as half of the total value of groundwater (Tsur 1993). During a 6-year drought in the late 1980s and early 1990s in California, economic impacts were minimal largely because farmers were able to shift from unreliable surface supplies to groundwater (Gleick and Nash 1991).

The reliability of groundwater underlies the tremendous demand for groundwater by millions of individual farmers that has emerged over recent decades in locations such as India. According to the Census of India, the country's population is approximately 70 percent rural and virtually all of those individuals depend on agriculture. Similarly, Shah et al. (2003) estimate that 55 to 60 percent of the total Indian population depends on groundwater. If these numbers are even approximately accurate, the role groundwater plays as a cornerstone of rural livelihoods in the region is clear.

Although supported by subsidies and government investment in electrification, most groundwater development in South Asia and other parts of the world has been through the initiative and investment of individual farmers. As Roy and Shah comment: "Groundwater exploitation and extraction is a function of predominantly *demand for irrigation* and has little to do with availability *per se*" (Roy and Shah 2003, p. 329). It is an outgrowth of rural demand and recognition on the part of rural populations of the high returns achievable through groundwater irrigation. Declines in poverty between the 1950s and 1990s across a variety of states in India are significantly correlated with levels of groundwater extraction (Moench 2001). While this could be an artifact of development (poverty declines enabling farmers to invest in groundwater rather than vice versa), logic suggests that access to groundwater has played a major role in its own right. Increases in yields and reductions in losses achieved through groundwater irrigation enable farmers to generate surpluses far more consistently than when they depend on rainfall or surface irrigation. Once generated, surpluses are available for saving or investment in health care, education, land, or other economic activity. Farmers become, to use entitlements terminology (Drez, Sen et al. 1995, Sen 1999), "entitled" to a much wider pool of assets than they would have been without access to groundwater. Furthermore, because losses are reduced, the capital stock held by groundwater users tends to accumulate over time and people who were once marginal farmers move out of poverty. In sum, a secure water supply represents the foundation for a pyramid of entitlements (physical, economic, and social assets) and helps break the cycle of poverty common in agricultural regions (Moench 2001).

This dynamic, recognized by millions of Indian farmers, underlies the explosion in groundwater irrigation and growth of the groundwater economy. A similar dynamic also probably underlies the massive explosion in groundwater extraction in the North China Plain. To quote Shah et al. again: "The Asian groundwater economy has emerged as a spontaneous response to a need felt by millions of farmers. As a result, it exists entirely within the private and informal sectors, with no, or very limited, regulation" (Shah et al. 2003). This is a key point—and one that directly relates to the larger question identified earlier: How accurately can emerging problems be quantified—and if they cannot be quantified—how can they be managed?

The Analytical Dilemma

Accurate data on groundwater availability and extraction are essential for any quantitative analysis of emerging overdraft and quality problems—but the structure of groundwater extraction in locations such as India and China greatly complicates such quantification. While it is, of course, possible to develop approximations, moving beyond this requires detailed information on well numbers, extraction rates, pumping hours, well depths, and a variety of other factors. In Europe and the West where most agricultural wells are fewer in number and larger in capacity, programs to monitor drilling activity and license wells generate some reliable information on well locations, capacities, depths, and other characteristics. In addition, since most wells are large and powered by electricity, it is possible to indirectly estimate pumping by monitoring power consumption. None of this applies to the tens of millions of small wells that supply agricultural users in Asia. The contrasting use situations highlighted in the data in Table 4.2 have fundamental implications for the ability to monitor groundwater use and quantify extraction. Furthermore, as discussed above in the context of the Aquastat and WRI databases, even in Europe and the West, comprehensive data on extraction or groundwater use within key sectors are often unavailable.

The lack of data creates a major analytical dilemma. Overdevelopment and quality declines are clearly emerging across large portions of Asia. The data required to quantify these problems or project their macro-level implications, however, either do not exist or are inaccessible. History is littered with compelling macro-analyses of environmental problems that have floundered on seemingly minor gaps in understanding and information. As argued elsewhere, this may well be the case with the emerging global debate over the implications of groundwater overdraft (Moench et al. 2003). While the impact of declining groundwater levels is often clear at the local level, groundwater conditions are highly variable and aquifer systems are, themselves, too poorly monitored and understood to quantify aggregate impacts. Moving forward requires recognition of fundamental data constraints and their relationship to management.

Exploring the Connection between Data, Analytical Approaches, and Management

Conventional concepts of groundwater management emphasize the balance between extraction and recharge as a key indicator of sustainability. While this may be true in some ultimate sense, it is highly misleading as an indicator of management needs. As previously noted, most social and environmental services are affected more by groundwater levels, pressure gradients, and quality changes than they are by changes in the amount of water in storage (Burke and Moench 2000). Furthermore, a variety of fundamental and local factors often constrain our ability to generate the types of data required to implement management approaches designed around sustained yield concepts. This is best illustrated in the context of specific management examples such as:

- The designation of critical overdraft areas for intensive management (as shown in parts of the United States, India, China, and elsewhere);

- Attempts to encourage or regulate groundwater extraction using regional economic incentives (e.g., India's attempts to use government credit support programs and restrictions on power supplies for wells as mechanisms to encourage or discourage groundwater pumping across broad regions);

- Establishment of volumetric allocation systems (implemented in parts of the Western United States and widely advocated as best practice elsewhere, including in the well registration process now underway in Mexico).

In each of these cases, the sustained yield concept is most valuable in a very limited number of areas where the need for management is high and aquifers are clearly defined and well monitored—but it breaks down as attempts are made to apply the concept more broadly. This is particularly evident with regard to the latter two of the management approaches. Let us look at the issues with each of these approaches sequentially.

Intensive Management

Where intensive management is concerned, identification of critical overdraft areas is generally stimulated first by falling water levels or declining water quality that affects users. The presence of such problems then triggers detailed hydrological studies needed to accurately quantify extraction-recharge balances and estimate sustained yield under different management scenarios. This, in turn, leads to the development of specific management responses designed either to increase supply or to decrease demand. This pattern is common from the early identification of overdraft areas in the western United States (Southern California and Arizona) to recent attempts to implement management in locations as diverse as India, Yemen, and Mexico.

A key point to emphasize, however, is that in all these situations accurate scientific evaluation (as opposed to rough estimation) of groundwater recharge and extraction is, in essence, a reactive analytical tool applied where problems are already evident. Regions almost never invest heavily in the type of hydrological data collection and analysis necessary to accurately quantify sustained yield within individual aquifers unless they already have evidence of problems. Furthermore, even where there is a clear need for management and regions do invest in hydrological analysis, sustained yield concepts are often inadequate as a basis for management. In the San Luis Valley of Southern Colorado, for example, interviews with regional hydrologists indicated that despite decades of research, monitoring, and analysis there is still a 30 percent gap between estimates of inflow and outflow. The complex nature of the regional hydrology (in particular the possibility of deep groundwater inflow) and difficult-to-quantify water demand components (evapotranspiration by native vegetation) make accurate quantification impossible even within this clearly defined sedimentary basin. Such problems expand when attempts are made to apply sustained yield concepts more broadly as a basis for regional management approaches or rights systems.

Regional Economic Incentives and Scientific Uncertainty

Overall, the conceptually straightforward idea of using sustained-yield estimates as a basis for identifying regional groundwater management needs runs into serious practical problems. Many groundwater problems are not related to the extraction-recharge balance. In addition, even where they are, aquifers are difficult to identify, data are difficult and expensive to collect, and estimates require numerous assumptions. When such estimates are used in politically sensitive management contexts they tend to shift debates away from practical responses and onto the uncertainties and political implications of the estimates themselves. The problem is even more acute where sustained-yield concepts are used as a basis for water rights or volumetric allocation systems.

India's attempts to use extraction and recharge estimates as a basis for allocating groundwater development support (credit for wells and power connections) on a regional basis illustrates the importance of scientific and data problems as well. For more than two decades, India has used the ratio between extraction and recharge estimates as the primary tool for classifying the groundwater development potential of local administrative units. Areas where extraction exceeds recharge are classified as "overexploited"; areas where extraction is greater than 85 percent of recharge are classified as "dark;" areas where extraction is between 65 percent and 85 percent of recharge are "grey;" and those where extraction is less than 65 percent of recharge are "white" (Ground Water Resource Estimation Committee 1997). This classification system is used as a tool for allocating development support funds—credit for well development is cut off in "dark" and "over-exploited" areas, only available on a limited basis in "grey" areas, and fully available in areas classified as "white".

Although this approach has been refined several times since it was initially developed, major questions remain over its scientific validity and effectiveness. Restrictions on credit and power connections appear to have done little to reduce pumping rates or the number of wells installed (The World Bank 1998). Scientific problems are more fundamental. While some of these problems, such as the use of administrative rather than hydrologic units as a basis for evaluation, can be corrected, others are inherent in the approach. Hydrologic conditions are highly variable making aquifer identification and monitoring both difficult and expensive. Although most states monitor more than 2,000 wells, individual administrative blocks often only contain 5 to 6 monitoring wells and there may be none within key aquifers.

To complicate matters, two-thirds of India is underlain by hard-rock formations, but the monitoring systems and analytical procedures do not differ greatly from those developed for and used in alluvial aquifers. This is inappropriate. Water table conditions in hard-rock regions can show tremendous variation between individual wells even within a single village. Furthermore, unlike alluvial aquifers, changes in storage in the vadose (unsaturated) zone may be the primary factor determining actual water availability in hard-rock areas (Narasimhan 1990). Even for the saturated zone, estimating changes in storage requires accurate assessments of specific yields from pumping tests.

Most analytical methods for interpreting pumping tests were initially devised for bore wells in alluvial aquifers with relatively simple geometric configurations. They do not apply to the large diameter wells and complex, heterogeneous, hydrologic conditions typical of the hard-rock aquifers extending throughout most of peninsular India (Moench 1996). As Narasimhan states, "indiscriminate fitting of hydraulic test data to available mathematical solutions will but yield pseudo hydraulic parameters that are physically meaningless" (Narasimhan 1990). Overall, for hard-rock regions, "a sound rational basis does not exist yet for quantifying resource availability and utilization" (Narasimhan 1990).

Similar problems apply in many alluvial areas. In numerous areas of the western portions of the Gangetic basin, for example, water levels and water quality have been declining for long periods of time. Technically, the basin can be viewed as one single interconnected aquifer system that stretches for thousands of kilometers from Bangladesh across most of North India and the Nepal Terai. According to the most recently available published data from the Central Ground Water Board, all districts in Uttar Pradesh, Bihar, or West Bengal, the main Indian states in the Gangetic Basin, have extraction levels below 60 percent of recharge (Central Ground Water Board 1995). The fact that the aquifer as a whole is not suffering from overdraft does not, however, minimize the impact pumping is having within many areas. Water-level declines or

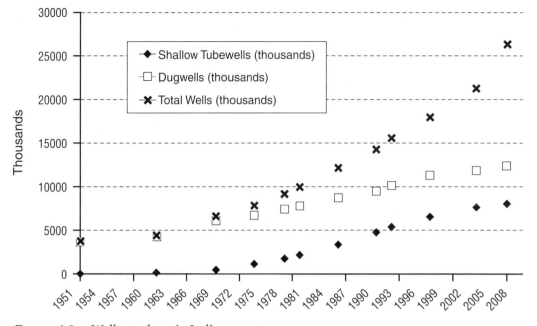

FIGURE 4.1   Well numbers in India.

Note: Also included in the total well numbers, but not plotted separately, are boreholes and public tubewells.
Numbers after 1998 are projections.

Source: The World Bank 1998.

increases in fluctuation are common and have major implications for key social and
environmental values including access to groundwater for the poor (whose wells may
not always be deep enough) along with the maintenance of instream flows and
wetlands. Equally important, water-quality changes, such as the problems with arsenic
(described above), may be related to extraction rates. Overall, the balance between
recharge and extraction gives little indication of the impact groundwater extraction
may be having on different classes of users or on key environmental values.

Another key scientific problem is the lack of real data on extraction. Two primary
methods are used to estimate groundwater extraction: estimates of irrigated area
supplied by groundwater and estimates based on well numbers and pumping rates.
Where the first approach is concerned, wells are known to be a major source of irrigation and buffer supply within the command of virtually all surface-irrigation systems.
Official data, however, generally classify irrigation based on the dominant source and,
as a result, probably underestimate the level of groundwater extraction for agriculture
(Roy and Shah 2003). Data on well numbers are equally problematic. Official estimates
clearly demonstrate the exponential growth in groundwater extraction that has
occurred over recent decades (see Figure 4.1).

By 2002, official sources indicated that there were approximately 22 million wells in
India, an increase of more than 21 million from the roughly 800,000 present in 1975
(Hydrology Project 2002). The numbers are, however, derived primarily from statistics
on the number of bank loans issued for well construction, not on actual surveys. Since
many farmers are not able to obtain such loans, the actual number of wells installed
since 1975 is likely to be even higher than official figures suggest. Operational wells are,
however, a different question. Water levels in many arid and hard rock sections of India
have declined dramatically and many, though generally uncounted, wells have gone
out of production. Although a partial census was carried out in 1995, the most recent

detailed information on abandoned wells for all states in India is the census of minor irrigation published in 1987—years before many areas began to experience water-level declines. In the absence of accurate information on operational wells or the contribution of groundwater to irrigated areas, any estimate of groundwater extraction will have a high level of uncertainty.

These scientific uncertainties have implications that go beyond the obvious difficulties in using them as tools to identify management needs. On a political level, the presence of substantial scientific uncertainty undermines India's ability to implement management. Groundwater recharge-extraction estimates can easily be manipulated and since these estimates are used as triggers for politically-sensitive management interventions, incentives for manipulation are often present (Moench 1994). Even where manipulation does not occur, estimates depend heavily on a variety of assumptions making legitimate disagreements between states and the central government (all of which operate separate monitoring systems) common.

A key indicator of the depth of such problems is the fact that, although the Central Ground Water Board estimates resource availability annually, the most recent official data were published in 1995 based on data from the late 1980s and early 1990s (Central Ground Water Board 1995). Another key indicator is the fact that Rajasthan, Haryana, and Punjab—some of the states with the most intense groundwater overdraft problems in India—have decided not to participate in the Hydrology Project, a major initiative to build a national database on water resources financed by The World Bank and the Government of the Netherlands. In essence, the tug of war over extraction recharge estimates has transformed groundwater data from an objective technical measure of resource availability into a politically charged debate.

Volumetric Allocation Systems

Volumetric allocation systems based on aquifer sustained yield are widely advocated as one of the conceptually best approaches to allocating available groundwater resources among users. This approach underlies regulatory allocation and water rights systems in locations as diverse as parts of the Western United States and Jordan and has been widely advocated for other regions including Mexico, India, and South Africa.

In a very similar manner to the viability of regional economic incentives, the viability of volumetric allocation systems depends heavily on the ability to monitor aquifer conditions and groundwater extraction rates. The approach is most effective in locations where well numbers are low, aquifers are well understood, and monitoring is relatively straightforward. The approach breaks down as technical uncertainties surrounding the relationship between extraction allocations and sustained yield become uncertain and as user/well numbers increase.

The situation in Colorado illustrates technical problems underlying allocation systems. Under Colorado law, groundwater is classified as either tributary (i.e., an integral part of) interlinked ground and surface hydrologic systems or non-tributary (i.e., isolated from) such systems. So-called "tributary" groundwater allocations are administered as part of the surface-rights system while "non-tributary" groundwater is allocated and administered separately. The technical problem is twofold: determining the difference between "tributary" and "non-tributary" groundwater, and determining the amount of water available for capture in both cases.

This technical challenge, inherent in most dynamic hydrologic systems, lies at the heart of ongoing litigation between groundwater and surface-water users in, for example, the South Platte river basin. Despite huge investments in monitoring and analysis, the technical issues inherent in quantifying the volume of water available for appropriation in interlinked groundwater and surface-water systems are not likely to be resolved soon. As previously discussed in the San Luis Valley case, fundamental uncertainties regarding the hydrological dynamics of regional systems make accurate quantification impossible in many situations. The inherent technical challenges often undermine the social and political acceptability of volumetric rights systems—if we cannot *prove* that additional water is not available for appropriation or that groundwater pumping is affecting surface water rights, it is very difficult for courts or politicians to place limits on new wells or the level of extraction from existing ones.

Inherent hydrological uncertainties are compounded as the number of wells and users increases. As Shah et al. document (see Table 4.2), there are roughly 200,000 major agricultural wells in the United States each pumping an average of 500,000 m^3/yr. At the other end of the spectrum in India, the number of wells (19 to 26 million), is orders of magnitude higher and the average water extraction per structure, some 7600 m^3/yr, is orders of magnitude lower (Shah et al. 2003). The huge number of wells and users in India greatly complicates monitoring and quantification.

Even in Mexico, where the number of wells is relatively low and the extraction per structure approaches United States levels, the process of just registering wells—which was started roughly a decade ago—is far from complete. While the establishment of aquifer management councils (COTAS—*Consejos Técnicos de Aguas*) is a major reform and has been portrayed as a vehicle for the implementation of innovative groundwater management approaches (Wester et al. 1999, Shah et al. 2003), actually quantifying pumping and establishing a licensing or rights system that would limit extraction to sustainable levels remains a distant and possibly unachievable vision. According to Patricia Romero Lankao, in Guanajato, for example, because the government lacks the capacity to monitor water use and farmers are unwilling to provide information, official data are unreliable. Mechanisms intended to reduce groundwater extraction, such as increasing electricity tariffs, imposing standards on the maximum level of water and energy consumption, imposing water fees, and modernizing irrigation systems, have not been implemented and most aquifers are overexploited. Estimates indicate that extraction by farmers is twice the maximum level allowed by regulatory bodies (Lankao 2003).

The situation is even more complicated in South Asia. While many recommend that Asia should follow Mexico's example: "it is easy to imagine how difficult it would be to enforce such a regime on 19 million tubewell owners given that Mexico is finding it difficult to enforce it on a mere 70,000 groundwater irrigators" (Shah et al. 2003). Most of the wells in India are small dispersed structures located on private lands. Some such wells are buried and capped when not in use. As a result, only the individual farm owner knows their location. In the South Asian context, simply locating the wells would, in itself, be a vast undertaking to say nothing of monitoring or regulating extraction. The situation is similar in China where a recent report by the Ministry of Water Resources states:

> "Currently there are thousands of local and personal databases storing key technical and licensing data in a very unsatisfactory manner. An absolutely fundamental need for effective groundwater management and protection is a comprehensive, publicly accessible, groundwater database (GDB). The complete lack of a GDB is seriously constraining the formulation and implementation of effective groundwater management through-

out China. The inability to access information, which at times is part of institutional secrecy, encourages inaction or incorrect decisions. GDBs are well established in almost every country where significant groundwater is used. The lack of such a database in China is surprising" (Ministry of Water Resources 2001, p. 231).

Even where groundwater databases have been established, fragmentation of, and access to the data often constrain policy. In sum, while conceptually simple, establishing groundwater rights or regulatory systems founded on extraction volumes appears impractical in many contexts.

A Way Forward: Simple Data as a Catalyst for Effective Management

Data availability and monitoring are directly linked to the practical applicability of management and water allocation concepts. In the groundwater case, managers often cannot develop effective management and allocation systems because they rely on concepts of sustained yield but cannot generate the basic scientific and monitoring data required to translate such concepts into practical tools for management.

Management approaches designed around extraction-recharge balances evolved over decades in locations, such as the Western United States, where there are few users and it is relatively easy to collect and analyze water resources data. Such approaches have proved successful only where society has sufficient financial and technical resources to eliminate or at least bound scientific uncertainties. This is not the case in much of the world. As Shah et al. note, "the per capita expenditure on groundwater management in the Netherlands is five times greater than the total per capita income in rural northern Gujarat" [India] (Shah et al. 2003, p. 139). Approaches to groundwater monitoring, allocation, and management that require intensive scientific analysis and detailed data on recharge and extraction simply are not viable in most locations, particularly in the developing world.

So what should be done? Addressing the situation calls for simplicity. While water balance (extraction-recharge) estimates have a nice conceptual link to sustainability, on a practical level changes in water levels mean much more to governments, water users, and the environment. Declining water levels directly affect energy requirements for pumping and the cost of establishing wells. These costs are critical for small poor farmers. As a result, water levels are closely linked to questions of poverty, agricultural economics, and equity. In addition, changes in groundwater levels are the primary factor influencing environmental impacts (baseflows in streams, maintenance of wetlands, changes in groundwater-dependent surface vegetation, etc.).

In most situations, the primary long-term groundwater data collected by water resource departments relate to water levels and water quality. These data are rarely analyzed or disseminated in any systematic manner. Such data could, however, serve as much better indicators of groundwater conditions than the current reliance on crudely estimated extraction and recharge balances.

Managers should also use some other simple indicators. These can include key groundwater quality parameters and operational well characteristics. When combined with existing hydro-geological information, they can provide a foundation for monitoring groundwater conditions at all levels from local regions or aquifers to global assessments. Furthermore, because each of these indicators is a direct measure, the level of uncertainty inherent in the measure itself is much lower than with extraction and recharge estimates. Changes in the indicators would, as a result, be much less subject to challenge or manipulation when used in socially or politically charged management contexts. The tangibility of these indicators along with their direct significance for economic and environmental use values could also serve as a much more effective catalyst for management than the current reliance on less tangible estimates.

As a final note it is important to emphasize that placing greater reliance on direct measures of groundwater conditions would not reduce the need for more detailed analyses. The nature of such analyses would, however, depend on local conditions and the specific types of problems emerging. In many cases, effective management is likely to require estimation of extraction and recharge within individual aquifers. In equally many cases, however, such information will be irrelevant.

Experts working on the implications of emerging groundwater problems face a fundamental conundrum: while they frequently encounter evidence of serious overdraft and quality problems at local levels, global and regional debates are dominated by qualitative, often anecdotal, arguments. The data in this chapter are no exception. Good quality data on groundwater conditions either do not exist or are inaccessible for much of the world. We must resolve this and create an information base necessary for management at local levels. This requires greater reliance on direct measures of groundwater conditions.

REFERENCES

Ahmed, N. 2001. *Water and fertilizers to support food production.* Rome:Food and Agriculture Organization.

Appleton, J. D., Fuge, R., and McCall, G. J. H. (editors). 1996. *Environmental geochemistry and health: With special reference to developing countries.* Geological Society Special Publication No. 113. London:The Geological Society.

Barraqué, B. 1997. Groundwater management in Europe: Regulatory, organisational and institutional change. Conference on how to cope with degrading groundwater quality in Europe. October 21–22. Stockholm.

Bhattacharya, P., Chatterjee, D., and Jacks, G. 1996. *Safeguarding groundwater from arseniferous aquifers.* WEDC Conference Proceedings (J. Pickford, editor), Reaching the unreached: Challenges for the 21st century, 258–261. New Delhi, WEDC, Loughborough University, Leicestershire, United Kingdom.

British Geological Survey and Mott Macdonald Ltd. 1999. *Groundwater studies for arsenic contamination in Bangladesh. Phase 1: Rapid investigation phase. Final report.* Dhaka and London: British Geological Survey and Mott MacDonald Ltd. in cooperation with the Government of the People's Republic of Bangladesh, the Ministry of Local Government, Rural Development and Cooperatives, the Department of Public Health Engineering, and the Department for International Development, United Kingdom.

Burke, J., and Moench, M. 2000. *Groundwater and society: Resources, tensions, opportunities.* New York:United Nations Publications.

California Department of Water Resources. 2003. California's groundwater. Bulletin 118-03. Sacramento, California.

Central Ground Water Board 1995. Groundwater resources of India. Faridabad, Central Groundwater Board, Ministry of Water Resources, Government of India.

Dhawan, B. D. 1989. *Studies in Irrigation and Water Management.* New Delhi:Commonwealth Publishers.

Edmunds, W. M., and Smedley, P. L. 1996. Groundwater chemistry and health: An overview. In *Environmental geochemistry and health with special reference to developing countries*, 91–105. Appleton, J. D., Fuge, R., and McCall, G. J. H., editors. Geological Society Special Publication No. 113. London:The Geological Society.

Freeman, D. M., Lowdermilk, M. K., and Early, A. C. 1978. Farm irrigation constraints and farmer's responses: Comprehensive field survey in Pakistan. Colorado State University, Fort Collins, Colorado.

Gleick, P. H. 2000. Arsenic in the groundwater of Bangladesh and West Bengal, India. *In The world's water 2000–2001*, 165–174, Gleick, P. H., editor. Washington, D.C.:Island Press.

Gleick, P., and Nash, L. 1991. *The societal and environmental costs of the continuing California drought.* Oakland, California:Pacific Institute for Studies in Development, Environment, and Security.

Ground Water Resource Estimation Committee. 1997. *Ground water resource estimation methodology–1997.* New Delhi:Ministry of Water Resources, Government of India.

Hernandez-Mora, N., Llamas, R., and Cortina, L. M. 2001. Misconceptions in aquifer overexploitation: Implications for water policy in Southern Europe. Workshop SAGA. FAIR-CT97-3673, Fondazione Enni Enrico Mattei, Milan.

Hydrology Project. 2002. Newsletter. New Delhi:The Hydrology Project.

Kahnert, F., and Levine, G. 1989. *Key findings, recommendations, and summary.* Conference proceedings, Groundwater irrigation and the rural poor: Options for eevelopment in the Gangetic Basin. Washington, D.C:The World Bank.

Kingdom of the Netherlands (KON). 1992. *India: Hydrological investigations for the Santalpur and Sami-Harij RWSS.* Kingdom of the Netherlands, Directorate General of International Cooperation; Government of India, Ministry of Agriculture; and Gujarat Water Supply and Sewerage Board. New Delhi.

Lankao, P. R. 2003. Mexican water reform, achievements and paradoxes. Draft Paper. Department of Politics and Culture, Metropolitan Autonomous University, Xochimilco. Mexico.

Lowdermilk, M. K., Freeman, D. K., and Early, A. C. 1978. Farmer irrigation constraints and farmer's responses: Comprehensive field survey in Pakistan. Water Management Technical Report Number 48-E, Volume V. Colorado State University. Fort Collins, Colorado.

Meinzen-Dick, R. 1997. Groundwater markets in Pakistan: Participation and productivity. Washington, D. C:International Food Policy Research Institute.

Ministry of Water Resources. 2001. China agenda for water sector strategy for North China. Beijing: Ministry of Water Resources.

Moench, M. 1994. Hydrology under central planning: Groundwater in India, *Water Nepal*, 4, issue 1:98–112.

Moench, M. 1996. Groundwater policy: Issues and alternatives in India. Colombo, Sri Lanka: International Irrigation Management Institute.

Moench, M. 2001. *Groundwater and poverty: Exploring the links.* Workshop on intensively exploited aquifers, Royal Academy of Sciences, Madrid.

Moench, M., Burke, J., and Moench, Y. 2003. *Rethinking the approach to groundwater and food security.* Rome:Food and Agriculture Organization of the United Nations.

Moench, M., and Matzger, H. 1994. Ground water availability for drinking in Gujarat: Quantity, quality and health dimensions. *Economic and Political Weekly*, XXIX, issue 13:Section A, 31–44.

Narasimhan, T. N. 1990. Groundwater in the Peninsular Indian Shield: A framework for rational assessment, *Journal of the Geological Society of India*, 36, 353–363.

Nordstrom, K. 2000. An overview of mass-poisoning in Bangladesh and West Bengal, India. Young and Courtenay (editors), *Minor elements 2000, Processing and environmental aspects of As, Sb, Se, Te, and Bi*, 21–30 Society for Min. Metall. Expl.

Perry, C. J., and Narayanamurthy, S. G. 1998. Farmer response to rationed and uncertain irrigation supplies. International Water Management Institute, Report No. 24. Colombo, Sri Lanka.

Postel, S. 1999. *Pillar of sand: Can the irrigation miracle last?* New York:Norton.

Renfro, R. Z. H. 1982. Economics of local control of irrigation water in Pakistan: A pilot study. Colorado State University, Fort Collins, Colorado.

Roy, A. D., and Shah, T. 2003. Socio-ecology of groundwater irrigation in India. In *Intensive use of groundwater: Challenges and opportunities*, ed. R. Llamas and E. Custodio, 307–336. A.A. Balkema, Lisse, the Netherlands.

Seckler, D., and Amarasinghe, U. 1999. Chapter 3, water supply and demand, 1995 to 2025: Water scarcity and major issues. IWMI web site. Colombo, Sri Lanka.

Seckler, D., Barker, R., and Amarasinghe, U. 1999. Water scarcity in the twenty-first century. *International Journal of Water Resources Development*, 15:issues 1/2, 29–42.

Shah, T. 1993. *Groundwater markets and irrigation development: Political economy and practical policy.* Bombay:Oxford University Press.

Shah, T., Molden, D., Sakthivadivel, R., and Seckler, D. 2000. *The global groundwater situation: Overview of opportunities and challenges.* International Water Management Institute, Colombo, Sri Lanka.

Shah, T., Roy, A. D., Wureshi, A. S., and Wang, J. 2003. Sustaining Asia's groundwater boom: An overview of issues and evidence. *Natural Resources Forum,* 27, 130–141.

Tsur, Y. 1990. The stabilization role of groundwater when surface water supplies are uncertain: The implications for groundwater development. *Water Resources Research,* 26, issue 5, 811–818.

Tsur, Y. 1993. *The economics of conjunctive ground and surface water irrigation systems: Basic principles and empirical evidence from Southern California,* Report P93-15, Department of Agricultural and Applied Economics, University of Minnesota. Minnesota.

United Nations. 1983–1990. *Regional groundwater reports.* Natural Resources Water Series. New York, UNDTCD. New York:United Nations Publications.

Wester, P., Pimentel, B. M., and Scott, C. A. 1999. *Institutional responses to groundwater depletion: The aquifer management councils in the State of Guanajato, Mexico.* Colombo, Sri Lanka: International Water Management Institute.

Wijdemans, R. T. J. 1995. Sustainability of groundwater for water supply: Competition between the needs for agriculture and drinking water. In *Groundwater availability and pollution: The growing debate over resource condition in India,* ed. M. Moench, 60. Ahmedabad, India:VIKSAT.

The World Bank and Government of India, Ministry of Water Resources. 1998. *India—water resources management sector review, groundwater regulation and management repor*t. Washington, D. C.:The World Bank, New Delhi:Government of India.

CHAPTER 5

Urban Water Conservation: A Case Study of Residential Water Use in California

Peter H. Gleick, Dana Haasz, Gary Wolff

Water conservation measures are real and practical, and offer enormous untapped potential. In fact, the largest and least expensive source to meet future needs for water in many regions is the water currently being wasted in each sector of the economy. The potential for conservation and improved efficiency is so large that it can often delay or eliminate the need for new sources of supply for a long period, even with expected growth in populations and economies. Moreover, capturing this water is often cheaper and more environmentally beneficial than any other alternative available.

What is the potential for water conservation and efficiency improvements? Remarkably, few governmental or private water organizations have ever made a comprehensive effort to find out. Yet this information is vital to decisions about meeting future needs, restoring the health of ecosystems, and reducing conflicts over shared water resources. Without information on the potential for water conservation, questions about industrial production, ecosystem restoration, immigration policy, land use, and urban growth will be much harder to answer, or worse, the answers provided will be wrong.

This chapter and the next summarize recent efforts to provide part of the missing information, in the context of a wealthy developed region—the State of California. In a report released in late 2003, the Pacific Institute quantified the potential for water conservation and efficiency improvements in California's urban sector, where around 8,600 million cubic meters (mcm) of water are annually used to satisfy commercial, industrial, institutional, and residential needs (Gleick et al. 2003). While this analysis is specific to California, the strong parallels in water use throughout the United States, and, indeed, in many parts of the world, strongly suggest that similar types, and perhaps degrees, of savings can be found elsewhere.

This research shows that most, if not all of California water needs in the coming years can be met by smart and thoughtful use of existing technology, revised economic and pricing policies, appropriate state and local regulations, and public education. Furthermore, the state's natural ecological inheritance and beauty do not have to be sacrificed to meet water needs for future economic development.

Capturing wasted water will require expanding existing conservation and efficiency programs, developing new programs and policies, and educating consumers and policymakers. Further technological advances will also help. Some of the needed improvements will be easy; some will be difficult. But there is no doubt that the path to a sustainable water future lies not solely with dams, centralized facilities, and pipelines but with the integration of this hard infrastructure with the soft infrastructure of local water management, smart small-scale technology, active community participation in decision making, and efforts of innovative businesses.

The Debate over California's Water

California has a long history of rancorous and contentious water debates. The sheer size of the state, the number and diversity of people, and the complexity of its natural climate and hydrology have led to the development of an expensive, sophisticated, and controversial water system to address the needs of competing interests and stakeholders. While California's population may increase by 25 percent in the next 20 years (CDOF 2002), financial, environmental, political, and social constraints will likely prevent any significant expansion of California's already well-developed, water-supply system.

Traditionally, western United States satisfied increasing water demands through centralized decision making and large infrastructure investments in dams, pipelines, and treatment plants. Much of this infrastructure was built at the expense of taxpayers from around the nation. But the most cost-effective water sources were developed decades ago, leaving only expensive, environmentally sensitive, and politically controversial sites available for future development. At the same time, California's water supply is likely to shrink due to a reduction in diversions from the Colorado River,[1] the return of water to natural ecosystems, and efforts to eliminate unsustainable groundwater overdraft.

During the twentieth century, California water policy revolved around the simple belief that regular additions to supply were the only viable options for meeting anticipated increases in demand. This belief led to the first pipelines to bring water to California towns and cities, followed by ambitious aqueducts and big reservoirs to capture and store water far from where the water was needed, culminating in the vision—now a reality—of the massive state and federal water projects that dominate today's landscape.

This classical approach to water policy, imitated around the world, led to enormous benefits to the state and its people. It permitted California to grow into the dominant economic power that it is today, with vibrant and dynamic industrial and agricultural sectors, and allowed the growth of large population centers where local water resources were inadequate. But this approach also came with high costs—costs largely unrecognized or ignored by those who created and implemented that vision. Those costs included the degradation and destruction of a significant part of California's ecological heritage, the growing mistrust of local communities toward state and federal water planners, and ultimate gridlock of water policy during the closing years of the twentieth century.

---

1. Due to high flows and unused water rights on the Colorado River in recent years, California has consistently had access to approximately 20 percent more water than its legal entitlement of 4.4 million acre-feet. A highly contentious process is underway now to reduce California's use of Colorado River water.

As we move into the twenty-first century, these costs can no longer be ignored. The old reliance on narrow definitions of supply can no longer be used to meet new needs. The failure of California's traditional water-planning process is slowly leading to new discussions, new ideas, and new participants. In the past decade, progress has been made in building bridges among competing water interests and in expanding directions for discussing and resolving disputes. In time, we hope that these efforts will lead to new ways of thinking and new ways of meeting California's diverse water needs in a sustainable and equitable manner. But the process of developing an alternative approach has not yet been completed.

One of the major new factors in California's long water debate is the first real discussion about how water is actually used and the potential for using the state's limited water resources more efficiently. The water community is slowly coming to the realization that current use of water is highly inefficient and wasteful. Rethinking needs for water and how those needs are met could go a long way toward reducing the pressure on the state's fixed water supply and perhaps provide a model for other states, regions, and nations to follow. Various terms have been used to describe this concept: conservation, water-use efficiency, demand management, water productivity, best management practices, and so on. Despite some subtle or not-so-subtle differences among these terms, they all refer to policies, technologies, and approaches that permit society to meet specific goals with less water.

Defining Water "Conservation" and "Efficiency"

The concept of conservation and improved management of water use goes back many decades. In 1950, the President's Water Resources Policy Commission published *A Water Policy for the American People*, which noted:

> "We can no longer be wasteful and careless in our attitude towards our water resources. Not only in the West, where the crucial value of water has long been recognized, but in every part of the country, *we must manage and conserve water* if we are to make the best use of it for future development."[italics added]

What does conservation mean? There are many different and sometimes contradictory definitions of conservation. Baumann et al. (1980) defined water conservation using a benefit-cost approach: "the socially beneficial reduction of water use or water loss." In this context, water conservation involves trade-offs between the benefits and costs of water-management options. The advantage of this definition is that it focuses on comprehensive demand-management strategies with a goal of increasing overall well-being, not curtailing water use. In the public eye, conservation sometimes seems to mean deprivation—simply cutting back use of a resource, even if that means cutting back the goods and services produced by using that resource. More recently, academics and water professionals have made a major effort to ensure that the term *water conservation* refers to reducing water use by improving the efficiency of various uses of water, *without decreasing services*.

Another term—*technical efficiency*—is sometimes used to refer to the ratio of output to inputs, such as dollars per unit of water used. Improving technical efficiency can be accomplished by either increasing output or reducing water inputs. While this term can be useful, if offers little guidance as to how much reduction in water use is enough (Dziegielewski et al. 1999). For some end uses, maximum technical efficiency for water could be infinite by cutting the water requirement to zero. For example, dry composting toilets or waterless urinals require minimal or even no water.

The concept of efficiency is also useful when put into the context of investment decisions. *Economic efficiency* offers insight into the level of conservation reached when the incremental cost of reducing demand is the same as the incremental cost of augmenting supply. Using this criterion, water utilities or individuals would invest in water conservation programs until the conserved water is as expensive as new supplies, taking into account all the costs and benefits of water conservation and supply augmentation, including environmental and other external factors.

For the purposes of the analysis done for California, we use several different terms: the most common, *conservation*, describes any action or technology that increases the productivity of water use. Collectively, we refer to these actions and technologies as conservation measures, demand management, or improving water productivity. We examine two broad types of conservation measures: improving *water-use efficiency* and, to a lesser degree, substituting reclaimed water for some end uses.[2] Improving water-use efficiency includes behavioral and managerial improvements, such as adjusting the schedule for watering lawns and gardens, and technological improvements. Technological improvements usually involve replacing water-using equipment with equipment that serves the same purpose with less water. Thus *improving water-use efficiency* means reducing the amount of water needed for any goal while still accomplishing that goal. We exclude from our analysis any options that limit the production of goods and services through deprivation or cutbacks in production.

Many technologies and policies are available for reducing water use. In this context, the theoretical maximum *water-use efficiency* occurs when society actually uses the minimum amount of water necessary to do something. In reality, however, this theoretical maximum efficiency is rarely, if ever, achieved or even computed because the technology is not available or commercialized, the economic cost is too high, or societal or cultural preferences rule out particular approaches. We have adopted the following additional terms and definitions to guide our analysis:

Best available technology (BAT): The best, proven commercial technology available for reducing water use. A good example is the composting toilet, capable of meeting all disposal needs without the use of water. These toilets are proven and commercially available. BAT is useful for quantifying a maximum savings technically available. This is an objective assessment of potential, independent of cost or social acceptability. Thus, the BAT for toilets uses no water.

Best practical technology (BPT): The best technology available for reducing water use that meets current legislative and societal norms. This definition involves subjective judgments of social acceptability but defines a more realistic estimate of maximum practical technical potential, independent of cost. Our assumption of the BPT for toilets in the United States is the ultra-low-flow toilet (ULFT) meeting existing national standards of 1.6 gallons per flush.[3]

---

2. The potential for substituting reclaimed water for a wide range of water needs is large in California and elsewhere. We address that issue for only a few end uses, primarily in the industrial sector (see Chapter 6).

3. We note here, and elsewhere, that "dual-flush" toilets, which use less water than the current United States standard of 1.6 gallons per flush (gpf), are the norm in Australia and Japan. We fully expect them to become more common in the United States over time, but for the purposes of this study, we use 1.6 gpf as the BPT.

Maximum available savings (MAS): For a given agency, region, or state, MAS is an estimate of the maximum amount of water than can be saved under full implementation of best available technology (BAT), independent of costs.

Maximum practical savings (MPS): For a given agency, region, or state, MPS is an estimate of the maximum amount of water that can be saved under full implementation of best practical technology (BPT), independent of current costs.

Maximum cost-effective savings (MCES): For a given agency, region, or state, we define the maximum cost-effective savings as the maximum amount of water that can be cost-effectively saved under full implementation of best practical technology (BPT). "Cost effectiveness" is defined as the point where the marginal cost (and benefits) of the efficiency improvements is less than or equal to the marginal cost of developing new supplies.

Current Urban Water Use in California

Like many western states and indeed, water-short nations, California faces a growing population but a fixed and limited water supply. Much of the state's population lives in urban centers along the coast and, increasingly, in the Central Valley. In 2003, the state estimated California's total population to be 35.6 million people (CDOF 2003). While the state's population may increase dramatically in the next 20 years (State of California 2001), financial, environmental, political, and social factors will likely prevent any significant expansion of California's water supply.

Urban water is used for residential, commercial, and industrial purposes, outdoor landscaping, and other miscellaneous uses. The Pacific Institute estimates urban water use in California in 2000 to be approximately 8,600 mcm (7 million acre-feet [MAF]) annually, with an uncertainty of at least 10 percent. This estimate is shown in Table 5.1 and Figure 5.1. Total residential use is around 4,600 mcm (3.75 MAF) per year. Commercial and industrial uses, addressed in the next chapter, are estimated to be around 2,300 mcm (1.9 MAF) and 860 mcm (700,000 AF) per year respectively, with governmental and institutional uses included in the commercial estimate. No independent estimate of unaccounted-for-water (UfW) was done—the best current estimate is 10 percent of all urban use (CDWR 1994b).

TABLE 5.1   California Urban Water Use in 2000

California Urban Water Use by Sector	2000 Water Use (mcm/yr)	Percent of Total Urban Use (%)
Residential indoor	2,800	33
Residential Outdoor	1,200 to 2,345[1]	21[2]
Commercial/Institutional	2,280	27
Industrial	860	10
Unaccounted for Water [3]	860	10
TOTAL	8,600 (+/− 10%)	100[4]

Notes:
1. We provide a range here given the uncertainties in the data.
2. Calculated using the average of this range is 1,800 million cubic meters (mcm).
3. No independent estimate of unaccounted for water was made in this report. We adopt the 10 percent estimate from the California Department of Water Resources.
4. Rounded.

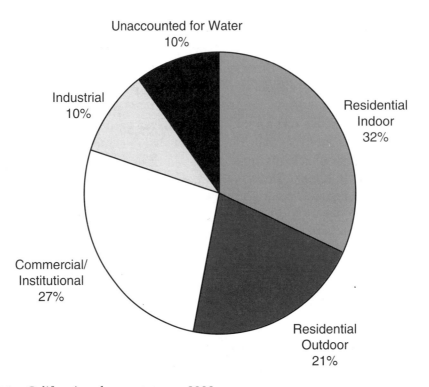

FIGURE 5.1   California urban water use, 2000.

We estimate indoor residential water use in 2000 was approximately 2,800 mcm. Table 5.2 shows our estimate of total indoor residential water use in 2000 by end use. Approximately a third of all indoor residential water goes to flush toilets. Other major uses are showers/baths, washing machines, and leaks. We estimate that leaks (which vary widely from house to house) average as much as 12 percent of total indoor water use.

Outdoor residential water use is highly uncertain since this use is not measured directly. Several approaches were taken to estimate this use at between 1,200 and 2,345 mcm per year in 2000, with an average of 1,800 mcm. Using this average, outdoor residential use is approximately 39 percent of total residential use.

TABLE 5.2   Estimated Indoor Residential Water Use in California (Year 2000)

Indoor End Use	Current Use (mcm/yr)	Fraction of Total Indoor Use (%)
Toilets	900	32
Showers	610	22
Washing Machines	410	14
Dishwashers	35	1
Leaks	350	12
Faucets	520	19
TOTAL	2,825	100

A Word About Agricultural Water Use

The vast majority of water used in many regions goes to the agricultural sector—an important part of the water "economy" not discussed in this chapter. Current estimates are that three-quarters of global applied water, and an even higher percentage of consumed water, is used for irrigation of food and fiber crops. This is also true in the western United States.

Water use in many parts of the agricultural sector is inefficient and wasteful, although some efforts are underway to address these problems. No comprehensive conservation and efficiency policy—indeed, no rational water policy—can afford to ignore inefficient agricultural water uses. If just 10 percent of this water can be saved with efficiency and conservation efforts—which we consider a highly conservative estimate given the available data and direct experience with on-farm efficiency programs in the United States and elsewhere—vast quantities of water would become available for alternative farming needs, ecosystem restoration, urban water use, or some combination. Obviously, a better, detailed assessment of the potential to improve efficiency of agricultural water use is urgently needed. We hope a future volume of *The World's Water* will address this issue.

Economics of Water Savings

Economics must play a fundamental role in helping to evaluate the relative merits of various water policy options and in implementing solutions to water problems. However, it is important to note limitations of economic analyses. Many economic data are uncertain. Water prices and rate structures vary over a wide range of values and designs. Humans respond to prices, but total water use is also determined by non-financial factors such as culture, preference, and tradition. And the costs of water-efficiency options change with time.

In order to address these uncertainties, assumptions and citations must be made explicit. (See Gleick et al. 2003 for a full examination of the uncertainties and inadequacies of the analysis.)

One way in which economic analysis is useful is that it provides rigorous methodologies for assessing costs of alternatives. For example, a thorough economic analysis can reveal to utility managers that interest, escalation, and delays associated with large capital-intensive water projects that lead to even slight forecasting errors can cause enormous increases in costs. It is an inherent characteristic of small-scale water-efficiency efforts that their lead times are substantially shorter than those of conventional big systems. Whether in development, distribution, installation, or repair, small and technically simple systems such as high-efficiency toilets, showerheads, or washing machines are faster than designing, permitting, financing, and constructing large-scale reservoirs. As Lovins (1977) noted for the energy industry, the industrial dynamics of this approach are very different—the technical risks are smaller and the dollars risked far fewer.

One of the reasons that efficiency approaches are difficult for traditional water agencies to adopt is that they shift the burden from engineering logistics to social ones. Traditional water agencies are often comprised of highly trained engineering experts who know how to design and build large structures that can serve a million people. But these same experts are unfamiliar with methods for designing and implementing conservation programs that reach a million individual customers.

The economic benefits of improving California water-use efficiency are substantial and compelling, and it is important to note that the popularity of conservation technologies should only increase in the future as competition for water grows, prices increase, and technology improves.

Our results also suggest that the benefits we have quantified understate the total benefits of these kinds of programs—perhaps substantially. Several kinds of benefits that we have not attempted to quantify, but that could have enormous additional advantages, are listed.

- Reductions in residential water use will lead directly to reductions in wastewater costs, both for treating wastewater as well as for building expensive new treatment facilities.

- Reductions in water use will lead to lower average peak water system loads—the most expensive kinds of water to provide.

- Reductions in water use will lead to lower average peak energy demands —the most expensive kind of energy to provide.

- Reductions in water use and subsequently in wastewater generation will lead to reductions in environmental damages from water withdrawals or wastewater discharges in sensitive regions.

- Investments in water-use efficiency leave money in local communities and create local jobs. Investments in distant new supply options usually take money from local communities and create distant jobs.

Data and Information Gaps

The "true" potential for water conservation technologies and programs will always be uncertain, because of wide variations in regional water use, prices, efficiency technologies, and many other factors. As a result, the estimates provided in these two chapters should be used with caution and an understanding that they are only as good as the assumptions and methods used to develop them. We have tried to be conservative in our estimates and explicit in describing our assumptions, and believe that the potential for cost-effective improvements in water use statewide most likely exceed the numbers reported here. But we urge that these kinds of estimates be done on local and regional levels as well, where uncertainties and data problems may be more readily resolved.

The lack of good data is a major constraint to comprehensive assessment of conservation potential. Data problems limit the ability of all researchers interested in water conservation and efficiency to evaluate potential savings and current success of conservation programs. We point out these data limitations throughout the chapter. But even when data were available, they often contained limitations that further affected the reliability of our estimates. Some data on efficiency programs were reported at the national level, which may be atypical. And when region-specific data were available, several factors often limited their usefulness.

In order to make intelligent decisions about water policy, gaps in the data urgently need to be filled, including the following examples (among many others):

- Residential and commercial landscape water use is poorly understood or measured.

- Distribution of residential water-using appliances, by type and use, is not well known.

- Economic costs of conservation options are sensitive to actual costs, lifetimes of conservation technologies, interest rates, and many other factors. Estimates of costs should be developed on a regional and utility basis.

- The water balance of major regions has not been adequately done.

- Rates of industrial water reuse are poorly reported.

- The implications for water quality of conservation options have not been explored analytically.

- Many benefits of water conservation are inadequately studied, poorly understood, or unquantified. These benefits include ecosystem improvements, reductions in wastewater treatment volumes, reduced need for investment for new facilities, and reductions in greenhouse-gas emissions from changes in energy use.

Indoor Residential Water Use

Efforts to reduce the wasteful use of water in homes have been underway for many years. Indeed, water conservation efforts have already made a big difference in improving the reliability of water resources in many regions, both reducing demand and freeing up new supply, reducing pressures to take any more water from overtapped rivers, lakes, and aquifers. Beginning in the early 1980s, Californians have participated in a range of programs to replace inefficient toilets, showerheads, and faucets, to audit heavy water users for leaks, and to reduce water use in gardens and other outdoor landscapes. We estimate that nearly 900 mcm (more than 700,000 acre-feet) per year of indoor savings have already been captured through a combination of smart regulation, improved technology, and educational programs. If used efficiently, this is enough water to meet the entire indoor residential needs of 17 million Californians each year.[4]

Among the first devices that water managers will chose for conservation programs are showerheads and toilets because they have a short payback period and are relatively simple to manage and install. In contrast, there has been little significant penetration of higher efficiency dishwashers (a relatively newly available technology) or reductions in leak rates (because of limited leak detection and prevention programs and inadequate data). In between these two extremes is the growing use of high-efficiency washing machines—these did not begin to appear in significant numbers in the United States until the late 1990s, but are now increasingly available and popular. For example, in 1999, an estimated 10,000 rebates were issued for high-efficiency washers in California (based on reporting data from the California Urban Water Conservation Council [CUWCC]); in 2002

---

4. One acre-foot currently satisfies the indoor residential needs of approximately 15 people in California. If currently available efficiency technology were used, one acre-foot could meet the indoor residential needs of 25 people. An acre-foot of water would cover one acre to a depth of one foot and would equal 326,000 gallons.

more than 24,000 rebates were awarded, and a total of 64,000 rebates have been awarded in the four years since 1999 (Dickinson, personal communications, 2003).

Figure 5.2 shows the indoor water savings that have already been achieved through current efforts and programs to replace inefficient toilets and showerheads. The top line is our estimate of what indoor residential water use in California would have been with no improvements in efficiency since 1980. The bottom line is our estimate of current indoor residential water use. As noted, we estimate that current use is around 900 mcm below what it would have been without existing conservation efforts.

Far more can be done to improve water efficiency, even with existing technology. The amount of water we estimate could be saved through comprehensive adoption of efficient technology and practices is presented in Figure 5.3. Table 5.3 summarizes the potential savings over current use for 2000 by specific end use. Although toilets have already had the single largest effect on indoor residential demand reduction, they still hold the greatest potential for savings. Leak reduction is also a worthwhile target for agencies' efforts. Reducing leaks usually requires adjustment of existing fixtures rather than complete replacement, which reduces overall costs. The savings potential of showers and washing machines is also relatively high, while that of dishwashers is modest. We estimate that full implementation of current conservation potential would cut current use by another 1,100 mcm (890,000 acre-feet)—approximately a further 40 percent reduction. This would have the effect of reducing current indoor residential use, on average, from around 230 liters per person per day (excluding some uses not evaluated here) to around 140 liters per person per day.

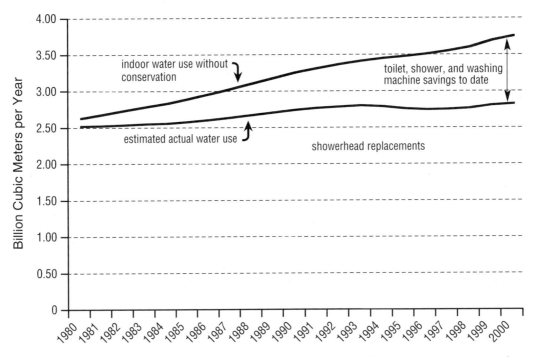

FIGURE 5.2   Total indoor residential water use with and without current conservation efforts (1980 to 2000).

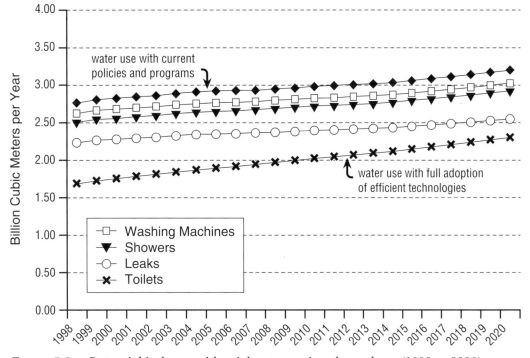

FIGURE 5.3 Potential indoor residential water savings by end use (1998 to 2020).

Note: Each line down from current use is the cumulative savings achievable by implementing each technology improvement noted.

TABLE 5.3 Indoor Residential Conservation Potential for 2000

Indoor Residential Water Use	Best Estimate of Additional Cost-Effective Water Conservation Potential (2000) (mcm/yr)	Conservation Potential: Percent Reduction Over Current Use
Toilets	520[1]	57
Showers	148[2]	24
Washing Machines	136[3]	33
Dishwashers	16	46
Leaks	284[4]	80
Faucets/Fixed Volume Uses	[5]	[5]
TOTAL Additional Indoor Savings	1,100	40

Notes:

1. For toilets, this requires full replacement of inefficient toilets with 1.6 gallon per flush models—the current United States standard.

2. For showers, this requires full replacement of showerheads with 2.5 gallon per minute models—the current United States standard (with actual flow rates averaging 1.7 gallons per minute).

3. For washing machines, these savings would result from the complete replacement of current models with the average (not the best) of the efficient machines currently on the market.

4. The 80 percent savings estimate comes from assuming that leak rates are reduced to the median value now observed. At the same time, CDWR (2003b) estimates that half of all leaks can be saved for less than $100 per acre-foot and 80 percent for less than $200 per acre-foot.

5. For faucets and other fixed volume uses such as baths, no additional "technical" savings are assumed.

For all indoor uses, additional temporary "savings" can be achieved during droughts by behavioral modifications (e.g., cutting back on frequency of actions like flushing, showering, washing). We do not consider these to be "conservation" or "efficiency" improvements.

Figure 5.4 summarizes both the water savings that have been achieved between 1980 and the present, and a projection of future potential indoor residential savings with both existing programs and all cost-effective savings to 2020, as a measure of the potential that remains. The top line is a projection of use if no conservation activities had been initiated in the state (i.e., using pre- 1980 conditions). The middle line is our "current use" projection (i.e., assuming the current mix of efficient and inefficient uses). The bottom line is our estimate of the further reduction in indoor residential water demand that is possible with all cost-effective savings using existing technology (for more detailed calculations see Gleick et al. 2003).

Indoor Residential Water Conservation: Methods and Assumptions

The first step in evaluating the savings potential of water-conservation options is to establish a reliable baseline of current water-use patterns. There are a number of different options for defining the baseline: water use by region, sector, household, individual, or by specific use. Typically the baseline is reported as water-delivery data (by water agencies or other institutions) but often, a good baseline is unavailable. There was none, for example, for the state of California as a whole. As a result, the Institute study developed one based on end uses of water. Looking at end uses allows one to evaluate the effect of

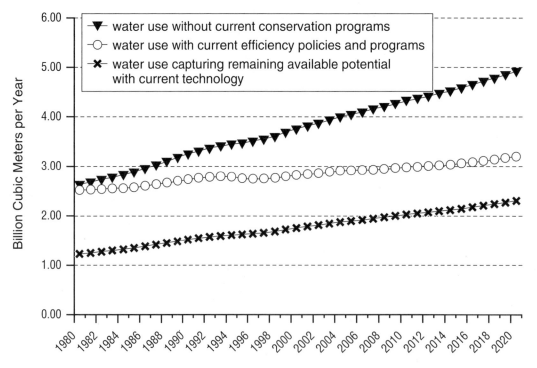

FIGURE 5.4   Indoor water use 1980 to 2020: the effect of conservation policies.

TABLE **5.4**   Summary of Estimated Water Use in California Residential Toilets: With No Efficiency Improvements, Current Use, and Maximum Practical Savings for 2000

Fixture	Water Use, No Efficiency Improvements (mcm)	Water Use, Estimated Current Use (mcm)	Water Use, Maximum Practical Savings (mcm)	Additional Percent Savings, Over 2000 Use
Toilets	1,415	905	385	57

Note: Assuming "maximum practical savings" is represented by 6 liter (1.6 gallon) per flush (6 gpf) models, "no efficiency" is represented by models using 22.7 liters (standard 6 gallon per flush) models, and "current use" represents the current mix of efficient and inefficient models.

improvements in end-use technology and management on water demand while maintaining the purpose for which the water is required.

The end uses examined for the indoor residential sector are: sanitation (flush toilets), bathing (showers and baths), washing dishes and clothes, faucet use, and water lost to leaks. The analysis of outdoor water uses evaluated improvements in water use in gardens and landscapes through technological changes, management efforts, and alternative landscape designs. The water demand of each indoor residential end use was modeled separately and assumptions and results are described in the full study.

Toilets

Flushing toilets is the largest single use of water inside the home in the United States. Estimates for toilet use range from 28 percent to almost 40 percent of total indoor use.[5] We estimate that 32 percent of current indoor residential water use goes to toilets. For this reason, improving the water efficiency of toilets has long been a high priority (A&N Technical Services 1994). Technical innovations in this field have made it possible to reduce the water used by toilets from six gallons per flush (gpf) to under two gpf and less. To tap this potential, federal and state water-efficiency laws now standardize flush volumes at a maximum of 1.6 gpf for all new toilets. Even more efficient toilets are now becoming available, but they were not included in the assessment.

The availability of more efficient toilets has already had a noticeable impact on the volume of water used by homes in California and the rest of the United States. We estimate current water used by residential toilets in California is around 900 mcm per year (in 2000), substantially below the 1,410 mcm that would have been used without the installation of any low-flow toilets. Yet if all the remaining inefficient toilets were replace statewide, current use would be less than 400 mcm per year—a potential further reduction of nearly 60 percent over current use (Table 5.4). The amount of water now used to meet the sanitation needs of 34 million people is less water than the state used for this purpose in 1980 to meet the needs of only 24 million people.

Emerging Technology Can Further Increase Efficiency

As noted earlier, full replacement with current ULFT technology does not represent the maximum technical savings. The current standard in the United States requires toilets

---

5. The lower estimates come from studies that include leaks in estimates of total indoor use.

that flush at 1.6 gallons, but more efficient technology has already been tested and installed extensively in other countries. The Save Water and Energy Education Program (SWEEP) in Oregon tested one example, the Caroma Caravelle 305, imported from Australia where dual-flush toilets are the norm. Dual-flush toilets have a two-button mechanism; one button is designated for liquid waste and flushes at about 3.5 liters; the one for solid waste flushes at the standard 6 liters (1.6 gallons). SWEEP found that the toilets performed well and that the liquid-flush mode was used about 65 percent of the time. Based on their sample, this design offers an additional 7,570 to 9,500 liters savings per home per year over the standard 6 liter (1.6 gallon) toilet (Sullivan et al. 2001). While these types of toilets are fairly common in other countries, they have yet to significantly penetrate the North American market.

Showers and Baths

Water used for showers and baths is typically the second or third largest category of indoor residential water use. Showers use 22 percent of all indoor home water in California. Federal legislation has already played a role in tapping the potential savings from showers and baths. The National Energy Policy Act of 1992 mandates that new faucets not exceed a flow rate of 2.5 gpm. Prior to that, the standard flow rate had been 5 gpm.

No savings were assumed to be possible from improving fixture efficiency when it comes to baths, which are a fixed-volume use, though temporary savings can be achieved during droughts by reducing the frequency of baths. This type of behavioral change represents a buffer for water agencies during periodic shortages. Water can, however, be saved by changing showerheads to low-flow units. Mayer et al.(1999) reports that households having all low-flow showerheads use on average about nine percent less water than households without these fixtures.

Replacing a 5.0-gpm showerhead with a 2.5-gpm model will save about 64 liters (17 gallons) per shower, or over 15,100 liters (4,000 gallons) per person per year (gpcy). Replacing a 3.5-gpm with a 2.5-gpm model will save about 32 liters (8.5 gallons) per shower, or about 7,570 liters per year. If no showerheads in California had been replaced with more efficient models, we estimate that water used for residential showers would be around 940 mcm per year (in 2000). Past conservation programs have managed to reduce this demand to around 612 mcm (in 2000), a reduction of 35 percent and a savings of around 325 mcm per year. If all showerheads today were high-efficiency models, they would save an additional 150 mcm/yr cutting demand for showers to around 465 mcm/yr —a reduction of 24 percent over current use.

Energy Savings of Efficient Showerheads

Switching to a low-flow showerhead also saves substantial amounts of energy by reducing the amount of water that requires heating. These energy savings are an important part of the cost effectiveness of replacing inefficient showerheads.

Washing Machines

Residential washing machines are one of the largest users of water in U.S. homes. The vast majority of residential washing machines in the United States are top-loading machines that immerse the clothes in water and spin around a vertical axis. Horizontal-axis

TABLE 5.5 Water Efficiency of Washing Machines in 1995 and 2000

Technology	Gallons per load	Water Factor	Source
Average machine in use (1995)	44	16.5	Kesselring and Gillman 1997
Average machine in use (1995)	37.5	14.3	CEE 1995, USDOE 1996
Average machine shipped (1995)	35.8	13.3	CEE 1995, USDOE 1996
Current generation efficient washers	24.2	9.1	CEE 1995; Kesselring & Gillman 1997

designs use a tumbling action where the washer tub is only partially filled with water, requiring far less water, energy, and detergent.[6] Horizontal-axis washing machines, long popular in Europe where they have captured over 90 percent of the market, have only recently been introduced to the United States.

In California, residential washers currently use around 410 mcm/yr (330,000 AF/yr) and significant savings can be achieved with new machines. Efficient machines can save a typical household up to 26,000 to 34,000 liters per year (7,000 to 9,000 gallons) of water (Bill Jacoby, personal communication, 2002; Consortium for Energy Efficiency 2003), cutting per-capita indoor use by 6 to 9 percent (Mayer et al. 1999), and these savings are accompanied by a wide range of secondary advantages.

In the past few years, increasing attention has been paid to the potential for efficient washing machines to reduce water and energy use. Rising pressure on water and energy resources nationwide has prompted detailed field and laboratory surveys evaluating savings from the use of more efficient washing machines (see Table 5.5, Kesselring and Gillman, 1997); Consortium for Energy Efficiency 1995, USDOE 1996, THELMA 1998). The High Efficiency Laundry Metering and Marketing Analysis project (THELMA) consisted of both a lab and field analysis of machines currently available on the market. Separately, the Department of Energy and the Oak Ridge National Laboratory conducted a 5-month field study in Bern, Kansas involving 103 machines and over 20,000 loads of laundry. Both studies yielded similar results: water savings of about 56 liters (15 gallons) per load.[7] Water savings from efficient machines are generally estimated to be between 40 and 50 percent (Hill et al. 1998, Pugh and Tomlinson 1999). This potential has encouraged many utilities nationwide to incorporate washing machine programs into their conservation programs.

In 1993 the Consortium for Energy Efficiency (CEE) launched a high-efficiency clothes washer initiative to accelerate the manufacture and sales of high-efficiency machines, recognizing the value of these machines in terms of reduced pollution, wastewater, energy, and water use. No federal water standard for washing machines has yet been passed into law. There has been more legislative success in California, which became the only state to adopt water-efficiency standards for washing machines with the passage of California Assembly Bill 1561, signed into law in the fall of 2002. The bill, which is supposed to take effect in 2007, requires newly manufactured home washers not to exceed a

---

6. For typical usage, 80 to 90 percent of the energy use attributed to clothes is used to heat water. The partial filling of the tub means less total water is required, less hot water, and less water-heating energy (DOE 1990 in http://www.ci.seattle.wa.us/util/recons/papers/p_sh1.HTM).

7. The two studies used a similar experimental design. The Bern study, however, examined only one efficient washing machine model while the THELMA study used three different H-axis models.

water factor of 9.5 (equivalent to current commercial standards). Currently, efficient washers have a water factor of 9.0 or even less, while the average washing machine sold in the mid-1990s has a water factor of 13.3.

On average, a medium-sized load requires 138 liters (36.4 gallons) in an inefficient machine and 98 liters (26 gallons) in an efficient machine. The savings assumed here are more conservative than some of the other estimates being applied; Seattle City Light uses 45 liters savings (12 gpl) in their analysis (Chin, personal communication, 2002) and the SWEEP study found average savings to be between 53 and 58 liters per load (14.1 and 15.2 gpl) (Sullivan et al. 2001). For maximum available savings we assumed that new machines averaged 91.6 liters (24.2 gallons) per load. We used the average for existing machines (136.6 liters) to estimate current conditions.

Information on the penetration of washing machines and frequency of use came from the 1995 American Housing Survey (United States Census Bureau 1995), which found that 73 percent of households in the United States have washing machines and calculate that there are just fewer than 9 million residential washing machines in California today.

Water Savings of Efficient Clothes Washers

In 2000, residential clothes washers in California used about 407 mcm/yr (330,000 AF), a reduction of around 86 mcm (70,000 acre-feet) over estimated use if no efficient machines were in use. If all current residential washing machines in California were as efficient as the average of the efficient models currently on the market, water use in California homes would be reduced by another 136 mcm/yr (110,000 acre-feet annually) —a 30 percent reduction.

As with showerheads, energy savings associated with efficient washers are extremely important for evaluating overall cost-effectiveness of this conservation choice. Studies show that these machines can reduce energy use for washing clothes by between 50 and 65 percent (Environmental Building News 2000).

Dishwashers

Dishwashers account for less than two percent of total residential water use in the United States (Mayer et al. 1999). Approximately 54 percent of United States housing units are equipped with dishwashers (United States Census Bureau 1995). This suggests that in 2000 there were approximately 6.3 million households in California with dishwashers. The typical rate of usage is 0.4 loads per household per day (AWWARF 1999, Mayer et al. 1999).

By integrating the distribution of fill volumes over the number of dishwashers in the state and multiplying it by the number of cycles per load, it is possible to estimate the total volume of water used by dishwashers.

As water becomes more of a concern, we expect there will be continued improvements in the water-use efficiency of newer models. The most efficient machine in the analysis used 17 liters (4.5 gallons) for a normal-sized load. By comparing similarly priced models, however, average efficient machines use about 20 liters (5.3 gallons) per load. Overall, dishwashers used almost 35 mcm (28,000 AF) of water in 2000 in California. If all of these dishwashers were to be replaced with the more efficient 20 liter (5.3 gpl) models, use in 2000 would have been reduced to under 18.5 mcm (15,000 AF).

Faucets

In 1992, the California Plumbing Code mandated that all faucets have a maximum flow rate of 2.2 gpm (note: 1 gallon per minute is the same as 3.785 liters per minute). This standard was replaced by the federal standard for faucets of 2.5 gpm enacted January 1, 1994. Prior to this, faucet flow rates ranged from 2.75 to 7.0 gpm. Faucet flow, like bath use, is largely volume based. For example, filling a pot will require the same volume of water regardless of flow rate of the fixture. The amount of water used for brushing teeth while leaving the faucet running, however, will be larger with a faucet that flows at a higher rate. Thus, a low-flow faucet may or may not reduce water needs, depending on the use and individual behavior.

Lack of consistent data on potential savings hinders making reliable estimates of conservation potential. Because of the uncertainties in this area, the California study did not estimate any savings from faucets. Technological options combined with change in users behavioral patterns do, however, have the potential to significantly affect faucet water use over time. One example is an automatic shutoff device that can be installed on any sink, such as a bar mounted in front of the sink at hip level that the user must press or lean against to turn on the faucet. When the user moves away, the faucet shuts off. This device also has a locking device and constant flow option.[8]

Leaks

Leaks within a home, including faulty faucets and toilets, are responsible for significant water losses. Leak repair, therefore, warrants substantial evaluation and investment—a conclusion reached by a number of studies (United States HUD 1984, Marin Municipal 1994, DeOreo et al. 1996, Steirer and Broder 1997, A&N Technical Services 1999, Mayer et al. 1999). The main difference between this measure and some of the ones previously discussed is that leak detection and repair generally do not require investment in new equipment and can often be performed by the homeowner with information and guidance from the utility.

Residential leak rates have been documented in a number of studies.[9] The early Housing and Urban Development (HUD) study (United States HUD 1984) estimated leakage to be 5 to 13 percent of total indoor water use. The REUW study (Mayer et al. 1999) found average leakage was 12.7 percent of indoor use, but with an unusual distribution: the 100 homes with the highest water use had leakage rates of 24.5 percent. In 5 of their 12 study regions, per-capita leakage rates exceeded total faucet water use. DeOreo et al. (1996) analyzed use for 16 single-family homes in Boulder County, Colorado and found that leaks averaged 11.5 percent of indoor water use, or around 80 liters per day (20.8 gpd) per account. In all these studies, toilets are the leading "leakers."

Leak rates are highly variable. In general, a small proportion of housing units accounts for the largest proportion of leaks. In Mayer et al. (1999), 10 percent of the homes were responsible for 58 percent of the leaks. The average leakage rate per household was

8. This is a fairly new product, so there has not been prolonged testing or extensive studies comparing water use. According to company estimates, this device can cut faucet water use in the kitchen and bathroom (excluding leaks) by about 83 percent. For more information, go to www.conservativeconcepts.com.

9. These studies do not differentiate between indoor and outdoor residential leaks. We include all leaks with indoor water use, presented as the percentage of indoor use. Note that by "leaks" we exclude "unaccounted for water," which is lost before reaching a home water meter.

83 liters per day, meaning that the top third of leaky households were more than doubling the average of the entire sample. In San Diego, Steirer and Broder (1997) found toilet leaks alone varied from 75 liters per day to an extreme of more than 15,000 liters per day.

The potential savings from reducing leaks are high. A&N Technical Services (1999) estimate that approximately 30 liters per day can be saved for each leaking toilet repaired and other household leak repairs can save an additional 47 liters per day. The HUD study of apartment buildings in Washington D.C. found that fixing leaking toilets saved 180 liters per day per unit, with two toilets per unit (in most units, both toilets were leaking). Fiske and Weiner (1994) estimate that leak detection and toilet repair can save about 76 liters per day per toilet and faucet leak repair can save 15 liters per day per leaking faucet.

This variability suggests that leak-reduction programs would be most effective if they were targeted at homes with the highest leakage rates. A sorting and filtering routine could permit utilities to identify accounts with dramatic increases in their use patterns. Audits can then be performed at these sites in order to identify the causes of the change. Targeting the high-end water users would make audits more cost effective to the utility. A number of utilities in California have been using this kind of targeted approach.

Although comprehensive audits and proper maintenance can reduce residential leaks to zero, in practice, there always will be a minimum level of lost water. Total water savings is estimated as the amount of water that is saved by reducing the distribution of residential water leaks down to the median level in California homes. If all homes reduced leakage rates to the average rate of 16 liters per day—total savings would be nearly 300 million cubic meters per year.

Indoor Residential Summary

Despite the significant and important progress that Californians have made in reducing indoor residential water use, substantial potential for conservation improvements remains untapped. At present, Californians use about 2,840 mcm (2.3 MAF) of water to meet indoor domestic needs annually, much less than the 3,700 mcm (3 MAF) per year that would have been necessary without past conservation programs. Indoor use could be reduced by approximately another 40 percent by replacing remaining inefficient toilets, washing machines, showerheads, and dishwashers, and by reducing the level of leaks, even without improvements in technology.

Outdoor Residential Water Use

Use of water indoors in homes is only a part of total residential water use. In many regions, especially arid and semi-arid regions of the western United States, a substantial amount of water is used outside to water lawns and gardens. While there are great uncertainties about the volume of total outdoor residential water use, estimates can be made for the purpose of evaluating the potential for improving efficiency through a wide variety of conservation options. These improvements have the potential to substantially reduce total and peak water demands.

There are additional benefits to such improvements as well. These include a reduction in energy and chemical use, mowings and other maintenance needs, and waste created. While these benefits are difficult to quantify and rarely evaluated, we describe them and urge that more work be done to understand and to quantify their scope. Given the magnitude of current outdoor residential water use, improved conservation programs, more data collection and monitoring, and better reporting by urban agencies should be top priorities for water policymakers and planners.

In California, substantial amounts of water are used in the outdoor residential sector, primarily for landscape irrigation, although great confusion accompanies estimates of actual use because of varying methods for calculation, lack of real data, limited metering, uncertainties about landscape area, and other variables. Watering gardens and lawns accounts for half of all residential water use, statewide, and as much as 70 percent of residential use in some parts of the state. Our study of California looked at several approaches to evaluating current and projected landscape water use in homes and quantified the potential to reduce that water use with existing technologies and cost-effective management approaches.

Many options are available for reducing residential landscape water use. Improving water use in gardens and landscapes could free up substantial quantities of water for new demands, ecological restoration, or other uses. And there are additional benefits from outdoor water conservation, such as reducing peak period demand. Outdoor water use rises to a maximum during the summer when California water supplies are most constrained. As a result, residential landscape use plays a large role in driving the need for increases in system capacity and reliability. Furthermore, much of this water is lost to evaporation and transpiration and is no longer available for capture and reuse, unlike most indoor use.

Efficient irrigation involves two things: proper design and proper landscape maintenance. Proper landscape maintenance requires that the homeowner be informed and diligent—difficult things for an agency to predict, control, or monitor. For example, planting a water-efficient landscape or installing a sophisticated irrigation system will not save water if the homeowner fails to match the irrigation schedule with plant needs. And a manual irrigation system on a traditional landscape can be efficient if it is properly maintained and used. In contrast, projecting the savings from an efficient toilet or showerhead program is relatively straightforward. When an agency decides whether to invest in a retrofit program, they can reliably calculate savings from switching their existing stock to efficient toilets and from that determine the costs and benefits of such a program. A similar evaluation of landscape programs is more difficult and is constrained by lack of data and consistency.

Farmers and, increasingly, large-lot landscape managers have been taking advantage of tools such as improved irrigation technologies, rebates, audits, and weather station data in planning and designing irrigation systems and schedules. While these tools are often available in the residential sector, homeowners are less likely to have the time, inclination, incentive, or expertise to adopt them. One challenge lies in educating, motivating, and in some cases requiring residential homeowners and managers of smaller residential lots to adopt proper irrigation scheduling and techniques.

Current Outdoor Residential Water Use

No satisfactory or consistent estimates of current outdoor residential water use are available for California. State agencies provide a variety of indirect estimates in different studies, mostly for a baseline of 1990. Five separate baseline estimates of outdoor residential water use were developed showing a range of use from 1,000 mcm to over 2,000

TABLE 5.6   Projections of Outdoor Residential Water Use in 2000

Estimate	Water Use (mcm/yr) in 2000
Low	1,210
High	2,350
Best Estimate	1,800

mcm per year—a factor of nearly two—showing the high uncertainties about actual outdoor residential water use (Table 5.6) (Gleick et al. 2003).

Lack of good data has greatly hindered progress in both capturing and measuring efficiency improvements in the residential landscape sector. There is agreement that the potential for saving water is substantial, but the tools to quantify and evaluate specific savings in specific landscapes are only beginning to develop. Most agencies know little about the characteristics of their residential landscapes. They do not always have reliable estimates of outdoor water use, let alone landscape acreage, type of plantings, or irrigation methods. Residential customers typically do not have dedicated irrigation meters, so site-specific information can be a challenge to collect. Because of the expense involved and because it is difficult for agencies to quantifying savings, outdoor water-use data collection and analysis has traditionally been considered low priority.[10]

Existing Outdoor Conservation Efforts and Approaches

Some efforts have been made at the regulatory level in California to improve landscape water use. California Assembly Bill 325, the Water Conservation in Landscaping Act of 1990, required that the Department of Water Resources develop a Model Water Efficient Landscape Ordinance. This Model Ordinance, the only residential landscape-specific state regulation, was adopted and went into effect January 1, 1993. The ordinance applies to all new and rehabilitated landscaping for public agencies and private-development projects that require a permit, and developer-installed landscaping of single-family and multi-family residential projects.[11] Landscapes must exceed 2,500 square feet to be subject to the ordinance. Cities and counties have the option of adopting the Model Ordinance, adopting their own ordinance, or issuing findings that no ordinance is necessary. If no action is taken, the Model Ordinance automatically goes into effect. By the late 1990s, the Model Ordinance or a similar water budget ordinance was being used in more than 250 jurisdictions. Turf limits or other approaches to water conservation had been adopted by nearly 200 jurisdictions. In a 1997 CDWR survey, 86 percent of communities questioned felt the ordinance was improving their landscape water-use efficiency. Most of those who felt the ordinance made no difference explained that their community was

---

10. There are a handful of California agencies, such as the EBMUD and IRWD, that have collected information on outdoor water use by landscapes. There has also been increased interest in obtaining this information and research into the most appropriate methods to do so. For these studies go to Landscape Area Measuring Study Final Evaluation Report, October 1999. Prepared for the United States Bureau of Reclamation by the Contra Costa Water District. http://watershare.usbr.gov. Annual Water Allocation and Methodology, Pilot Project Executive Summary. May 1998. Prepared for MWDOC, MWDSC, USBR, and the Moulton Niguel Water District by Psomas and Associates.

11. For more information on the Model Water Efficient Landscape Ordinance see http://wwwdpla.water.ca.gov/cgi-bin/urban/conservation/landscape/ordinance

small or nearly built-out and very few projects were in the development phase (http://-www.dpla.water.ca.gov/urban/land/itworks.html).

The concept behind California Assembly Bill 325 is that by establishing a water allowance based on limiting evapotranspiration losses, landscapes will be maintained to ensure water efficiency. To ensure proper irrigation, the ordinance requires documentation for each landscape that includes a calculation of maximum applied water allowance, applied water use, total water use, and an irrigation design plan. The ordinance's greatest weakness is lack of enforcement and monitoring, according to a statewide implementation review (Bamezai et al. 2001). Few developers and contractors interviewed were even aware of the ordinance. Only two among the 66 agencies responding to the survey had ongoing outreach programs. The reviewers concluded that the key to improving the success of the ordinance is more education, economic incentives (pricing), and better integration of enforcement efforts between land-use agencies and water suppliers.

Outdoor Residential Water Conservation: Methods and Assumptions

There are a large number of options available to the homeowner for reducing the amount of water used for landscape purposes. The options range from relatively simple and inexpensive practices such as maintaining a proper irrigation schedule to more demanding practices such as retrofitting an irrigation system with new efficiency options or changing landscape design. Efficiency options can be separated into four general categories: management practices, hardware improvements, landscape design, and policy options. Existing field studies, audit results, technical reports, and related published literature on these options help us quantify the potential water savings. The following are examples of studies and programs in the residential landscape sector as well as the potential savings that can be achieved.

Management Practices

Proper management of outdoor water use is the most effective way to reduce water waste. Without it, no amount of investment will make an irrigation system efficient. Proper management practices can stand on their own as efficiency measures by ensuring that plants are being watered according to their needs or, ideally, they can be used to enhance the savings from other options. Efficient landscape management practices include ET-based irrigation scheduling, regular system maintenance (such as checking for leaks and fixing broken or misaligned sprinkler heads), and proper horticultural practices (such as fertilization and soil aeration).

Successful management involves an understanding of the irrigation system, an ability to recognize problems with the system, and an ability to adapt landscape needs to various conditions. These practices are not difficult but, because they are so dependent on individual behavior, they are difficult to quantify or predict.

Table 5.7 lists some of the various management options analyzed here and their potential savings, assuming no change in landscape area or design. Savings can vary widely depending on climate, geography, and behavioral patterns among other things, but these estimates help to define and bracket the potential options.

TABLE 5.7   Management Options for Reduction of Landscape Water Use

Reduction options	Potential savings %	Source
Turf maintenance[1]	10	SPUC 1998, 1999
Turf maintenance, irrigation system maintenance, irrigation scheduling	20	WPR 1997
Mulching in ornamental gardens	20	SPUC 1998, 1999
Soil amendments (compost)	20	SPUC 1998, 1999
Irrigation scheduling	~25	Steirer and Broder, SPUC 1998, 1999
Irrigation/soil maintenance	65–75	Pittenger 1992
Allow lawn to go dormant	90	SPUC 1998, 1999

Note:
1. Includes thatching, aerating, over-seeding, and top-dressing.

Hardware Improvements

Hardware devices that reduce water use in outdoor residential gardens vary widely in cost and sophistication. For example, a hand-held probe that measures soil moisture may cost around US$12. At the other extreme, home plumbing systems can be redesigned and a "gray-water" system installed, which permits replacing potable water use in gardens with household water that has been used once for some other purpose. Savings from devices also range widely, from about 10 percent for automatic rain shut-off devices, to 50 percent for drip-irrigation systems, to gray water systems, which can potentially eliminate use of all potable water for landscape needs (Table 5.8). (For more detailed information on irrigation systems and devices see Vickers, 2001, and other hardware-specific sources.)

Installing water-saving devices alone does not ensure that less water will be applied to the landscape. The landscape can be just as easily overwatered with a sophisticated drip irrigation system as with a traditional sprinkler. Effectiveness depends on the homeowner knowing to how to use their irrigation system, reset run times as the season warrants, and match water application to water needs. Similarly, soil probes are useful only if the homeowner properly uses the results to design a scheduling system.

To ensure that water-saving technologies meet their full potential, conservation programs must address behavioral variations. Some tackle the problem by trying to make the technology as independent of the homeowner as possible. A pilot study of irrigation controllers that are linked to local weather information stations and automatically respond to weather changes was recently conducted in Orange County. This allows the landscape to be irrigated according to its climate needs without requiring any involvement from the homeowner. The pilot program resulted in a 24 percent reduction in outdoor use (Hunt et al. 2001). Other conservation programs emphasize proper use of the available tools through public policy programs. These programs can include public education, outreach, rebates, loans, and rate structures, among other things. Using these tools alone, the Irvine Ranch Water District reduced overall landscape water use by about 27 percent (Lessick, personal communication, 2002, Wong 1999). They later included soil probes and irrigation software (which they continued to support with a public education program) and succeeded in reducing use to 50 percent of baseline.

TABLE 5.8   Hardware Improvement Options for the Reduction of Landscape Water Use

Reduction Options	Potential Savings %	Source
Auto rain shut off	10	SPUC 1998, 1999
Soil moisture sensors; soil probes	10–29 Allen 1997; Lessick 1998; Wong 1999	SPUC 1998,1999;
Improve performance[1]	40	SPUC 1998, 1999
Drip/bubbler irrigation	50	SPUC 1998, 1999
Gray water[2]	Up to 100	SPUC 1998, 1999
Rain barrel catchment	Up to 100	SPUC 1998, 1999

Notes:
1. This includes repair, removal, or adjustment of in-ground system components.
2. This option is used to reduce the volume of potable water used; it does not affect the total volume of water used.

Landscape Design

One of the most reliable ways of eliminating variability in effectiveness of outdoor conservation options is to modify the design of gardens and landscapes. The California study did not develop estimates of statewide potential on this approach, because of the fundamental assumption that there be no change in the "service" provided by water, even though we believe that xeriscaping and reduction in turf area produces perfectly acceptable, and sometimes even improved, garden aesthetics. Nevertheless, the potential for significant reductions in outdoor water use is high, and we discuss that potential here as an option available to all homeowners.

There are two aspects to landscape design: the choice of plants and the physical layout of the landscaped area. Water needs of different plant species vary considerably and some vegetation is better equipped to withstand the hot, dry regions and periods of parts of California than others. Water requirements for vegetation commonly found throughout the state range from up to 1.0 ET_o for cool season grasses (Kentucky bluegrass, rye, tall fescue, red fescue, etc.), 0.7 ET_o for warm season grasses (Bermuda, Zoysia, etc.), 0.5 ET_o or less for groundcovers, to 0.2 ET_o for shrubs and trees (http://www.owue.water.ca.gov/docs/wucols00.pdf) (CDWR 2000). Proper landscape layout involves controlling the area and perimeter of turf, minimizing narrow paths or steep areas that cannot be irrigated efficiently, and grouping plants with similar irrigation needs.

A limited number of studies have quantified savings from xeriscape practices (typically defined as water-efficient landscaping) (Table 5.9). The North Marin Water District conducted a series of such studies and found that proper choice of plants and careful landscape design could reduce water use by up to 54 percent (Nelson 1986, 1987). Less water use was not the only benefit—the water demands of the xeriscape landscapes were more level throughout the growing season and lacked the dramatic peak demands common to traditional landscapes. The Southern Nevada Water District compared the water use of traditional landscapes with those that had been converted to xeriscape. They found that relatively few properties in each group used vastly more water on a per-unit area basis than the bulk of the rest of the sample. Mean monthly household consumption dropped an average of 33 percent following conversion. The xeriscaped landscapes

TABLE 5.9   Potential Water Savings from Landscape Design Improvements

Reduction Options	Potential Savings %	Source
Landscape design[1]	19–54	Nelson 1986, 1987, CDWR 2000
Turf reduction[2]	19–33	Nelson 1994, Sovocool and Rosales 2001
Choice of plants[3]	30–80	CDWR 2000

Notes:
1. Based on minimizing turf area and perimeter.
2. Non-turf areas are not necessarily comprised of low water use plants.
3. Savings based on ET_0 range of 0.2 to 1.0 and a current ET_0 of 1.0

consumed, on average, 20 to 25 percent as much water as the traditional landscapes. These savings took place in the year following conversion and remained stable during the following three years of analysis (Sovocool and Rosales 2001).

Rate Structures, Outreach

Properly designed rate structures can be a valuable tool to help homeowners improve the efficiency of their water use. There are few agencies in California that effectively employ rates to encourage conservation, but some innovative utilities successfully use rates to encourage efficient water use. One of the most well-known examples is the Irvine Ranch Water District (IRWD) (see Wong 1999, Owens-Viani et al. 1999). In 1991, IRWD replaced its flat rate-per-unit charge with an increasing block rate structure (Table 5.10). These rates are designed so that conservation is rewarded and unreasonable use is penalized. The point at which rates go up to the next block is based on a percentage of initial allocation provided each customer. The new rate structure was combined with a well-developed public outreach and education program that allowed the district to help customers identify why they might fall into more expensive blocks and how they can reduce their use to save money.

The base allocation is based on the number of household residents, landscape area, actual daily weather, and ET. Customers receive a fixed allotment for indoor use based on the number of residents (75 gallons per person per day), while the landscape allotment is calculated as a function of landscape area, cool-season ET for grasses, the crop coefficient, and irrigation efficiency.

TABLE 5.10   Summary: Ascending Block Rate Structure for Residential Customers at IRWD

Tier	Water Use (as percent of base allocation)	Price per Unit Use in Each Tier
Low Volume Discount	0–40	3/4 Base Rate
Conservation Base Rate	41–100	Base Rate
Inefficient	101–150	2x Base Rate
Excessive	151–200	4x Base Rate
Wasteful	201 and above	8x Base Rate

Sources: Wong 1999; Lessick, personal communications, 1998, 2002.

IRWD coupled the new budget-based rate structure with an aggressive education and outreach program. During the first two years following implementation of the rate structure (drought years), water use fell by 19 percent from the pre-program baseline. Water use rebounded slightly after the drought in the late 1980s and early 1990s, but remained below pre-program levels. On average, use has remained about 12 percent below 1990–1991 levels. In 1997–1998, targeted audits and soil probes were added to the program.

Residential Outdoor Water Use Summary

Outdoor residential water conservation and efficiency improvements have the potential to significantly reduce total water demand in California and improve supply reliability by reducing both average and peak demand. Savings will result from improved management practices, better application of available technology, and changes in landscape design away from water-intensive plants. There are great uncertainties in total water currently used in the outdoor residential sector, with best estimates ranging from between 1,230 and 2,470 mcm per year (one and two million acre-feet per year) and averaging 1,800 mcm in 2000. The California case study estimated that 25 to 40 percent of this water could economically be saved through proven approaches, a reduction of 444 to 715 mcm per year (360,000 to 580,000 AF/yr) or even more.

There are additional benefits to such improvements as well. While we have not quantified these benefits, we describe them briefly below and urge that more work be done to understand and to quantify their scope:

- Moller at al. (1996) found that precisely managing turf water applications with moisture sensors reduced vegetative growth by 73 percent, thus reducing the number of mowings required, energy expended, and waste created. They also saw water-quality benefits—the correct placement of water and fertilizer through continuous monitoring and irrigation scheduling minimized leaching below the root zone and into groundwater sources, waterways, and estuaries.

- Studies by Nelson (1986, 1987, 1994) not only showed water savings of 54 percent, but found that xeriscapes decreased resource requirements in general. The efficient landscapes studied reduced labor needs by 25 percent, fertilizer use by 61 percent, fuel use by 44 percent, and herbicide use by 22 percent. These reductions make investment in xeriscape more economically attractive and offer improvements in both water and air quality.

- In the SNWA study, savings of both time and money of more than 30 percent were realized in sites converted to xeriscape. The xeric sites required 2.2 hours/month less to maintain than the traditional sites and cost $206 per year less to maintain on top of savings in the water bill (Sovocool and Rosales 2001). Added benefits include savings on wastewater disposal and a decrease in the amount of lawn-care chemicals in garden runoff.

Better estimates of both total outdoor water use and the conservation potential in this sector are needed. Given the magnitude of current outdoor residential water use in California, improved data collection, monitoring of outdoor use, and reporting by urban agencies should be top priorities for water policymakers and planners.

Conclusions

Residential water use satisfies a wide range of needs, from providing drinking water and sanitation services to cooking and cleaning to maintaining attractive and desirable gardens. The water provided for these services in the United States and many other countries is often extremely high-quality, potable water, obtained and purified at great economic and ecological expense. In many regions, obtaining more water for domestic uses is increasingly costly and controversial. As a result, there has been growing interest in evaluating our ability to continue to meet our needs for water, while using less of it, by increasing the efficiency of use. This chapter summarizes an extensive research assessment on the potential to improve the efficiency of residential water use in the State of California and evaluates the cost-effectiveness of that potential. More than 30 percent of California's current residential water use can be saved with existing technologies such as efficient washing machines, dishwashers, and toilets, by reducing leaks, and by smarter and more careful use of outdoor landscape water. These savings remain to be captured, even after more than a decade of attention to conservation and efficiency in the region. This strongly suggests that inefficient use of water may be far more widespread and that efficiency improvements represent a potential new "reservoir" of new supply waiting to be tapped.

Abbreviations and Acronyms

AF: acre-feet
AF/yr: acre-feet per year
AWWARF: American Water Resources Association Research Foundation
BMPs: Best Management Plans
ccd: hundred cubic feet
CDOF: California Department of Finance
CDWR: California Department of Water Resources
CEE: Consortium for Energy Efficiency
CUWA: California Urban Water Agencies
CUWCC: California Urban Water Conservation Council
EBMUD: East Bay Municipal Utilities District
ET_0: reference evapotranspiration
gpcd: gallons per-capita per day
gpcy: gallons per-capita per year
gpd: gallons per day
gpf: gallons per flush
gpl: gallons per load
gpm: gallons per minute
HE: high efficiency
HUD: United States Department of Housing and Urban Development
kWhr: kilowatt hours
kWhr/yr: kilowatt-hours per year
LRMC: Long-run marginal cost

MAF: million acre-feet

MAF/yr: million acre-feet per year

MCC: marginal cost of avoidable capacity investment

MWD: Metropolitan Water District of Southern California

REUW: Residential End-Use of Water Study (see Mayer et al. 1999)

rpm: revolutions per minute

SRMC: short-run marginal cost

TAF: thousand acre-feet

THELMA: The High Efficiency Laundry Metering and Marketing Analysis Project

UfW: unaccounted for water

ULFT: ultra-low-flow toilet

USDOE: United States Department of Energy

USHUD: United States Department of Housing and Urban Development

UWMPs: urban water managements plans

REFERENCES

A&N Technical Services. 1994. *Public facilities toilet retrofits, evaluation of program outcomes and water savings.* Encinitas, California.

A&N Technical Services, Inc. 1999. Guide to data and methods for cost-effectiveness analysis of urban water conservation best management practices. Working draft prepared for the California Urban Water Conservation Council.

Allen, R. G. 1997 *Demonstration of potential for residential water savings using a soil moisture controlled irrigation monitor.* Department of Biological and Irrigation Engineering, Logan, Utah: State University.
http://www.engineering.usu.edu/bie/research/papers/smcim/cons96p.html

American Water Works Association Research Foundation (AWWARF). 1999. *Residential End Uses of Water.* Denver.

Bamezai, A., Perry, R., and Pryor, C. 2001. *Water efficient landscape ordinance (California Assembly Bill 325): A statewide implementation review.* A Report submitted by Western Policy Research to the California Urban Water Agencies. Sacramento, California.

Baumann, D. D., Boland, J. J., and Sims, J. H. 1980. *The problem of defining water conservation.* The Cornett Papers, 125–134. Victoria, Canada:University of Victoria.

California Department of Finance (CDOF). 2002. Demographics department. Sacramento, California. http://www.dof.ca.gov/HTML/DEMOGRAP/repndat.htm

California Department of Finance (CDOF). 2003. E-4 Population estimates for cities, counties and the state, 2001–2003, with 2000 DRU benchmark. State of California. Sacramento, California.

California Department of Water Resources (CDWR). 1994a. *California water plan update.* CDWR Bulletin 160-93, October. Sacramento, California.

California Department of Water Resources (CDWR). 1994b. *Urban water use in California.* CDWR Bulletin 166-4. Sacramento, California.

California Department of Water Resources (CDWR). 2000. *Estimated irrigation water needs of landscape plantings in California.* University of California Cooperative Extension. Department of Water Resources. Sacramento, California.
http://www.owue.water.ca.gov/docs/wucols00.pdf

California Department of Water Resources (CDWR). 2003b. Water use efficiency: Leak detection fact sheet. September. http://www.owue.water.ca.gov/leak/faq/faq.cfm

Consortium for Energy Efficiency (CEE). 1995. *High efficiency clothes washer initiative: Program description.* Boston:Consortium for Energy Efficiency.

Consortium for Energy Efficiency (CEE). 2003. Fact sheet: Residential clothes washers. September. http://www.ceeformt.org/resrc/facts/rwsh-fx.php3

DeOreo, W. B., Heaney, J. P., and Mayer, P. W. 1996. Flow trace analysis to assess water use, *Journal of the American Water Works Association,* Vol. 88:79–90.

Dziegielewski, B., Davis, Y. W., and Mayer, P. W. 1999. Existing efficiencies in residential indoor water use. In Proceedings of the Conserv99 Annual Conference, American Water Works Association, Monterey, California, January 31–February 3.

Environmental Building News. 2000. http://www.buildinggreen.com/products/washers.html

Fiske, G. S., and Weiner, R. A. 1994. *A guide to: Customer incentives for water conservation.* Barakat and Chamberlin, Inc., prepared for California Urban Water Agencies, California Urban Water Conservation Council, and United States Environmental Protection Agency. Oakland, California.

Gleick, P. H. 1999. The power of good information: The California irrigation management information system (CIMIS). In *Sustainable uses of water: California success stories*, Owens-Viani, L., Wong, A. K., and Gleick, P. H., editors. Oakland, California:Pacific Institute for Studies in Development, Environment, and Security.

Gleick, P. H., Haasz, D., Henges-Jeck, C., Srinivasan, V., Wolff, G., Cushing, K. K., and Mann, A. 2003. *Waste not, want not: The potential for urban water conservation in California.* Oakland, California:Pacific Institute for Studies in Development, Environment, and Security. November.

Hill, S., Pope, T., and Winch, R. 1998. THELMA: Assessing the market transformation potential for efficient clothes washers in the residential sector. http://www.ci.seattle.wa.us/util/recons/papers/p_sh1.HTM

Hunt, T., Lessick, D., Berg, J., Weidman, J., Ash, T., Pagano, D., Marian, M., and Bamezai, A. 2001. Residential weather-based irrigation scheduling: Evidence from the Irvine "ET Controller" Study. Irvine, California. http://www.irwd.com/Welcome/FinalETRpt.pdf

Kesselring, J. R., and Gillman. 1997. Horizontal-axis washing machines, *The ERPI Journal*, 22:no. 1:38–41.

Lessick, D. 1998, 2002. Irvine Ranch Water District. Personal communication.

Lovins, A. B. 1977. *Soft energy paths: Toward a durable peace.* Friends of the Earth. Cambridge, Massachusetts:Ballinger Publishing Company.

Marin Municipal Water District (MMWD). 1994. *Water conservation baseline study.* Final report and technical appendices. Oakland, California:Demand Management Company.

Mayer, P. W., DeOreo, W. B., Opitz, E. M., Kiefer, J. C., Davis, W. Y., Dziegielewski, B., and Nelson, J. O. 1999. *Residential end uses of water.* Final report. Denver:AWWA Research Foundation.

Mayer, P. W., DeOreo, W. B., and Lewis, D. M. 2000. *Results from the Seattle home water conservation study. The impact of high efficiency plumbing fixture retrofits in single-family homes.* Boulder, Colorado:Aquacraft Inc. Water Engineering and Management.

Moller, P., Johnston, K., and Cochrane, H. 1996. Irrigation management in turfgrass: A case study from Western Australia demonstrating the agronomic, economic, and environmental benefits. Presented at the Irrigation Association of Australia, National Conference, Adelaide, Australia. May 14–16. Agrilink Water Management Services. http://members.iinet.au~agrilink/turf.html

Nelson, J. O. 1986. Water conserving landscapes show impressive savings. Paper presented at the American Water Works Association Annual Conference and Exposition. Denver. June 24.

Nelson, J. O. 1987. Water conserving landscapes show impressive savings. *Journal of the American Water Works Association*, 79, no. 3:35–42, March.

Nelson, J. O. 1994. Water saved by single family xeriscapes. Paper presented at the American Water Works Association National Conference, June 22. New York.

Owens-Viani, L., Wong, A. K., and Gleick, P. H., editors. 1999. *Sustainable uses of water: California success stories.* Oakland, California:Pacific Institute for Studies in Development, Environment, and Security.

Pittenger, D. R., Hodel, D. R., Shaw, D. A., and Holt, D. B. 1992. *Determination of minimum errigation of Needs non-turf groundcovers in the landscape.* Berkeley, California:University of California Water Resources Center.

Pugh, C. A., and Tomlinson, J. J. 1999. High efficiency washing machine demonstration, Bern, Kansas. CONSERV99 Conference, Monterey, California.

Seattle Public Utilities Commission (SPUC). 1998. Water conservation potential assessment, Final project report. http://www.ci.seattle.wa.us/util

Seattle Public Utilities Commission (SPUC). 1999. Water conservation potential assessment. http://www.cityofseattle.net/util/RESCONS/CPA/default.htm

Sovocool, K. A., and Rosales, J. L. 2001. A five-year investigation into the potential water and monetary savings of residential xeriscape in the Mojave Desert. 2001 AWWA Annual Conference Proceedings, June. Southern Nevada Water Authority, Nevada. (Working paper supported by the Southern Nevada Water Authority and the United States Bureau of Land Management.) http://www.snwa.com/assets/pdf/xeri_study.pdf

State of California. 2001. *Interim County Population Projections: Estimated July 1, 2000, 2005, 2010, 2015, and 2020.* Sacramento, California:Demographic Research Unit June

Steirer, M. A., and Broder, M. I. 1997. *Residential water survey program final report for fiscal year 1995–1996 and supplemental data tables for July 1, 1996 through December 31, 1996.* Prepared by The City of San Diego Water Department Water Conservation Program. San Diego, California.

Sullivan, G. P., Elliott, D. B., Hillman, T. C., and Hadley, A. R. 2001. *To save water and energy education program: SWEEP.* Water and energy savings evaluation. Prepared for the United States Department of Energy Office of Building Technology State and Community Programs. Washington, D.C. http://www.pnl.gov/buildings/download_reports.html

THELMA. 1998. *The high-efficiency laundry metering and marketing analysis.* A joint venture of the Electric Power Research Institute, United States Department of Energy, United States Bureau of Reclamation, and two dozen electric, gas, water, and wastewater utilities. EPRI final report, 1998. Palo Alto, California.

United States Census Bureau. 1995. American housing survey. AHS-N data Chart Table 2-4. http://www.census.gov/hhes/www/housing/ahs/95dtchrt/tab2-4.html

United States Department of Energy (USDOE). 1996. Energy conservation program for consumer products: Test procedure for clothes washers and reporting requirements for clothes washers, clothes dryers, and dishwashers. *61 Federal register 17589.* Washington, D.C.

United States Department of Housing and Urban Development (USHUD). 1984. *Residential water conservation projects—summary report.* Report No. HUD-PDR-903. Prepared by Brown and Caldwell Consulting Engineers for the Office of Policy Development and Research, Washington, D.C.

Vickers, A. 2001. *Handbook of water use and conservation.* Amherst, Massachusetts:Waterplow Press.

Water Resources Policy Commission. 1950. *A water policy for the American people.* Washington D.C.: United States Government Printing Office.

Western Policy Research. 1997. *Efficient turfgrass management: Findings from the Irvine Spectrum Water Conservation Study.* Project designed and implemented by D. D. Pagano, Inc., Irrigation Consultants and James Barry M.S., Environmental Consulting for the Metropolitan Water District of Southern California. Los Angeles, California.

Wong, A. K. 1999. Promoting conservation with Irvine Ranch Water District's ascending block rate structure. In *Sustainable uses of water: California success stories,* Owens-Viani, L., Wong, A. K., and Gleick, P. H., editors. 27–35. Oakland, California:Pacific Institute for Studies in Development, Environment, and Security.

Urban Water Conservation: A Case Study of Commercial and Industrial Water Use in California

Peter H. Gleick, Veena Srinivasan, Christine Henges-Jeck, Gary Wolff

California's urban water use can be divided into residential use (discussed in Chapter 5) and commercial and industrial use. The commercial and industrial sectors of California generate hundreds of billions of dollars of economic goods annually and use approximately 3,000 million cubic meters (mcm) per year (2.5 million acre-feet [MAF] of water annually),[1] or about one-third of all the water used in California's urban areas. This chapter summarizes the results of a first-ever statewide assessment of the potential savings in the commercial, institutional, and industrial sectors (CII sectors) from conservation and improving water-use efficiency (Gleick et al. 2003). We found that total estimated conservation savings potential in the CII sector ranges from 860 mcm/yr to 1,600 mcm/yr (700,000 to 1.3 MAF per year), with a best estimate of 1,200 mcm/yr—about 39 percent of current use.

Two broad types of conservation measures were evaluated—improving water efficiency and substituting reclaimed water. Improving water efficiency includes behavioral improvements, such as adjusting a watering schedule, and technological improvements, such as on-site reuse of water or implementing point-of-use reduction technologies. On-site reuse of water includes reusing water in the original process, such as recycling water in cooling towers, or recovering process water for use in alternative applications, such as irrigation. Point-of-use reduction involves implementing fixtures, such as ultra-low-flow-toilets (ULFTs) or autoshutoff valves that reduce the amount of water used to accomplish a certain task. (See Figure 6.1)

The potential conservation measures described in this chapter are "technically achievable" savings. How much of this potential can be realized depends on how favorable the economics of doing so are and the ability to overcome other barriers, as described in Chapter 5. Long-term conservation is an alternative to developing new sources of water supply, and is cost-effective as long as the cost per unit of conserved water is less than

---

1. Institutional water use—primarily schools, universities, and government buildings—is included here in the analysis of commercial water use.

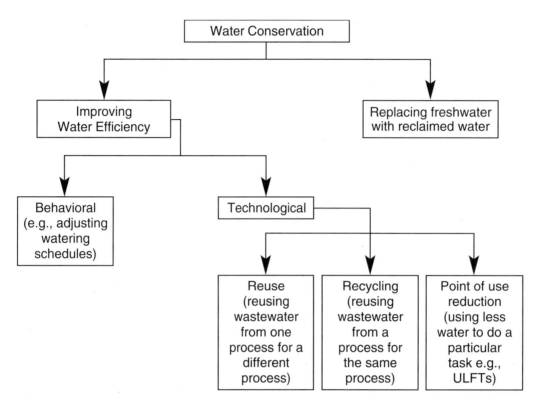

FIGURE 6.1   Defining CII water conservation.

the true cost of the cheapest alternative source of water. Unfortunately, firms do not apply the same criteria as water agencies to judge cost-effectiveness. They instead often look for paybacks of two years or less—a criterion that we show to be excessively stringent.

Most of the measures discussed here are cost-effective (as discussed in Gleick et al. 2003). No attempt was made to determine the specific regional or sectoral cost-effective potential since this depends on rates of individual water agencies. It is important to note that as water becomes scarcer and the cost of water increases, the economically achievable potential will increase. In California, and more widely, the popularity of conservation technologies should only increase in the future as competition for water grows, prices increase, and technology improves.

Background to CII Water Use

Definitions of the commercial, institutional, and industrial sectors vary widely. For the purposes of this work, "commercial water" use refers to water used in private facilities providing or distributing a product or service, such as hotels, restaurants, or office buildings. Institutional facilities include public facilities dedicated to public service including schools, courthouses, government buildings, and hospitals. Industrial facilities are those that manufacture or process materials as defined by the Standard Industrial Classification (SIC) code numbers 2000 through 3999.[2]

---

2. The SIC system was recently replaced by the North American Industrial Classification System (NAICS). We use the SIC code system because our largest single data set available, the CDWR's industrial survey data (CDWR 1995a), is classified by SIC code.

Studies of CII water use often group commercial and institutional users of water together for analytical purposes, since the distinction between what is considered commercial (i.e., a private school) and what is considered institutional (i.e., a public school) is somewhat arbitrary (Sweeten, personal communication, 2000). This approach was used in the analysis.

Current California Water Use in the CII Sectors

Calculating water conservation potential requires knowing how much water various industries in the CII sectors use annually. Although the California Department of Water Resources has estimated CII water use by sector at the state level, and a few other studies calculated water use by industry in specific regions, no statewide estimate of water use by industry exists. Therefore, the first step in calculating water conservation potential involved estimating baseline CII water use by sectors and end use. Table 6.1 summarizes current water use in California's CII sectors in 2000. All together more than 3,000 mcm/yr (2.5 MAF/yr) were used for these purposes—about 30 percent of all urban water use.

Within the CII sectors, water use varies among individual users in both quantity and purpose. Because of these differences in use, conservation potential varies from one industry to the next. Detailed analyses were done for industries that account for about 70 percent of total CII water use. Insufficient data were available for detailed studies of the remaining sectors and more general estimates were made. Table 6.1 lists the industries examined in detail and their estimated water use in 2000.

End Uses of Water

Although individual industries use water differently, nearly all of them use some water for "common" purposes. Through examining water use in the industries shown in Table 6.1, water use in all industries could be classified into six broad end uses: sanitation (restroom); cooling; landscaping; process; kitchen; and laundry. With the exception of process water use, the end uses (i.e., toilet flushing or dishwashing) are very similar among industries. For example, although a hospital and dairy plant use process water for very different

TABLE 6.1   Estimated 2000 Water Use in California's CII Sectors, (mcm/yr)

Commercial		Industrial	
Schools	310	Dairy Processing	21
Hotels	37	Meat Processing	19
Restaurants	201	Fruit and Vegetable Processing	86
Retail	189	Beverage Processing	70
Offices	418	Refining	104
Hospitals	46	High Tech	93
Golf Courses	283	Paper	27
Laundries	37	Textiles	38
		Fabricated Metals	25
Unexamined Commercial	766	Unexamined Industrial	340
TOTAL	2,285	TOTAL	821

Note:

1. Commercial water use, as reported herein, includes both commercial and institutional uses.

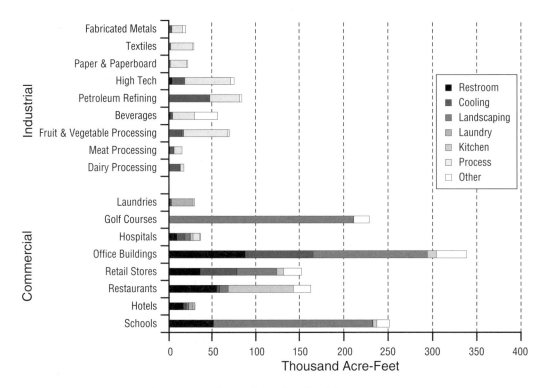

FIGURE 6.2 . Estimated water use by end use for the CII sector, 2000.

Source: See Gleick et al. 2003, Appendices C and D for derivations of use by industry.
http://www.pacinst.org/reports/urban_usage

purposes, they both use landscape water for irrigating turf and other vegetation and restroom water for flushing toilets and running faucets.

The mix of end uses and quantity of water used varies widely by industry type. Industrial facilities tend to use water mostly for processes, although they do use (relatively) small amounts of water for common end uses. Commercial facilities tend to use water almost exclusively for common end uses. Figure 6.2 shows the estimated breakdown of CII water use into these six end uses.

Landscaping uses more water than any other end use in the CII sectors. Other significant end uses include restrooms, cooling, and process, which, when combined, comprise close to fifty percent of total water use. The smallest end uses, in terms of total use, are kitchens, laundries, and other. See Figure 6.7.

Process

Process water use includes any water uses unique to a particular industry for producing a product or service (Scaramelli and Cohen 2002). In the food-processing industry, for example, any water used in the production of canning tomatoes, whether for cleaning the equipment or cooking the tomatoes, counts as process water.

Process end uses vary tremendously among industries. While hospitals use process water for x-ray machines, sterilizers, and vacuum pumps, beverage production plants use water for cleaning equipment and bottles and as part of the final product. Even within specific industries, process water use can vary greatly. In food producers who make tomato salsa, for example, plants that produce salsa from preprocessed tomatoes use water very differently from plants that produce salsa from whole tomatoes.

The analysis shows that process water use comprised approximately 18 percent (550 mcm) of California CII use in 2000. Nearly all of this water use took place in the industrial sector with the high tech, beverage, and food and vegetable industries using the most process water of the examined industries. In the commercial sector, only the hospital industry used significant amounts of process water (see Figure 6.3).

Restroom

In restrooms, water is used for toilet and urinal flushing and faucets and, in hospitals and hotels, it is used for showers. Our estimates indicate that toilets consumed nearly three-quarters of restroom water use (see Figure 6.4).

Approximately 15 percent (444 mcm) of total California CII water use in 2000 was used in restrooms. Restroom water use is ubiquitous across all industries, but it is most significant in the commercial sector, particularly hotels, where it represents as much as 55 percent of total water use. In the industrial sector, restrooms often use a very small percentage of total water relative to process and cooling uses. For some of these industries, therefore, restroom water use is combined with landscaping and kitchen into the generic category of "other."

Cooling

Cooling involves using water either as part of the production process or for air conditioning units. In the production process, water either directly cools heated equipment or components (contact cooling) or cooling towers chill the water, which then runs through heat exchangers to cool hot fluids or air (non-contact cooling). Cooling as part of the production process generally occurs in the industrial sector and is particularly significant in the petroleum refining and dairy industries. Water use by air conditioning units is common in both industrial and commercial industries.

Kitchen

Water is used in kitchens for a number of purposes including pre-rinsing and washing dishes and pots, making ice, preparing food, and cleaning equipment (MWRA 2002). As illustrated in Figure 6.5, over 50 percent of "kitchen" water use goes to cleaning dishes and pots.

In 2000, approximately 6 percent (185 mcm) of total CII water use occurred in kitchens. While restaurants provide the most obvious and significant example of kitchen water use, most industries use some kitchen water, whether in the cafeteria of a hospital, factory, or school or in the kitchenette of an office or deli of a retail store. In some industries, the amount of kitchen water use relative to the amount of water used in processing is so small that it is rarely counted separately. In these cases, we assume that it falls in the category of "other."

Landscaping

Landscaping includes water used for irrigating turf and shrubs. Most landscaping water goes to turf irrigation because it is both more dominant and water intensive than other vegetation used in landscaping. Figure 6.6 shows the breakdown between turf and other vegetation water use.

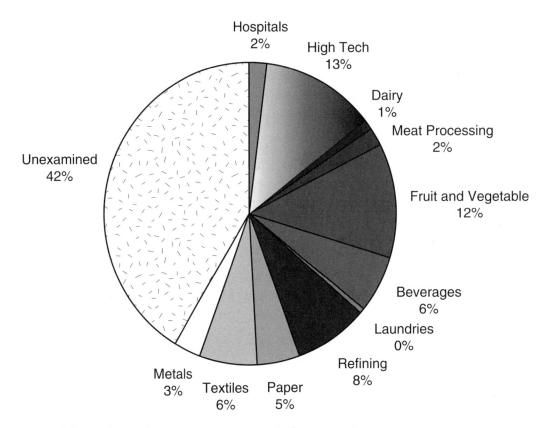

FIGURE 6.3 Estimated process water use by industry, 2000.

Source: See Gleick et al. 2003, Appendices C and D for derivations of use by industry.
http://www.pacinst.org/reports/urban_usage

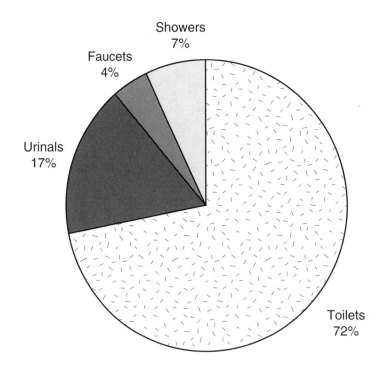

FIGURE 6.4 Water used in restrooms in the CII sector, 2000.

Source: See Gleick et al. 2003, Appendix C for a detailed description of how restroom water use was estimated.
http://www.pacinst.org/reports/urban_usage

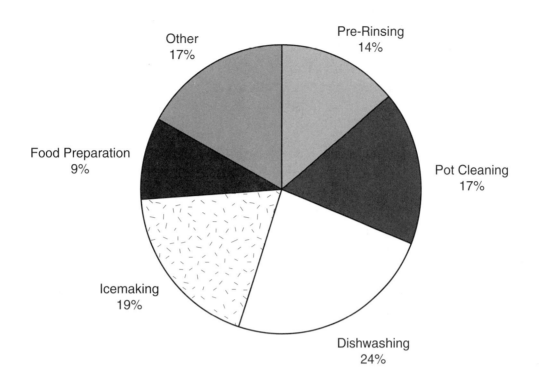

FIGURE 6.5   Water used in kitchens in the CII sector, 2000.

Source: See Gleick et al. 2003, Appendix C for a detailed description of how kitchen water use was estimated. http://www.pacinst.org/reports/urban_usage

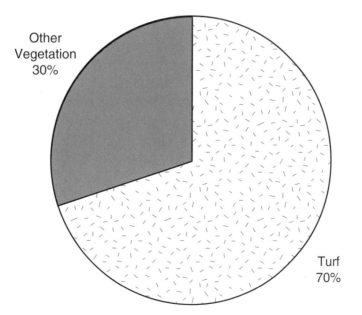

FIGURE 6.6   Landscape water use in the CII sector, 2000.

Note: This ratio is derived by averaging different California regional CII data set on turf and vegetation extent.
Source: City of Santa Barbara, 1996a,b, Contra Costa County 1996, Haasz 1999.

Many commercial and industrial facilities in the state use substantial amounts of water for landscaping. In 2000, 38 percent (1,190 mcm) of CII water use went to landscaping statewide, according to our estimates. In many industrial facilities, water use for landscaping is so small relative to other uses that it is counted as "other," whereas landscaping generally comprises a sizable portion of water use in the commercial and institutional sectors, particularly in schools and office buildings.

Laundry

Laundry water use includes water used to wash clothing and other fabrics in standard and commercial washers. Laundries use almost all of their water in the washing process (we classify it as process water use). Many establishments such as hotels, nursing homes, and universities offer coin laundry facilities. Some hotels and hospitals (about 5 percent) have in-house laundries, but increasingly they are outsourcing their laundry to commercial laundries. In individual establishments that do have in-house washing machines, laundry often represents a major percentage of water use, although laundry use may only represent a small percentage of water use for the industry as a whole, because of outsourcing.

Other

"Other" includes uses that do not fall in the end uses listed above or uses that represent such a small percentage of total water use that they are consolidated into one category. In the industrial sector, where almost all water is used for process purposes, "other" may describe all non-process uses and include restroom, kitchen, cooling, and landscaping uses. In both the industrial and commercial sectors, "other" often captures miscellaneous uses such as water use in janitorial closets in schools and hospitals or leaks in any type of industry.

Estimated CII Water Use in California in 2000

Baseline data were obtained from a comprehensive survey of industrial users developed by the California Department of Water Resources in the mid-1990s,[3] water-use surveys by sector as reported by nearly 150 water districts in 1995 and 2000, and a few studies based on surveys of water use primarily in southern California's commercial sector (CDWR 1995a,b, Davis et al. 1988, Dziegielewski et al. 1990). A further valuable source of information was the *Commercial and Institutional End Uses of Water* study published by the American Water Works Association Research Foundation (AWWARF) (Dziegielewski et al. 2000, Mayer et al. 1999).

When estimating water use in the CII sectors in 1995, two independent approaches were used and then crosschecked against other published estimates. These methods involved using the CII survey data and the public water-supply delivery data reported to the CDWR by 147 water agencies across the state (CDWR 1995b).[4] Once publicly supplied water use was calculated, self-supplied water use was estimated based on survey data and on information from USGS analyses (Solley et al. 1998).

---

3. More than 2,600 firms responded to this survey. Although the survey data were from 1994, the data were updated for 1995 and 2000 to produce a baseline estimate.

4. Over 470 agencies were listed in the CDWR file, but most of these agencies did not differentiate between commercial, institutional, and industrial uses and, therefore, could not be included in this analysis.

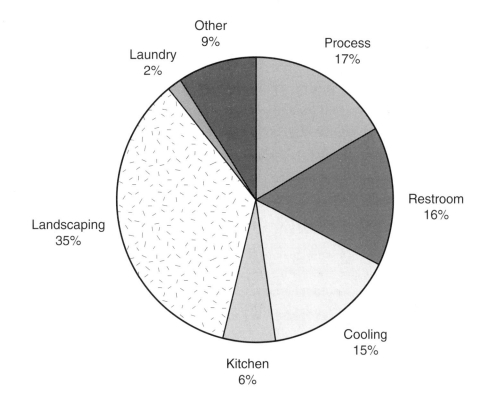

FIGURE 6.7 Estimated water use in the CII sectors by end use, 2000.

Note: Consolidation of water use estimates by end use. These estimates were calculated in each of the industries examined here by applying end-use percentages (from multiple sources) to GED estimates of total water use.

Source: See Gleick et al. 2003, Appendices C and D for these calculations.
http://www.pacinst.org/reports/urban_usage

Data Challenges

Truly accurate estimates of current total CII water use cannot be developed without better information, an improvement in reporting methods, and more detail on regional and agency variations in water use and conservation. While some water agencies currently break out their urban water use into residential and non-residential sales and the more advanced water agencies further classify their non-residential sales into commercial, institutional, and industrial sectors, even these data are not always comprehensive. A handful of California water agencies, such as Sacramento, East Bay Municipal Utility District, and Torrance, break up their sales by user type, but this is not common, and is even less common outside of California. Unfortunately, agencies that do classify their sales by sector or user type do not use a standard classification system, making comparisons difficult. Many agencies also fail to accurately report the population served, at either the local or hydrological region level.

If water agencies implemented a standard customer classification system, however, calculating CII water use by industry would simply require adding up the water delivered by customer categories. Creating such a system would require water agencies to add a few extra fields to each customer record including the NAICS code and facility description (office building, educational, manufacturing, restaurant, hospital, parking lot, etc.) in addition to refining their population counts. Standardized database maintenance could be encouraged through numerous means, including adding such requirements to Urban Water Management Plans or BMP reporting.[5]

The addition of another field to each record—the number of employees/residents at the customer facility—could further improve the reported information. While the 1995 CDWR Water Survey was an excellent attempt to collect such data, the routine collection of employment data and its entry into a central CDWR database would allow CDWR to better spend its funds in collecting more detailed water-use surveys.

The Potential for CII Water Conservation and Efficiency Improvements: Methods and Assumptions

Improving the efficiency of water use in the CII sectors can be accomplished with a broad range of technologies and actions that decrease water use without affecting production. The water manager often has several options to choose from when improving water efficiency and these technologies and actions vary in their potential water savings, cost, and payback period. Industries, which use varying quantities of water for different purposes, have historically implemented conservation measures at different rates, giving each industry a unique conservation potential. Conservation potential also varies among regions because of differences in industrial concentrations and in the extent of past efforts to improve water-use efficiency.

Through literature and audit reviews, discussions with equipment manufacturers, and meetings with water managers, we identified the most common conservation measures that apply to the different end uses, including process use by the various industries.[6] Most of these measures are point-of-use reduction measures, although several involve on-site reuse. The potential savings from these technologies depends on their specific water-saving characteristics, economic factors, and other barriers to implementation.

Very few measures identified involve water reclamation or behavioral modifications.[7] For purposes of the report, only a few behavioral modifications that were judged to be long-term measures, such as switching from turf to other vegetation, were included. Short-term measures that are usually instituted in response to drought situations, such as lawn-watering restrictions, were excluded. This conservative assumption also means that these kinds of responses are still available during drought periods.

Potential Water Savings Summary

The total amount of water that these measures can save in the CII sectors varies tremendously by industry and end-use. Estimates of savings also vary within industries because different sources report different or vague penetration rates[8] and potential savings. To

---

5. Accuracy of data entry would also have to become a priority because, as suggested by Sweeten (2002), in districts that currently categorize users, errors often exist due to low prioritization of this task.

6. See Gleick et al. (2003) Appendix C and D for a complete glossary of all of the technologies examined here.

7. Even though behavioral or reclaimed water measures were mentioned very few times, they can still save significant quantities of water. Indeed, if all potable water currently used in California golf courses were replaced with reclaimed water, more than 280 million m^3/yr. could be saved.

8. The rate at which conservation technologies have already penetrated a market.

TABLE 6.2   Estimated Potential Savings in California's Commercial Sector for 2000

Commercial	Potential Savings (mcm/yr)		
	Low	High	Best
Schools	113	153	143
Hotels	11	14	12
Restaurants	54	63	59
Retail Stores	51	83	69
Office Buildings	125	190	164
Hospitals	14	21	18
Golf Courses	69	261	101
Industrial Laundries	14	22	18
Unexamined Industries	228	407	295
TOTAL	678	1213	880

Note: The Commercial Sector includes California's institutional water use (government buildings, schools, and universities).

Source: See Gleick et al., 2003, Appendices C and D for details.
http://www.pacinst.org/reports/urban_usage/

TABLE 6.3   Estimated Potential Savings in California's Industrial Sector for 2000

Industrial	Potential Savings (mcm/yr)		
	Low	High	Best
Dairy Processing	2.5	8.6	6.2
Meat Processing	2.5	6.2	4.9
Fruit and Vegetable Proc.	8.6	30.8	22.2
Beverages	7.4	12.3	11.1
Petroleum Refining	48.1	96.2	76.4
High Tech	23.4	45.6	35.8
Paper and Pulp	3.7	12.3	8.6
Textiles	11.1	16.0	13.6
Fabricated Metals	6.2	11.1	8.6
Unexamined Industries	81.4	170.2	133.2
TOTAL	194.8	409.4	320.6

address these differences, potential savings were calculated as "best" (what we judge to be the most accurate estimate based on source of the data, age of the data, and/or sample size), "low" (assuming low penetration of the conservation technologies), and "high" (assuming high penetration of the conservation technologies). Overall, the range of potential savings is between 875 mcm/yr (710,000 AF/yr) and 1,600 mcm/yr (1.3 MAF/yr) over current use. Our best estimate of potential savings in the CII sectors is about 1,200 mcm/yr (975,000 AF/yr), or 39 percent, of total current annual water use (see Tables 6.2 and 6.3).

Using our best estimates of potential savings as a guide, the greatest percentage of water savings could be realized in traditional heavy industries, such as petroleum refining, which could potentially save nearly three-quarters of its total current water use (see Figure 6.8). Other industries that could save a large percentage of their total water use include paper and pulp (40 percent), commercial laundries (50 percent), and schools (44 percent).

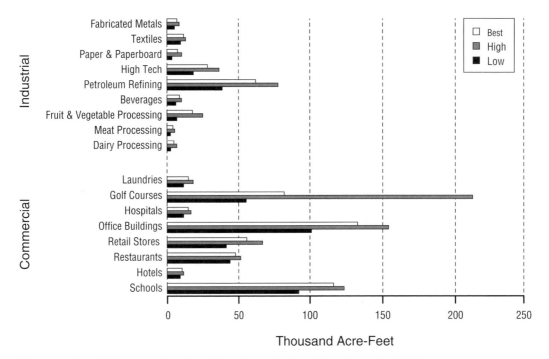

FIGURE 6.8   Estimates of potential savings in the CII sectors for 2000.
Source: See Gleick et al. 2003, Appendices C and D for derivations.
http://www.pacinst.org/reports/urban_usage

Although many of the largest percentages of water savings relative to use appear in the industrial sector, our findings suggest that the largest *quantities* of water could be saved in California's commercial sector because commercial facilities use more water overall. The best estimate shows, for example, that office buildings and schools could each save more than 140 mcm/yr if all recommended conservation measures were implemented. In contrast, potential savings for the petroleum refining industry, which has the highest potential savings in the industrial sector, are about 76 mcm/yr (62,000 AF/yr).

Approximately half of these total savings would come from reductions in landscaping water use, which could be cut by 50 percent with the conservation measures recommended here (see Figure 6.9). Implementing the recommended conservation measures could also reduce restroom and laundry water use by approximately 50 percent. The potential savings for restrooms (196 mcm/yr) is much higher than laundries (32 mcm/yr), however, because restrooms comprise a larger percentage of total CII water use than laundries. And we estimate the potential savings in kitchens and cooling at approximately 20 percent of their total use, which would total over 120 mcm annually.

Conservation by Region

Because California is such a diverse state geographically and climatically, conservation potential also varies by region. For water-planning purposes, state water agencies divide California into 10 hydrological regions that approximately correspond to the state's major drainage basins (CDWR 1998).

California's regions have implemented conservation measures at different rates depending on the reliability and adequacy of the regional water supply. Problems with water-supply reliability often manifest themselves in terms of increased water rates, poor service,

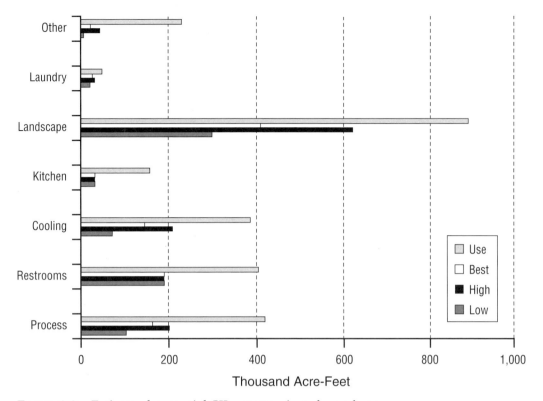

FIGURE 6.9 Estimated potential CII water savings, by end use.
Source: See Gleick et al. 2003, Appendices C, D and E for details.
http://www.pacinst.org/reports/urban_usage

accelerated implementation of conservation measures relative to other regions, or the development of new supplies.

In many regions of California, the population continues to grow, but options for increasing supply remain limited, leaving these regions susceptible to shortages, especially in times of drought. This situation has encouraged some water agencies to raise water rates and promote the implementation of conservation measures to improve efficiency and reduce demand. Water agencies in the coastal regions appear to be more aggressive in implementing conservation measures than those in the interior regions. Specifically, the report shows that the North Coast and the South Coast regions are implementing more comprehensive conservation measures than the Central Valley and Colorado regions. Given these results, the state's interior regions have the greatest remaining conservation potential as a fraction of total use, though overall remaining savings may be higher in coastal regions.

Methods for Estimating CII Water Use and Conservation Potential

Calculating water conservation potential in California's CII sectors requires taking account of differences in how individual industries use water. Because time, resource, and data limitations prevented us from calculating conservation potential in every industry, a sub-group of industries was examined in detail.

After selecting a group of industries to represent both the commercial and industrial sectors, we looked closely at each industry. We first determined how much water was utilized by each end use, crosschecked these estimates when possible, and then listed the conservation measures corresponding to each end use before calculating the potential savings. Finally, potential savings from each end use were summed to get an overall potential savings for each industry.

Differences among commercial and industrial facilities required that different methodologies be used for computing conservation potential. While commercial facilities use water mostly for common end uses, industrial facilities use water mostly for processing products, in boilers to generate steam, or in process cooling.

Since commercial facilities use water primarily for common end uses, it was easier to identify general conservation measures for this sector. In contrast, potential savings at each industrial facility must be examined individually. For example, California's 500 fruit and vegetable plants use water for diverse purposes ranging from peach canning to producing tomato paste. Such differences usually require a detailed site audit followed by an economic analysis to identify what technologies are cost-effective for each facility.

Commercial

For the commercial sector, the end uses were grouped by type rather than by examining SIC code water use. For example, psychiatrists' offices, engineering firms, and banks use water in similar ways, even though they belong to completely different SIC codes. Conversely, psychiatrists' offices and nursing homes are classified under SIC code 80, even though the nursing homes use water more like a multi-family residential complex.

To avoid these inconsistencies, we selected the top five commercial groups from the AWWARF study of commercial and institutional end uses of water (Dziegielewski et al. 2000), and other commercial groups with reliable and relatively comprehensive data sets. In total, the groups we selected accounted for 73 percent of commercial water use.

Industrial

To select the industrial groups, the most water-intensive industries at the two-digit SIC code level were identified. Then, within each of these two-digit SIC codes, we examined how the individual industries at the more detailed three-digit SIC code level used water. For some industries, water use at the three-digit SIC code level was similar enough that the entire two-digit SIC code was included in our analysis. In the case of the textiles industry (SIC code 22), for example, the three-digit sub-classification was based on the type of fabric being processed and the water-intensive processes such as dyeing, printing, and finishing that were common to all fabrics. Given this similarity in process water use, SIC code 22 was selected as one industry group. The same grouping was made for the high-tech sector (SIC codes 35, 36, and 38) (Allen and Hahn 1999, Boyko 2000).

In other industries, however, processing varied greatly among the three-digit SIC codes and only certain sub-industries were included. The paper and pulp industry (SIC code 26), for example, includes paper mills, pulp mills, and paperboard production. While paper and pulp mills use very water-intensive processes to convert raw fibrous material into a finished product, paperboard and converted paper products industries

(SIC codes 264 and 265) merely cut and assemble boxes out of raw paperboard and use no process water. Because these differences in use were so great, we included only the water-intensive industries (SIC codes 261, 262 and 263) in our analysis. A more extreme example occurred in the chemical (SIC code 28) industry, which is one of California's more water-intensive industries. Because this industry includes sub-industries as diverse as pharmaceutical drugs, industrial resins, petrochemicals, and fertilizers, we could not conduct a detailed analysis of how water is used in the general chemical industry.

Once we selected industries for more detailed assessment, data were gathered on water use and conservation in these industries in order to identify conservation technologies that are currently being implemented or are in the development stage (in either research or pilot testing); to identify the typical magnitude of savings for each technology as a percentage of total or process water use; and to determine the penetration rates of each technology.

Data reported in Tables 6.4 and 6.5 consist of surveys of specific sectors and best estimates from conservation and efficiency experts. A few important sectors were omitted due to the lack of data.

Calculation of Conservation Potential

Upon selecting the industries, we quantified how much water each industry used for specific end uses, such as restroom or kitchen use. The first step involved reviewing case studies (e.g., Anderson 1993, Asnes 1984, Black and Veatch 1991a, 1991b, 1995, NRTS 2001), a summary of the Metropolitan Water District's (MWD) CII audit data (MWD 1992, 1996, 2002), technical papers (e.g., CIFAR 1995a,b, Chiarello 2000, and CII water-conservation materials (e.g., Carawan and Sheldon 1989, City of San Jose 1992a–j, MnTaP 1994a,b, North Carolina DNRCD 2002 and NCDENR 1998, Koeller and Mitchell 2002), to determine the average breakdown of water use, by end use, for each industry. These percentages were then multiplied by the industry's total water use to calculate the quantity of water going to each end use. These results were crosschecked against additional sources, including industry statistics, case studies, and calculations from end-use studies.

Potential savings for each end use were estimated from a variety of information on existing conservation measures. We calculated the conservation potential for each common end use and then applied the potential savings to all of the industries. Due to the diverse nature of process-related end uses, the potential process savings were calculated separately for each industry.

The first step of these calculations involved breaking down each end use into sub-end uses and identifying existing conservation technologies (and their savings) corresponding to the sub-end uses. We used a number of sources, including case studies of individual facilities, technical industry papers, summary results from detailed surveys from the MWD, published audit summary results, and manufacturers specifications to determine which technologies could be used to save water for each sub-end use.

Upon identifying the conservation technologies, we estimated their current penetration rates throughout California using existing penetration information that we collected from various sources listed in Table 6.6.

While these sources provided fairly complete information on penetration rates for some technologies, important gaps remained. For some technologies and/or indus-

TABLE 6.4   Estimated Water Use in the Commercial and Institutional Sectors in 2000

Commercial	2000 (mcm/yr)	SIC Codes
Schools	309	821, 938
Hotels	37	701
Restaurants	201	58
Food and beverage stores	43	54
Other retail stores	145	53, 55, 56, 57, 59
Office buildings	418	60–67, 86
Hospitals	46	806
Golf courses	422	7992
Coin laundries	6	7215
Industrial laundries	37	721 (except 7215)
Unexamined commercial	619	
TOTAL	2,284	17–19, 41–99
Percentage water use selected	73	

Note: Total may not add up precisely due to rounding.

TABLE 6.5   Estimated Water Use in the Industrial Sectors in 2000

Industrial	2000 (mcm/yr)	SIC Codes
Food processing		
Dairy	21	202
Meat	18	201
Fruit and vegetable	86	203
Beverages	74	208
Petroleum refining	104	291
High tech		28
Semiconductors	18	3674
Other high tech	74	358, rest of 36, 38
Paper and paperboard mills	27	261, 262, 263
Textiles	36	22
Fabricated metals	25	34
Unexamined industrial	337	Rest of 20–39
TOTAL	820	20–39
Percentage water use selected	59	

Note: Total may not add up precisely due to rounding.

tries, little or no penetration rate data existed. And even where the data were available, the descriptions of penetration were often qualitative, using terms such as "very few" or "several" to describe the number of facilities using such measures. When actual penetration rates were unavailable, estimates were made based on the age of the technology or, more commonly, on available qualitative data. These values must be considered estimates only.

Upon identifying the appropriate conservation technologies, the savings from implementing them, and their penetration rates, total conservation potential due to the tech-

TABLE 6.6   Sources of Market Penetration of Conservation Technologies

Data Source	Industry/End Use	Geography
Surveys from Industry Associations	Food Processing; Coin Laundry- Golf Courses; Metal Finishing and Semiconductor	California; Southwestern United States
Surveys from the U.S. EPA	Industrial Laundries	United States
Reclaimed Water Data, State Water Resources Control Board	Schools; Golf Courses; Textiles; and Refining	California
Assumptions: Industry Experts	Various	Various
Interviews: Consultants and Industry Officials	Cooling; Textiles; Kitchens; and Paper and Pulp	United States; California
Individual Facility Data	Refineries	Various
Summary of MWD Survey Results	Restrooms; Landscaping; All Industries	South Coast Region, California
Survey/Audit Results: Water Agencies	Various	Various

nology could be identified. Based on the type of information available for a particular technology and sub-end use, we used one of two methods to calculate conservation potential.

The first method involved the "best case" scenario and we used it when comprehensive data were available. The information required includes: water use per unit or per event by the efficient and inefficient technology (e.g., gallons per flush for toilets, gallons per minute for showerheads, gallons per rack for dishwashers, or gallons per load for clothes washers [see, for example, Hagler Bailly 1997, Pope et al. 2002]); the penetration rate of efficient and inefficient models; and the total number of units/events per year for the industry (i.e., total number of toilet flushes per year, total loads of laundry per year, or total minutes of showering per year). With this information, current water use by the efficient and inefficient models could be estimated. The difference between the current use and the most efficient use (assuming 100 percent penetration of the efficient model) yields an estimate of the technical potential available (see Box 6.1).

In many cases several technologies can be applied *simultaneously* to a particular end-use in an industry. For instance, case studies on low-flow nozzles and autoshutoff valves suggest that each have savings potentials of 50 percent and can be simultaneously implemented at the same facilities. Clearly, the savings are not additive because, if we implement both, water use does not decrease by 100 percent. We describe technologies as *complementary* if they can be simultaneously implemented at one facility.

Another situation occurs when technologies are mutually *exclusive* and either one or the other is applicable depending on some specific characteristic of the facility. An example of this type of a situation is in landscape water use where different technologies apply to turf and shrubs. In this case it is necessary to know how much of the total water use is by turf and shrubs respectively.

A third situation occurs when technologies apply to only a *component* of the water use. For instance, kitchen water use includes dishwashing, pre-rinsing, and icemakers. Different technologies apply to each of these components of water use such as efficient dishwashers, low-flow pre-rinse nozzles, and efficient icemakers respectively (e.g., Pike et al. 1995, Pike 1997, ASHRAE 1994).

Data Constraints and Conclusions

As has been noted elsewhere, data constraints affect final estimates of conservation potential in the CII sectors. These constraints are encountered when calculating current water use by specific end uses, penetration rates, and potential water savings. The primary data constraint is a fundamental lack of key information. At the most basic level, reliable end-use analyses for a few industries in the industrial sector, such as textiles, are unavailable. The penetration rates of conservation technologies are also rare or unavailable, forcing one to estimate potential savings and adding another level of uncertainty to estimates of untapped conservation potential.

Even when data are available, they often contain limitations that further affected the reliability of conservation estimates. Often, only average penetration data are available, making regional conservation potential estimates unreliable. Performance of some technologies is poorly reported (Vickers 2001). When region-specific data are available, several factors often limit their usefulness. Details of how savings are realized are omitted from many studies, particularly for the industrial sector. For example, the literature may report potential savings for process water in a given industry, but it often does not report the amount of this water being saved from the sub-end uses of process-

---

Box 6.1 Definitions

Technology Savings: Percentage of water saved by implementing a particular technology, assuming service provided remains the same

e.g., Technology
savings from ULFTs $=$ $\dfrac{\text{Water use in 3.5 gpf—Water use in 1.6 gpf toilet}}{\text{Water use in 3.5 gpf toilet}}$

$= (3.5\text{-}1.6)/3.5 = 54.3\%$

Measure of Technology Penetration: Percentage of the total number of potential sites using the efficient technology

e.g., Penetration
of ULFTs $=$ $\dfrac{\text{Number of ULFTS}}{\text{Total population of toilet}}$

Conservation Potential Percentage: Percentage of the total water used for a particular purpose that can be eliminated

e.g., Conservation potential from replacing all toilets by ULFTs

$=$ $\dfrac{(1\text{- Penetration of ULFTs}) * \text{Technical Savings from ULFTs}}{(1\text{- Penetration of ULFTs} * \text{Technical Savings from ULFTs})}$

ing, such as rinsing and sterilizing. Without these breakdowns, crosschecking estimates of potential savings is more difficult. In the commercial sector, problems arise when data are in varied and variable formats such as: gallons/employee/day; gallons/square foot/day; or gallons/meal served/day. Each conversion of these numbers into a comparable figure risks introducing uncertainty.

When data on specific conservation technologies are available, they are sometimes out of date. In the area of water conservation, technologies are continuously changing (see Box 6.2) and a water-savings technology may become obsolete a few years after implementation when an even better technology is introduced. Trying to sort out the mix of several existing technologies, and the potential for new technologies, further complicates calculating conservation potential.

Despite these data constraints, working through the water-conservation potential provides a framework for further discussion and improvements. The "modular" approach employed in the California assessment allows agencies with better information to update

---

Box 6.2 Evolution of Water Conservation Technologies

Water conservation technologies are constantly evolving. The technologies that were adopted in the 1970s and early 1980s were the easiest and cheapest to implement—the low hanging fruit. In this period, the water conservation literature focused on preventing waste and typical water conservation measures implemented included autosensors to turn off water when production lines were not in use, elimination of single-pass cooling, reuse of non-contact cooling water, and replacement of inefficient toilets. Most of these measures were fairly low technology and repaid quickly.

In the late 1980s and 1990s, there were further improvements in water-efficient equipment (clothes washers, dishwashers, toilets, and pre-rinse nozzles). More recently, the focus has been on reducing overall freshwater demands by reusing treated wastewater streams. A detailed analyses of every waste stream of every industry was beyond the scope of our report (Gleick et al. 2003), but the broad steps include: segregating effluent streams; identifying the characteristics of each waste stream; identifying processes that can potentially use water of a lower quality; and treating effluent streams with chemicals and membrane filtration to increase its quality for reuse.

This trend is expected to continue in the future, with more and more fresh water being substituted with treated internal waste streams or reclaimed water from local water recycling plants. Indeed, some industries, such as paper and pulp, industrial laundries, and metal finishing, are beginning to develop "closed-loop" systems where all the wastewater is reused internally, with only small amounts of fresh water needed to make up for water incorporated into the product or lost in evaporation.

penetration rates or other components of conservation potential to reflect status in their service area. Similarly, industry associations with better information on conservation potential in process water use can adjust these figures without changing the conservation estimates for cooling or restroom use. And, most importantly, the process provides the first overview of the conservation measures in each industry and illustrates which measures will produce the most savings.

Recommendations for CII Water Conservation

Developing recommendations for industrial and commercial water users is increasingly critical to encouraging actions, and agencies, states, and local groups are working to identify how best to move to capture conservation potential (Schultz Communications 1999, SPUC 1999).

Encourage Conservation Through Proper Water Pricing, Including Wastewater Charges

Incentives for improving water efficiency and conservation are always greater when the price of water accurately reflects its true costs. All water providers should charge appropriate prices for water, including charging for wastewater separately, by volume of water. When wastewater charges fall below the cost of pollutant disposal, industries often choose to use extra water to dilute their wastewater streams until the pollutant levels reach acceptable levels. Wastewater charges can be adjusted to discourage this practice.

Encourage Conservation Through Wastewater Permitting

When an industry wants to expand its operations, it usually undergoes a wastewater permitting process. Several water districts have successfully incorporated water conservation requirements into this process so that as companies grow, and their demand for water increases, they increase their level of water conservation.

Encourage Smart Management Practices at the Industry Level

Often, industry managers will introduce conservation measures, but differences in management and worker goals can prevent the full implementation of these measures. For example, not budgeting additional worker time for implementing water conservation technologies contributes to poor implementation rates and may even increase water use. If managers take such worker concerns into consideration, however, they can achieve more long-term results. Managers also need to remember that, like all equipment, conservation devices have regular maintenance or replacement requirements. Facilities must be encouraged to incorporate checking water-efficient fixtures as a part of routine maintenance.

Budgeting practices also frequently contribute to a poor conservation ethic. At large facilities, individual departments may not know how much water they use, much less how to conserve it, when a central office handles their accounts. Managers should provide the appropriate incentives to individual departments, such as deducting the utilities bill from the department's budget or ensuring that the facility's maintenance department receives a copy of the water bill.

Educate Industry Decision Makers and the Public About Hidden Conservation Opportunities

Industries sometimes choose less efficient technologies because they are operating with incomplete information. Discussions with the Champion dishwasher company, for example, revealed that sales of an inefficient dishwasher model (UH-150B) far exceeded sales of the efficient model in the same range (UH-200B) because the efficient model costs about 10 percent more than the less efficient model. The customers were unaware that an efficient commercial dishwasher pays back in about six months.

Other hidden conservation opportunities exist when an industry does not own its water-using equipment, but rents from an independent rental agency that charges a monthly or a use-based fee. In the case of some dishwashing rental companies, for example, the rental company makes most of its margin selling cleaning chemicals that require more water for rinsing. In this arrangement, these companies have a perverse incentive to lease inefficient dishwashers and the customer pays for more chemicals and water.

Water agencies should also encourage the implementation of new technologies that are not intended to achieve reductions in water use but do so anyway. Occasionally, shifts to water-conserving equipment have occurred for reasons unrelated to water conservation. In hospitals, for example, water-ring vacuum pumps were historically installed because flammable gases were used as anesthetics. Once the flammable gases were discontinued, hospitals slowly shifted to oil-based pumps. Similarly, digital x-ray film processors that use no water are gaining market share for their superior ability to process, transmit, and manipulate x-ray images.

Give Industries an Opportunity to Tout Their Conservation Achievements

Programs such as promoting the most efficient water users in local newspapers or other media outlets during a drought or instituting green-certification programs often encourage industries to conserve water out of a desire to improve their public image. Instituting water-efficiency certification programs for industry groups such as hotels, restaurants, or hospitals can reinforce this trend.

Promote Reclaimed Water as a Secure Source for Water Supply

The desire for a guaranteed water supply during drought conditions has driven some petroleum refineries to switch to reclaimed water for their cooling needs. Even if water is not a major cost component, an interruption of water supply can cause shutdowns in many industries and result in lost income. Promoting reclaimed water as a secure supply may encourage some industries to invest in the necessary infrastructure for using this water.

Implement Financing Schemes That Encourage Conservation

Many conservation technologies are cost-effective for the water agency, but not for individual industries. When we considered cost-effectiveness in the original analysis, we use the weighted average cost of capital, which is about 7 to 10 percent for most private companies, to calculate the $/unit water cost. A technology is cost-effective for a firm if the $/unit water cost is less than the current price of water.

Realistically (but unreasonably), however, most companies expect a payback period of two years or less. This translates to a discount rate of 40 to 50 percent, depending on the lifetime of the equipment. This is a major reason for the difference between the economically achievable conservation potential and what actually gets implemented. Experience from successful energy-efficiency programs could be used as models to address this problem. Financing schemes, such as shared savings programs or leasing of efficient equipment, would require little or no capital to be invested up front by the customer.

Data Issues

As highlighted throughout this report, problems with data influenced research and results. Although we calculated water use and conservation potential estimates in the CII sectors with the information available, increasing the accuracy of future estimates requires water users, suppliers, and managers at all levels to increase the reliability and accessibility of water use and conservation data.

Currently, data are neither collected nor reported in standard formats. This lack of standardization affected the reliability of our estimates because it prevented us from cross-checking some of our calculations and accessing key background data that were often lost in the reporting process. Privacy concerns also limited our access to data while uncertainties about differences in data and various reporting units further affected our estimates. And, finally, the absence of certain data from the literature—such as end use breakups of water use in certain industries—required that we estimate certain findings based on very general information.

Recommendations for addressing these problems, and thus increasing the accuracy of future estimates of water use and conservation potential, are presented below.

Definitions of water-related terms should be standardized.

Currently, various agencies define water-related terms in the CII sectors differently. For example, one water agency may define a nursing home as a multi-family residential establishment while another agency classifies it as a commercial establishment. Until such terms are standardized, comparing data will remain difficult.

Water agencies should develop standard formats for water-use audits.

Every water agency and consulting firm uses a unique reporting form for data collection practices, reporting methods, and data categories. Standard forms should include fields that capture background information about each establishment, such as the area of the establishment, the number of employees, and other relevant facts that may vary by industry. Including these data would make comparisons of audits administered by different agencies more accurate.

Audits should also include a wide variety of data that audit administrators already collect for their final estimates, but that get lost somewhere between the field and the final report. Examples of such data include recording the amount of water used by the dishwasher, the sink, and the icemaker in kitchens rather than merely reporting "kitchen use." Similarly, an audit should capture information on specific conservation technologies in place, rather than simply report "process savings." Including these data would decrease confusion about what is included in each calculation and would consequently increase the accuracy of estimating penetration rates and, ultimately, conservation potential.

Reporting mechanisms currently used in the CII sectors must be further standardized.

Examples of reporting mechanisms include Urban Water Management Plans (UWMPs) that are developed by California water agencies, "best management practices" reports, and water-use data that the state water agency collects from local water agencies. While these mechanisms can be useful, differences in defining terms, calculating results, and other factors often limit their usefulness. Perhaps the best method for standardizing these reports would involve creating detailed manuals on what to report and how to report it. Although some guidelines currently exist, strict requirements about which units are used, the definitions of specific terms, such as what a survey is, and the best way to obtain specific information are not always explicitly outlined. Adherence to these guidelines must also be enforced somehow beyond what currently occurs. If such standards could be reached, the data provided through these reporting mechanisms would increase in accuracy and, thus, reliability and usefulness.

Water agencies should store customer records and audit results so that they can be shared with independent researchers while the privacy of the customer is protected.

Access to data is often limited by privacy concerns. The simplest way to overcome such barriers may involve assigning an identification number to each record, rather than just the customer name. If an identification number was used on audit forms, for example, researchers could access the raw data contained within the forms without concerns about privacy violations. Access to these raw data would increase the amount of information available, which would in turn have increased the accuracy of our findings. This practice would prove particularly helpful if the format of audits was standardized, as suggested above.

Collect additional data.

Perhaps the most obvious—and labor intensive—solution to increasing the accuracy of future estimates of water use and conservation potential in the CII sectors involves the collection of additional data. While the standardization of data, increased access to data, and reductions in the reporting inconsistencies of water agencies would certainly generate more useful and accessible data, some types of data are simply not collected reliably. Data on self-supplied water, for example, was very limited. Two other key pieces of information that we could not uncover were end-use breakdowns for several industrial users and the penetration rates of certain conservation technologies. Because these factors are central to accurate estimates of water use and conservation potential, we recommend improving current audits or using additional audits to collect this information.

Conclusions

The good news is that water users in the commercial and industrial sectors can save very substantial amounts of water with existing technologies and modest changes. We estimate that in 2000, the commercial, institutional, and industrial sectors in California used more than 3,000 mcm and that more than 1,200 mcm of this water can be saved through existing cost-effective strategies and technologies. Much of this savings comes from improving efficiency in outdoor watering, bathroom, and kitchen use—thus the same technologies that have proven so useful in the home can also cheaply save water in the CII sectors. But changes in the way water is recycled and modifications to specific CII end-use processes also show considerable potential, despite the progress that has already been made to improve efficiency and reduce waste.

REFERENCES

Allen, S., and Hahn, M. R. 1999. *Semiconductor wastewater treatment and reuse.* Semiconductor Fab Tech, ninth edition, Sunnyvale, California:Microbar Incorporated.

Anderson, B. 1993. *Commercial and industrial program case studies.* American Water Works Association, Conserv93 Proceedings, 485–515.

ASHRAE Refrigeration Handbook. 1994. Cited in Bose, James E., Smith, Marvin D., and Spitler, Jeffrey D. 1998. *Icemakers, coolers and freezers, and GX.* Washington, D.C.:Geothermal Heat Pump Consortium Inc.

Asnes. H. 1984. *Reduction of water consumption in the textile industry.* IFATTC Conference. London.

Black and Veatch (B&V).1991a. Water conservation reports: Unpublished restaurant audit report. B&V Project 17559. Prepared for the Los Angeles Department of Water and Power.

Black and Veatch (B&V). 1991b. Water conservation reports: Unpublished hospital audit report. B&V Project 17559. Prepared for the Los Angeles Department of Water and Power.

Black and Veatch (B&V).1995. *Norwood hospital case study, Norwood, Massachusetts.* B&V Project 24239. Prepared for the Massachusetts Water Resources Authority (MWRA). Boston, Massachusetts.

Boyko, M., Lelic, F., and Liberator, T. 2000. *A water efficiency program for the high-tech industry.* Report prepared by the City of Portland's BIG Water Conservation Program for Electronics and Microelectronics Programs. Portland, Oregon.

California Department of Water Resources (CDWR). 1995a, 2000. Unpublished water use reported to CDWR by water agencies. Sacramento, California.

California Department of Water Resources (CDWR). 1995b. Unpublished survey of industrial water use. Database available from CDWR. Sacramento, California.

California Department of Water Resources (CDWR). 1998. *California water plan update.* DWR Bulletin 160-98, November. Sacramento, California.

California Institute of Food and Agricultural Research (CIFAR). 1995a. *Membrane applications in food processing. Volume 1: fruit and vegetable processing industries.* Davis, California. July.

California Institute of Food and Agricultural Research (CIFAR). 1995b. *Membrane applications in food processing. Volume 2: Algin fiber, pasta, dairy, fruit, and wine processing industries.* Davis, California. September.

Carawan, R., and Sheldon., A. 1989. *Systems for recycling water in poultry processing.* North Carolina Agricultural Extension Service. CD-27. http://es.epa.gov/techinfo/facts/nc/poltry5.html

Carawan, R. E., Jones, V. A., and Hansen, A. P. 1979. Water use in a multiproduct dairy. *Journal of Dairy Science.* 62:no. 8.

Chiarello, R. P. 2000. Rinse optimization for reduction of point-of-use ultra-pure water consumption in high technology manufacturing. Report prepared for the Silicon Valley Manufacturing Group and the Silicon Valley Pollution Prevention Center. February. San Jose, California. http://www.ci.san-jose.ca.us/esd/PDFs/SVMGfinalreport.pdf

City of San Jose. 1992a. *Water conservation guide for cooling towers.* Environmental Services Department. City of San Jose, San Jose, California.

City of San Jose. 1992b. *Water conservation guide for the metal finishing industry.* Environmental Services Department. City of San Jose, San Jose, California.

City of San Jose. 1992c. *Water conservation guide for office buildings and commercial establishments.* Environmental Services Department. City of San Jose, San Jose, California.

City of San Jose. 1992d. *Water conservation guide for schools.* Environmental Services Department. City of San Jose, San Jose, California.

City of San Jose. 1992e. *Water conservation guide for new construction.* Environmental Services Department. City of San Jose, San Jose, California.

City of San Jose. 1992f. *Water conservation guide for hotels and motels.* Environmental Services Department. City of San Jose, San Jose, California.

City of San Jose. 1992g. *Water conservation guide for printed circuit board manufacturers and metal finishers.* Environmental Services Department. City of San Jose, San Jose, California.

City of San Jose. 1992h. *Water conservation guide for computer and electronic manufacturers.* Environmental Services Department. City of San Jose, San Jose, California.

City of San Jose. 1992i. *Water conservation guide for restaurants.* Environmental Services Department. City of San Jose, San Jose, California.

City of San Jose. 1992j. *Water conservation guide for hospitals and health care facilities.* Environmental Services Department. City of San Jose, San Jose, California.

City of Santa Barbara. 1996a. Water facts. Santa Barbara, California.

City of Santa Barbara. 1996b. *City of Santa Barbara urban water management plan.* March. Santa Barbara, California.

Contra Costa County. 1996. *Urban water management plan.* California.

Davis, W. Y., Rodrigo, D. M., Opitz, E. M., Dziegielewski, B., Baumann, D. D., and Boland, J. J. 1988. *IWR-MAIN water use forecasting system, version 5.1—Users manual and system description consultant report.* Carbondale, Illinois.

Dziegielewski, B., Rodrigo, D., and Opitz, E. M. 1990. *Commercial and industrial water use in Southern California.* Final report submitted to the Metropolitan Water District of Southern California by Planning and Management Consultants, Ltd. March.

Dziegielewski, B., Kiefer, J. C., Opitz, E. M., Porter, G. A., Lantz, G. L., DeOreo, W. B., Mayer, P. W., and Olaf, J. Nelson. 2000. *Commercial and institutional end uses of water.* American Water Works Association (AWWA) Research Foundation and the American Water Works Association. Denver, Colorado.

Gleick, P. H., Haasz, D., Henges-Jeck, C., Srinivasan, V., Wolff, G., Cushing, K. K., and Mann, A. 2003. *Waste not, want not: The potential for urban water conservation in California.* Oakland, California:Pacific Institute for Studies in Development, Environment, and Security. November.

Haasz, D. 1999. Reducing water use in residential, industrial, and municipal landscapes. In *Sustainable use of water: California success stories.* Owens-Viani, L., Wong, A. K., Gleick, P. H., editors. 49–67. Oakland, California:Pacific Institute for Studies in Development, Environment, and Security.

Hagler Bailly Services, Inc. 1997. *The CII ULFT savings study.* Final report. Sponsored by the California Urban Conservation Council. Boulder, Colorado. August.

Koeller, J., and Mitchell, D. 2002. Commercial dishwashers: A new frontier in energy and water conservation. American Water Works Association, Conserv 2002 Proceedings.

Massachusetts Water Resources Authority (MWRA). 2002. Water efficiency and management for restaurants. http://www.mwra.state.ma.us/water/html/bullet3.htm

Mayer, P. W., DeOreo, W. B., Opitz, E. M., Kiefer, J. C., Davis, W. Y., Dziegielewski, B., and Nelson, J. O. 1999. *Residential end uses of water.* Final report. AWWA Research Foundation. Denver, Colorado.

Metropolitan Water District of Southern California (MWD). 1992. Justa Restaurant. Unpublished restaurant audit report. Metropolitan Water District. Los Angeles, California.

Metropolitan Water District of Southern California (MWD). 1996. Unpublished hospital audit report. Metropolitan Water District. Los Angeles, California.

Metropolitan Water District of Southern California (MWD). 2000. *The regional urban water management plan for the metropolitan water District of Southern California.* December. Los Angeles, California.

Metropolitan Water District of Southern California (MWD). 2002. Unpublished data based on site surveys conducted between 1992 and 1996. Los Angeles, California.

MnTaP. 1994a. Water conservation opportunities for a printed circuit board manufacturer. Minnesota Technical Assistance Program. http://www.mntap.umn.edu/intern/Aci-it15.htm

MnTaP. 1994b. Schroeder milk saves $400,000 through product savings and water conservation. Minnesota Technical Assistance Program. http://www.mntap.umn.edu/food/cs80.htm

North Carolina Department of Environment and Natural Resources (NCDENR). 1998. *Water efficiency manual for commercial, industrial, and institutional facilities.* North Carolina Division of Pollution Prevention and Environmental Assistance, North Carolina Division of Water Resources, and Land-of-Sky Regional Council—WRATT Program. Asheville, North Carolina.

North Carolina Department of Natural Resources and Community Development (NCDNRCD). 2002. Pollution prevention tips: Water conservation for textile mills. http://es.epa.gov/techinfo/facts/nc/tips12.html

Pike, C. W. 1997. *Some implications of the California Food Processor Survey.* Sacramento, California:California Department of Water Resources.

Pike, C. W., Fierro, S., and Sheradin, H. L. 1995. Efficient water appliances for restaurants. American Water Works Association (AWWA) 1995 National Conference, Anaheim, California.

Pope, T., Fernstrom, G., Horowitz, N., and Maddaus, L. 2002. Commercial clothes washer standards: Highlighting the water-energy connections. American Water Works Association, Conservation 2002 Proceedings.

Scaramelli, A., and Cohen, R. 2002. Breakthrough in water conservation: Take it to the next level— Look at your process equipment. Paper presented at The American Society of Healthcare Engineers Conference. June 21–25, 1999. Philadelphia, Pennsylvania.

Schultz Communications. 1999. *A water conservation guide for commercial, institutional and industrial users.* Prepared for the New Mexico Office of the State Engineer. Albuquerque, New Mexico.

Seattle Public Utilities Commission (SPUC). 1999. Water Conservation Potential Assessment. http://www.cityofseattle.net/util/RESCONS/CPA/default.htm

Solley, W. B., Pierce, R. R., and Perlman, H. A. 1998. *Estimated use of water in the U.S. in 1995.* USGS Circular 1200. United States Geological Survey. Reston, Virginia.

Sweeten, J. 2000 and 2002. Metropolitan Water District of Southern California. Personal communication.

Vickers, A. 2001. *Handbook of water use and conservation.* Amherst, Massachusetts:Waterplow Press.

CHAPTER 7

Climate Change and California Water Resources

Michael Kiparsky, Peter H. Gleick

Global climate change is playing an increasing role in scientific and policy debates over effective water management. In recent years, the evidence that global climate change will affect water resources has continued to accumulate. More than 3,000 peer-reviewed scientific articles on climate and water have now been published, addressing everything from improved regional forecasting by general circulation models to adapting reservoir operations to new conditions.[1]

At the same time, however, regional water managers have not fully understood the state-of-the-science and taken account of climate changes in planning and managing complex water systems. California water planners and managers have been among the first in the United States to consider these issues, though most of their efforts have been modest and informal. Initial research on climate risks facing California water resources began in the early 1980s. By the end of the decade, state agencies such as the California Energy Commission had prepared the first assessments of state greenhouse gas emissions and of possible impacts of global warming on a wide range of sectors (CEC 1991). The official California Water Plan (Bulletin 160) briefly addressed climate change in the 1993 and 1998 versions (CDWR 1993, 1998), but without details or substantive recommendations.

Since that time, the scientific consensus has strengthened even further that climate changes will be the inevitable result of increasing concentrations of greenhouse gases. There is also strong scientific evidence that various anthropogenic climate impacts are already appearing worldwide (e.g., for Karl and Trenberth 2003). Impacts on California's hydrologic system have also begun to appear. Water agencies are beginning to consider the implications of climate change for the reliability and safety of water systems, and

---

1. To improve access to and information on these articles, the Pacific Institute has compiled a searchable electronic bibliography at www.pacinst.org/resources (see Water Brief 3).

professional water organizations have begun urging managers and planners to integrate climate change into long-term planning. In 1997, the American Water Works Association issued a committee report concluding that "Agencies should explore the vulnerability of both structural and nonstructural water systems to plausible future climate changes, not just past climatic variability," and "Governments at all levels should reevaluate legal, technical, and economic approaches for managing water resources in light of possible climate changes" (AWWA 1997).

Many uncertainties remain. Responsible planning, however, requires that the water community work with climate scientists and others to reduce those uncertainties and to begin to prepare for those impacts that are well understood, already appearing, or likely to appear. This chapter reviews efforts to integrate climate change science with water management in California, as a case study for other regions and water-resource areas around the world.

The State of the Science

Climate change is a scientific reality. The broad consensus of the scientific community is that greenhouse gases emitted by human activities are accumulating in the atmosphere and that these gases will cause a wide range of changes in climate dynamics (IPCC 1996, 2001, NRC 2002). Some of the most significant impacts will be on water resources—impacts that are of special concern to regions like California where water scarcity, quality, and allocation are already of great concern (Gleick and others 2000, Wilkinson and others 2003). As concentrations of these gases continue to increase, greater amounts of terrestrial radiation will become trapped, temperatures will rise further, and other impacts will become more significant.

Climatic impacts at the international and national level have been well researched, but far less information is available on regional and local impacts. This chapter, an outgrowth of work done for the Government of the Netherlands and state agencies in California, summarizes some of the consequences of climate change for water resources and water systems in California as well as some of the efforts that policymakers have begun to take. A more comprehensive assessment, supported by multiple state agencies and including the participation of a wide range of stakeholders, would be a valuable tool for policymakers and planners. We urge such an assessment be undertaken in the near future in water basins around the world.

Projecting regional impacts of climatic change and variability relies on a wide range of tools, ranging from large-scale general circulation models of the climate (GCMs) to small-scale analyses of ecological sensitivity to specific climatic parameters. GCMs develop global scenarios of changing climate parameters, usually comparing scenarios with different carbon dioxide (CO_2) concentrations in the atmosphere. This information is typically at too coarse a scale to make accurate regional assessments, and GCMs must have their scale reduced, or downscaled, statistically or using regional climate models. The resulting finer-scale output can then be analyzed for given watersheds, ideally with the incorporation of other hydrologic parameters such as local evaporation, transpiration, snowpack, and groundwater.

Models are typically calibrated by comparing model runs over historical periods with observed climate conditions. It should be emphasized that these model results are not intended to be specific predictions, but rather are scenarios based on the potential climatic variability and change driven by both natural variability and human-induced factors. Nonetheless, they are useful for assessing potential possible future conditions.

Temperature

Modeling results from GCMs are consistent in predicting increases in temperatures globally with increasing concentrations of atmospheric CO_2 from human activity. Temperature as a driver of hydrologic variables is particularly important in California because of the state's vast agricultural activity and its dependence on snowmelt for water supply. Recent work by Snyder et al. (2002) has produced some of the finest-scale temperature and precipitation estimates to date. Resulting temperature increases for a scenario of doubled CO_2 concentration are 1.4 to 3.8 degrees C throughout the region. This is consistent with the global increases predicted by the Intergovernmental Panel on Climate Change (IPCC 2001). In a regional model of the Western United States, Kim et al. (2002) project a climate warming of from 3 to 4 degrees C. Of note in both studies is the projection of uneven distribution of temperature increases. For example, the greatest projected warming effects are in the Sierra Nevada mountains, with implications for snowpack and snowmelt (Kim et al. 2002, Snyder et al. 2002).

Precipitation

In general, while modeling of projected temperature changes is broadly consistent across most modeling efforts, there is far less agreement about precipitation estimates among different GCM studies. Regional modeling efforts conducted for the western United States generally indicate that overall precipitation will increase (Giorgi et al. 1994, Kim et al. 2002, Snyder et al. 2002), but this result is considered uncertain. The most recent regional modeling efforts agree broadly that precipitation increases will be centered in Northern California (Kim et al. 2002, Snyder et al. 2002). Kim et al. also project precipitation increases, mainly in the winter months. Further work is in progress to extend and improve these modeling efforts, and to make use of these data in watershed-scale hydrological models that will be of more direct use to planners. Until better precipitation forecasts are available, uncertainties about runoff timing, the impacts of storms and droughts, and water supply and demand will remain.

Evaporation and Transpiration

Evaporation and transpiration affect climate, plant growth, and distributions, and water demand and use. Increasing average temperatures generally lead to an increase in the potential for evaporation, though actual evaporation rates are constrained by the water availability on land and vegetation surfaces and in the soils. Atmospheric moisture content can limit evaporation rates, so changes in humidity are relatively important. Vegetative cover is also important because plants intercept precipitation and transpire water back to the atmosphere. Different vegetation types play different roles in evaporation; so evaluating the overall hydrologic impacts of climate change in a region requires some understanding of current vegetation patterns and of the ways in which vegetation patterns may change.

Transpiration, the movement of water through plants to the atmosphere, is affected by variables including plant cover, root depth, stomatal behavior, and the concentration of carbon dioxide in the atmosphere. Investigations of the impacts of increased carbon dioxide concentrations on transpiration have yielded conflicting results—some assessments suggest reductions in overall water use while others indicate that some plants acclimatize to increased CO_2 levels, limiting improvements in water-use efficiency (Field et al. 1995, Korner 1996, Rötter and Van de Geijn 1999). Multiple factors related to climate change can have more complex effects when taken together, including suppressing gains in plant growth (Shaw et al. 2002). While reproducible generalizations for evapotranspiration (ET) are not yet available, climate models have consistently projected that global average evaporation would increase in the range of 3 to 15 percent for an equivalent doubling of atmospheric carbon dioxide concentration. The greater the warming, the larger the anticipated increases (IPCC 2001).

Snowpack

By delaying runoff from winter months when precipitation is greatest, snow accumulation in the Sierra Nevada acts as a massive natural reservoir for California. As early as the mid-1980s and early 1990s, climate models, regional hydrologic models, and theoretical studies of snow dynamics projected that higher temperatures would lead to a decrease in the extent of snow cover in mountainous regions such as the Sierra Nevada of California (e.g., Gleick 1986, Gleick 1987a,b, Lettenmaier and Gan 1990, Lettenmaier and Sheer 1991, Nash and Gleick 1991a,b, Dettinger and Cayan 1995, Cayan 1996, Hamlet and Lettenmaier 1999).

These changes, in turn, would affect the timing and magnitude of runoff in California. Higher temperatures will have several major effects: they will increase the ratio of rain to snow, delay the onset of the snow season, accelerate the rate of spring snowmelt, and shorten the overall snowfall season, leading to more rapid and earlier seasonal runoff. Indeed, over the past two decades, this has been one of the most persistent and well-established findings on the impacts of climate change for water resources in the United States and elsewhere.

A few broad assessments have simulated the effects of climate change on snowpack in the United States (McCabe and Legates 1995, Cayan 1996, McCabe and Wolock 1999). McCabe and Wolock (1999) evaluated the links between climate conditions and snowpack for over 300 different snow sites in the western United States, including the Sierra Nevada and the Colorado basin. They used long-term historical records to develop a snow model that used altered climatic information from GCMs. For most of the sites, strong positive correlations were found between precipitation and snowpack; strong negative correlations were found between temperature and snowpack. These correlations indicate that the supply of winter moisture is the best predictor of snowpack volume, while temperature is the best predictor of the timing of snowmelt and the overall nature of the snow season. This correlation breaks down only for those high-altitude sites where mean winter temperatures are so cold that the ratio of rain to snow is not affected.

The climate projections from a wide range of GCMs lead to large decreases in April 1 snowpack for California (Gleick and others 2000, Snyder et al. 2002, Kim et al. 2002, Knowles and Cayan 2002). In some of the extreme cases, model snowpack is completely eliminated by the end of the next century, although some snowfall and snowmelt would certainly continue in high-altitude sites. Recent work with a detailed regional scale shows snow accumulation in February will be reduced by up to 82 percent in a $2xCO_2$ scenario,

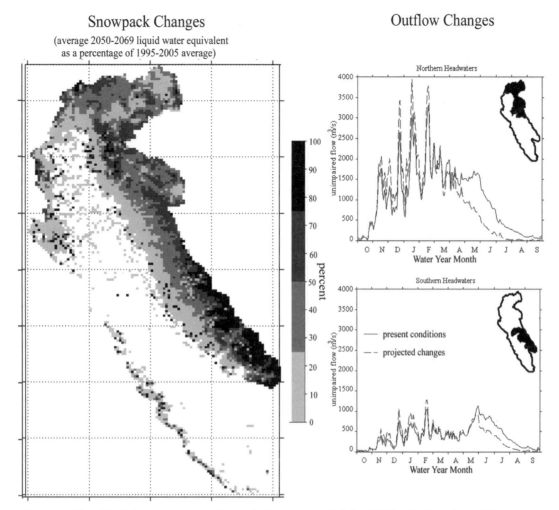

Snowpack Changes

(average 2050-2069 liquid water equivalent as a percentage of 1995-2005 average)

Outflow Changes

FIGURE 7.1 Global warming: projected changes in California's hydrology by 2060.
Sources: Kim et al. 2002, Knowles and Cayan 2002.

with an almost complete melting by the end of April (Snyder et al. 2002). Figure 7.1 shows other modeling efforts projecting that decreased snowfall and enhanced winter snow-melt could deplete most of the snow cover in California by the end of the winter (Kim et al. 2002, Knowles and Cayan 2002).

Recent field surveys provide evidence that snowpack is already diminishing in accord with these projections. Snow cover over the Northern Hemisphere land surface has been consistently below the 21-year average (1974 to 1994) since 1988 (Robinson et al. 1993, Groisman et al. 1994), with an annual mean decrease in snow cover of about 10 percent over North America. In the mountainous western United States, snow cover has signifi-cantly retreated during the latter half of the twentieth century, leading to related shifts in seasonal discharges (Dettinger and Cayan 1995).

Variability, Storms, and Extreme Events

Much of the analysis of climate and water impacts looks at how changes in various averages (such as mean temperatures, precipitation patterns, sea level, and so on) will affect water and water systems. But some of the most important impacts will result not from changes in averages but from changes in extremes. Variability is a natural part of any climatic system, caused by processes that will continue to exert an important influence on the climate system even as changes induced by rising concentrations of greenhouse gases are felt. Efforts to understand how natural patterns of variability, such as hurricanes, intense rainstorms, and El Niño/La Niña events affect California's water resources help to identify vulnerabilities of existing systems to hydrologic extremes (McCabe 1996, Vogel et al. 1997, Piechota et al. 1997, Cayan et al. 1999).

Large climatic variability has been a feature of California's past. Paleoclimatic evidence from tree rings, buried stumps, and lakebed sediment cores suggests that the past 200 years have been relatively wet, and relatively constant when compared with longer records (Michaelsen et al. 1987, Hughes and Brown 1992, Earle 1993, Haston and Michaelsen 1997, Meko et al. 2001, Benson et al. 2002). These longer records reveal greater variability than the historical record, in particular in the form of severe and prolonged droughts (Stine 1994). In spite of this evidence, water resource planning and operation are generally based on the historical climate record since 1900, which may not be representative of past or future conditions.

Models produce various pictures of increased storminess, but increased storm intensity is consistently forecast, whether or not its frequency also increases (Carnell and Senior 1990, Hayden 1999, Lambert 1995, Frei et al. 1988). Some modeling studies suggest that the variability of the hydrologic cycle increases when mean precipitation increases, possibly accompanied by more intense local storms and changes in runoff patterns (Noda and Tokioka 1989, Kothavala 1997, Hennessy et al. 1997). In addition, another long-standing model result points to an increase in drought often resulting from a combination of increased temperature and evaporation along with decreased precipitation (Haywood et al. 1997, Wetherald and Manabe 1999, Meehl et al. 2000, Lambert 1995, Carnell and Senior 1998, Felzer and Heard 1999).

The frequency of El Niño events may increase due to greenhouse warming. Timmermann et al. (1999) used a high-resolution global climate model to simulate the El Niño/Southern Oscillation phenomenon (ENSO) under conditions of warming. Their model indicated that the tropical Pacific climate system would undergo systematic changes if greenhouse gas concentrations doubled. In particular, their results suggest a world where the average condition is like the present-day El Niño condition and events typical of El Niño will become more frequent. Their results also found more intense La Niña events and stronger interannual variability, meaning that year-to-year variations may become more extreme under enhanced greenhouse conditions. More frequent or intense El Niños would alter precipitation and flooding patterns in the United States in a significant way.

In a study that analyzed 20 GCMs currently in use worldwide, extreme events intensified over the next century. The study concludes that the West Coast will probably be less affected because of its heavier rainfall and more moist soil (Meehl and Easterling 2001). In a study that reviewed several GCM scenarios, an increased risk of large storms and flood events was shown for California (Miller and Dettinger 1999). Conflicting conclusions about storms support the need for higher-spatial-resolution models with better cloud and precipitation processes.

Water managers and planners are especially interested in extreme events and how they may differ with climate change. Unfortunately, this is one of the least-well understood categories of impacts. Hydrological fluctuations impose two types of costs on society: the costs of infrastructure to provide more even and reliable flows, and the economic and social costs of disruptions such as floods and droughts that occur in spite of these investments. Major floods on California's rivers are produced by slow-moving Pacific storm systems, which sweep moist subtropical air from a southwesterly direction into the state. In modeling by the California Department of Water Resources on the American River basin, increased storm temperatures of three degrees Celsius increased storm runoff by about 10 percent (personal communication, Roos 2003). The 1986 flood on which these experiments were based had the highest 3-day average flow on record for the American River, claimed 15 lives, and caused more than a billion dollars in property damage (http: //www.news.water.ca.gov/1997.spring/quest.html). Since existing flood-control facilities in the Central Valley and elsewhere can barely accommodate such a large flood event, even a modest increase caused by climate warming could pose problems.

Ironically, some regions could be subjected to both increases in droughts and increases in floods. Even without increases in variability, both problems may occur in the same region. In California, where winter precipitation falls largely as snow, higher temperatures will increase the ratio of rain to snow, shifting peak runoff toward the period of time when flood risk is already highest as noted earlier. At the same time, summer and dry-season runoff will decrease because of a decline in snowpack and accelerated spring melting. This produces more frequent extremes of both kinds, even without fundamental changes in storm patterns.

Large-Area Runoff

Runoff is directly affected by changes in precipitation and temperature. However, runoff in actual watersheds is rarely explicitly evaluated in GCMs because their resolution is insufficient to include other critical watershed characteristics. Estimates of changes in runoff over large areas are thus often relatively simple evaluations of changes in large-scale precipitation and evapotranspiration patterns (Arnold et al. 1998, Arnold et al. 1999, Srinivasan et al. 1993). Despite remaining uncertainties in precipitation patterns, especially, Brown et al. (1999) concluded that altered precipitation and increases in evapotranspiration should be considered in any analysis of future regional or national water supply and demand as should changes in seasonality of the hydrologic cycle.

Several different conclusions about runoff can be drawn from a review of the literature. First, the great differences in results show the difficulty of making accurate "predictions" of future runoff—these results should be viewed as sensitivity studies and used with considerable caution. Second, runoff is extremely sensitive to climate conditions. Large increases in precipitation will probably lead to increases in runoff: such increases can either worsen or lessen water-management problems. Third, far more work is needed, on a finer scale, to understand how climate will affect local water resources. Until GCMs get better at evaluating regional temperature and precipitation, their regional estimates of future runoff must be considered speculative and uncertain.

Regional Runoff

River runoff reflects multiple climatic factors, which makes it an important indicator of climatic variability and change. Runoff also integrates numerous human influences such as flow diversions for irrigation and municipal use, natural streamflow regulation by dams and reservoirs, and baseflow reduction by groundwater pumping. Detecting a climate signal in the midst of these complicating factors would be strong evidence of human-induced change, but such detection can be difficult (Changnon and Demissie 1996).

Sierra Nevada Rivers

By using anticipated, hypothetical, or historical changes in temperature and precipitation and models that include realistic small-scale hydrology, modelers have consistently projected significant changes in the timing and magnitude of runoff in California resulting from quite plausible changes in climatic variables. In California, runoff is extremely sensitive to rainfall: a small percentage change in rainfall can produce a much larger percentage change in runoff. Considerable effort has been made to evaluate climate impacts in particular river basins, including the Sacramento, the San Joaquin, the Carson/Truckee, and others. Even in the absence of changes in precipitation patterns, higher temperatures resulting from increased greenhouse gas concentrations lead to higher evaporation rates, reductions in streamflow, and increased frequency of droughts (Schaake 1990, Rind et al. 1990, Nash and Gleick 1991a,b 1993). In such cases, increases in precipitation would be required to maintain runoff at historical levels.

For California, one of the most important results for planners has also been one of the most consistent research conclusions. Warming leads to changes in the timing of streamflow, including both the intensity and timing of peak flows. A declining proportion of total precipitation falls as snow as temperatures rise, more winter runoff occurs, and remaining snow melts sooner and faster in spring (Gleick 1986, 1987a,b, Lettenmaier and Gan 1990, Nash and Gleick 1991b, Miller et al. 1992). In some basins, spring peak runoff may increase; in others, runoff volumes may significantly shift to winter months. With warming, snow levels in the mountains will rise on average, and the average amount of snow-covered area and snowpack will decrease. A reasonable estimate is about 500 feet of elevation change for every degree Celsius rise (M. Roos, personal communications, 2003).

For a doubling of atmospheric carbon dioxide (equivalent), only about one-fourth of the snow zone of the Sacramento River basin would remain with an estimated decrease of 5 MAF of April-through-July runoff (Cayan 1996, Knowles and Cayan 2002, Miller and Dettinger 1999). The impact would be less in the higher elevation of southern Sierra. For example, in the San Joaquin/Tulare Lake region, about seven-tenths of the snow zone would remain. Nevertheless, these are dramatic change in the hydrologic system.

Under current operating rules, less spring snowmelt could make it more difficult to refill winter reservoir flood control space during late spring and early summer of many years, thus potentially reducing the amount of surface water available during the dry season. Lower early-summer reservoir levels also would adversely affect lake recreation and hydroelectric power production, with possible late-season temperature problems for downstream fisheries. Not all river systems would be equally affected; much depends on the existing storage capacity and on how the system is controlled and operated.

Lins and Slack (1999) looked at historical trends in monthly mean flow across broad regions of the United States, finding statistically significant increases in California. Lettenmaier et al. (1994) evaluated trends using monthly mean discharge and also found significant increases in western streamflow from 1948 through 1988. During 1948 through 1991, snowmelt-generated runoff came increasingly early in the water year in many basins in northern and central California. A declining fraction of the annual runoff was occurring during April to June in middle-elevation basins (as previously described) and an increasing fraction was occurring earlier in the water year, particularly in March (Dettinger and Cayan 1995). Gleick and Chalecki (1999) observed this same basic pattern in an analysis of the Sacramento and San Joaquin Rivers over the entire twentieth century.

Colorado River

The Colorado River supplies water to nearly 30 million people and irrigates more than one and a half million hectares of farmland in Wyoming, Colorado, Utah, New Mexico, Arizona, Nevada, California, and the Republic of Mexico. Spanning 2,300 kilometers and eventually running through Mexico to the Sea of Cortez, the river is the only major water supply for much of the arid southwestern United States and the Mexicali Valley of Mexico, and it plays a special role in California's water situation (see Figure 7.2).

Colorado River basin water supply, hydroelectricity generation, reservoir levels, and salinity are all sensitive to both the kinds of climate changes that are expected to occur and to the policy options chosen to respond to them. Because of concerns about these

FIGURE 7.2   Map of the Colorado River Basin.
Source: United States Bureau of Reclamation.

issues, some of the very first river basin climate studies examined the impacts of climatic changes on the Colorado River basin and several of its major tributaries.

The earliest studies used historical regression approaches to evaluate the impacts of hypothetical temperature and precipitation changes (Stockton and Boggess 1979, Revelle and Waggoner 1983). Both of these studies suggested that modest changes in average climatic conditions could lead to significant changes in runoff. Revelle and Waggoner concluded that a 2 degree Celsius (C) increase in temperature with a 10 percent drop in precipitation would reduce runoff by 40 percent. Stockton and Boggess' results were similar, with a projected 35 to 56 percent drop in runoff.

By the late 1980s, researchers began to use physically based models capable of evaluating climatic conditions outside of the range of existing experience and hydrologic statistics. Schaake (1990) used a simple water-balance model to evaluate the elasticity of runoff in the Animas River in the upper Colorado River basin. That study suggested that a 10 percent change in precipitation would lead to a 20 percent change in runoff, while a 2 degree C increase in temperature would reduce runoff by only about 2 percent. More significant, however, was the finding that changes in temperature would have significant seasonal effects on snowmelt, a finding in agreement with the earlier conclusions of Gleick (1987) for the Sacramento River.

Nash and Gleick (1991a,b, 1993) analyzed the impacts of climate change on the Colorado basin using conceptual hydrologic models coupled with the United States Bureau of Reclamation Colorado River Simulation System (CRSS) model of the entire water-supply system of the river. They evaluated hypothetical temperature and precipitation scenarios as well as the equilibrium GCM scenarios available at the time. A GCM transient run was done as well with one of the first models to use transient greenhouse gas inputs. River flows were found to be very sensitive to both precipitation and temperature, though less sensitive than the earlier regression studies. As with earlier studies, major changes in the seasonality of runoff resulted from the impacts of higher temperature on snowfall and snowmelt dynamics.

One of the most important qualitative findings was that the effects of climate changes on water supplies were dependent on the operating characteristics of the reservoir system and the institutional and legal rules constraining the operators. The variables most sensitive to changes in runoff were found to be salinity, hydroelectric generation, and reservoir level. This study also evaluated the possible utility of increased storage capacity to address the impacts of climate changes and concluded that additional storage would do nothing to alleviate potential reductions in flow. Only if climatic changes were to increase streamflow variability without decreasing long-term supply might additional reservoirs in the Upper Colorado River Basin have any benefits.

Another comprehensive assessment of the Colorado Basin's systems of reservoirs was done for the Colorado River Severe Sustained Drought study (CRSSD) (Lord et al. 1995). That analysis focused on a scenario of long-term drought, rather than a single climate change scenario, and concluded that the "Law of the River" as currently implemented would leave ecosystems, hydropower generation, recreational users, and Upper Basin water users vulnerable to damages despite the extensive infrastructure.[2] A related study also found that water reallocation through marketing had the power to reduce drought damages (Booker 1995).

---

2. "The Law of the River" refers to the combined set of agreements, legal water rights, international treaties, environmental constraints, and other "laws" that determine how the entire Colorado River system is to be operated. Many of these conflicts with each other—contributing to ongoing debate and dispute.

Eddy (1996) looked at extreme events in the Colorado Basin and evaluated the impact of an increase or decrease in precipitation of 10 percent on the duration of wet and dry periods. Eddy concluded that changing average precipitation would not change the number of consecutive wet or dry years by more than one year, but that about once every 20 years, some groupings of stations would experience a dramatic change in consecutive extreme years. If several portions of the Upper Colorado Basin experienced these major wet or dry periods simultaneously, "an episode of crisis proportions could occur."

Soil Moisture

Soil moisture—a measure of the water in different depths of soil—helps define vegetation type and extent, influences agricultural productivity, and affects groundwater recharge rates. The amount of water stored in the soils is influenced by vegetation type, soil type, evaporation rates, and precipitation intensity. Any changes in precipitation patterns and evapotranspiration regime directly affect soil-moisture storage. Decreased precipitation or increased temperature can each lead to decreases in soil moisture. Soil-moisture response has important implications for crop yield and irrigation demand (Brumbelow and Georgakakos 2000).Where precipitation increases significantly, soil moisture is likely to increase, perhaps by large amounts.

GCM results suggest large-scale regional soil drying in summer from higher temperatures. Drying in California could have significant impacts on agricultural production and on the supply of and demand for water. One consequence is an expected increased incidence of regional droughts as measured by soil-moisture conditions, even where precipitation increases, because of the increased evaporation (Vinnikov et al. 1996).

Early modeling of the Sacramento Basin identified reductions in summer soil moisture of 30 percent or more resulting from a shift in the timing of runoff from spring to winter, a decrease in snow, and higher summer temperatures and evaporative losses (Gleick 1986, 1987a,b). Similar results are seen for the Colorado River basin, where large increases in precipitation were found to be necessary in order to simply maintain soil moisture at present historical levels as temperatures and evaporative losses rise (Nash and Gleick 1991b, 1993).

Water Quality

Water quality depends on a wide range of variables, including water temperatures, flows, runoff rates and timing, and the ability of watersheds to assimilate wastes and pollutants. Climate change could alter all of these variables. Higher winter flows of water could reduce pollutant concentrations or increase erosion of land surfaces and stream channels, leading to higher sediment, chemical, and nutrient loads in rivers. Changes in storm flows will affect urban runoff, with attendant water-quality impacts. Lower summer flows could reduce dissolved oxygen concentrations, reduce the dilution of pollutants, and increase zones with high temperatures. Less directly, changes in land use resulting from climatic changes, together with technical and regulatory actions to protect water quality, can be critical to future water conditions. The net effect on water quality for rivers, lakes, and groundwater in the future, therefore, depends not just on how climatic conditions might change but also on a wide range of other human actions and management decisions (Earhart et al. 1999).

In a review of potential impacts of climate change on water quality, Murdoch et al. (2000) conclude that significant changes in water quality are known to occur as a direct result of short-term changes in climate. They note that water quality in ecological transition zones and areas of natural climate extremes is vulnerable to climate changes that increase temperatures or change the variability of precipitation. They also note the importance of non-climatic factors such as changes in land and resource use and recommend long-term monitoring of water quality in order to identify severe impacts. Additionally, they recommend taking action now, such as developing appropriate management strategies for protecting water quality in the light of future changes.

Moore et al. (1997) note that increased water temperatures enhance the toxicity of metals in aquatic ecosystems and that increased periods of biological activity could lead to increased accumulation of toxics in organisms. Ironically, increased bioaccumulation could decrease the concentration of toxics in the water column, improving local water quality. Similarly, higher temperatures may lead to increased transfer of chemicals from the water column to sediments. However, increases in air temperature, and the associated increases in water temperature, are likely to lead to adverse changes in water quality, even in the absence of changes in precipitation.

Decreased water flows can exacerbate temperature increases and increase the concentration of pollutants, flushing times, and salinity (Schindler 1997, Mulholland et al. 1997). Decreased surface-water volumes can increase sedimentation, concentrate pollutants, and reduce non-point source runoff (Mulholland et al. 1997). Increases in water flows can dilute point-source pollutants, increase loadings from non-point source pollutants, decrease chemical reactions in streams and lakes, reduce the flushing time for contaminants, and increase export of pollutants to coastal wetlands and deltas (Jacoby 1990, Mulholland et al. 1997, Schindler 1997). Higher flows can increase turbidity in lakes, reducing UVB penetration. These studies point to potentially important impacts, but do not yet provide a clear picture of the overall sensitivity of water quality to climate change. More work on these issues specific to California is needed.

Lake Levels and Conditions

California has an amazing diversity of lakes—high-altitude, deep lakes, closed saline lakes, desert oases, lowland seasonal lakes, and more. Work is needed to identify threatened lakes in California and projected impacts of climate change on inflows and outflows and water quality. Although only limited California-specific work has been done, lakes are known to be sensitive to a wide array of changes in climatic conditions. Variations in temperature, precipitation, humidity, and wind conditions can alter evaporation rates, the water balance of a basin, ice formation and melting, and chemical and biological regimes (McCormick 1990, Croley 1990, Bates et al. 1993, Hauer et al. 1997, Covich et al. 1997, Grimm et al. 1997, Melak et al. 1997). Closed (endorheic) lakes are extremely sensitive to the balance of inflows and evaporative losses. Even small changes in climate can produce large changes in lake levels and salinity (Laird et al. 1996).

Other effects of increased temperature on lakes could include higher thermal stress for cold-water fish, higher trophic states leading to increased productivity and lower dissolved oxygen, degraded water quality, and increased summer anoxia. Decreases in lake levels coupled with decreased flows from runoff and groundwater may exacerbate temperature increases and loss of thermal refugia and dissolved oxygen. Increased net evaporation may increase salinity of lakes. Hostetler and Small (1999) also note that climate

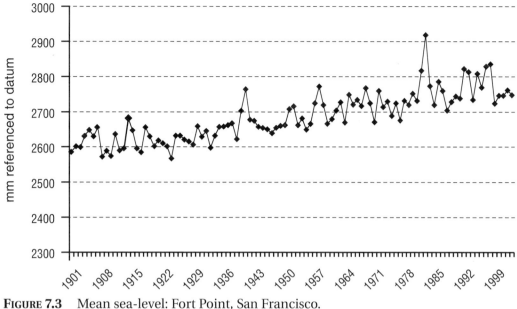

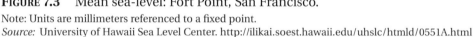

FIGURE 7.3   Mean sea-level: Fort Point, San Francisco.

Note: Units are millimeters referenced to a fixed point.
Source: University of Hawaii Sea Level Center. http://ilikai.soest.hawaii.edu/uhslc/htmld/0551A.html

variability may amplify or offset changes in the mean state under climate changes and may ultimately be more important than changes in average conditions. Some non-linear or threshold events may also occur, such as a fall in lake level that cuts off outflows or separates a lake into two isolated parts.

Sea Level

Sea-level rise, caused by thermal expansion of ocean waters and melting of ice from land surfaces, will affect groundwater aquifers and coastal ecosystems. Mean sea-level (msl) data for stations along the coast of California show msl rising. Figure 7.3 shows the increase as measured at Fort Point in San Francisco over the past 100 years. Early studies of the impacts of sea-level rise in California show that estuarine impacts of sea-level rise will be felt in the San Francisco Bay and the Sacramento-San Joaquin River delta in northern California (Williams 1985, 1987, SFBCDC 1988). Among the risks will be threats to levee integrity and tidal marshes, the salinity of water in the Delta region, and intrusion of salt water into coastal aquifers.

Delta levees protect transportation systems, agriculture, and homes in the region. Williams projected that levees would fail at a higher rate, sediment movements would be changed, mudflats and salt marshes would experience more erosion, and ecosystem impacts could be substantial (Williams 1985, 1987). In addition, tidal marshes in parts of the San Francisco Bay would be submerged by a one-meter sea-level rise (SFBCDC 1988). One analysis showed that only a 15-centimeter (6 inch) rise would transform the current 100-year high tide peak in San Francisco Bay into about a 10-year event (Gleick and Mauer 1990). Severe high tides could become a more frequent threat to the delta levees and their ability to protect land and water systems there.

Williams (1985, 1987) also concluded that the average salinity level could migrate roughly 15 kilometers upstream in the Sacramento-San Joaquin river system, impacting the state's water-supply infrastructure. Greater reservoir releases, reduced pumping, or some other operational or infrastructure solution would have to be implemented to push the increased salinity intrusion caused by the sea-level rise back towards the delta and away from the state water-project facilities.

Groundwater

Groundwater withdrawals in California in the mid-1990s were estimated to be nearly 18 billion cubic meters annually, nearly 20 percent of all the groundwater withdrawn in the entire United States (Solley et al. 1998). In typical years, groundwater accounts for around 30 percent of all urban and agricultural water use in the state. In some areas, current levels of groundwater use are already unsustainable, with pumping rates exceeding natural recharge. Groundwater overdrafts in California in the drier years of the 1990s averaged 1.8 billion cubic meters per year (CDWR 1998, CDWR 2003).

Little work has been done on the impacts of climate changes for specific groundwater basins, or for general recharge characteristics or water quality. Changes in recharge will result from changes in effective rainfall as well as a change in the timing of the recharge season, previously described. Increased winter rainfall, expected for some mid-continental, mid-latitude regions could lead to increased groundwater recharge. Higher temperatures could increase the period of infiltration where soils freeze. Higher evaporation or shorter rainfall seasons, on the other hand, could mean that soil deficits persist for longer periods of time, shortening recharge seasons (Leonard et al. 1999). A significant portion of winter recharge comes from deep percolation of precipitation below the rooting zone, whether of native vegetation or farmland. Warmer winter temperatures between storms would be expected to increase evapotranspiration, thereby drying out the soil between storms. A greater amount of rain in subsequent storms would then be required to wet the root zone and provide water for deep percolation.

Pumping from some coastal aquifers in California already exceeds the rates of natural recharge, resulting in saltwater intrusion into the aquifers. Sea-level rise—a highly probable impact from climate change, as noted above—could increase saltwater intrusion into coastal aquifers. Oberdorfer (1996) used a simple water-balance model to test how changes in recharge rates and sea level would affect groundwater stocks and flows in a California coastal watershed. While some sensitivities were identified, the author notes that the complexity of the interactions among the variables required more sophisticated analysis.

Warmer, wetter winters would increase the amount of runoff available for groundwater recharge. However, this additional runoff in the winter would be occurring at a time when some basins, particularly in Northern California, are either being recharged at their maximum rate or are already full. Conversely, reductions in spring runoff and higher evapotranspiration because of higher temperatures could reduce the amount of water available for recharge.

The most recent California groundwater report from the Department of Water Resources (CDWR 2003) notes that these possible changes may require more sophisticated conjunctive management programs in which the aquifers are more effectively used as storage facilities. They also recommend that water managers consider evaluating their systems to better understand existing snowpack-surface water-groundwater relationships, and identify opportunities that may exist to optimize groundwater storage or reduce risks under new hydrologic regimes that may result from climate change (CDWR 2003).

Ecosystems

The direct effects of climate change on ecosystems like those found in the western United States and California will be complex. Previous assessments have established a wide range of possible direct effects, including changes in lake and stream temperatures, lake levels, mixing regimes, water residence times, water clarity, thermocline depth and productivity, invasions of exotic species, fire frequency, permafrost melting, altered nutrient exchanges, food web structure, and more (for reviews see Gleick and others 2000, Wilkinson and others 2003).

Increased concentrations of greenhouse gases have been observed to both increase and decrease plant growth, depending on species and the availability of other key growth conditions (Field et al. 1995). Availability of water at a critical time of the plant life will determine actual plant growth. Predicted drier summers might adversely affect drought sensitive plants. Further research has to be done in translating possible increased plant growth to increases in yield.

Humans are dependent upon ecosystem processes to supply essential goods and services, such as primary productivity and inputs from watersheds, fish for commercial and recreational purposes, decomposition and biological uptake, and water purification. The health and dynamics of ecosystems are fundamentally dependent on a wide range of climate-sensitive factors, including the timing of water availability, overall water quantity, quality, and temperature. All of these factors may be altered in a changed climate. Freshwater systems are rich in biological diversity, and a large part of the fauna is threatened in California—150 species of animals are listed as endangered or threatened under state and federal law, and more than 200 species of plants are facing similar threats (http://www.dfg.ca.gov/hcpb/species/t_e_spp/tespp.shtml).

A changing climate may intensify these threats in many ways, such as accelerating the spread of exotic species and further fragmenting populations (Firth and Fisher 1991, Naiman 1992). Experience with ecosystem dynamics strongly suggests that perturbing ecosystems in any direction away from the conditions under which they developed and thrive will have adverse impacts on the health of that system (Peters and Lovejoy 1992, IPCC 2001).

The ecological response to a modification in natural flow regime resulting from climate change depends on how the regime is altered relative to historical conditions (Meyer et al. 1999). It is likely that the ecosystems at greatest risk from climate change are those that are already near important thresholds, such as where competition for water is occurring, where water temperatures are already near limits for a species of concern, or where climate change will act with other anthropogenic stressors such as large water withdrawals or wastewater returns (Meyer et al. 1999, Murdoch et al. 2000).

There will be both positive and negative direct effects of increasing temperatures on aquatic and terrestrial ecosystems. In general, while many uncertainties remain, ecologists have high confidence that climatic warming will produce a northward shift in species distributions, with extinctions and extirpations of temperate or cold-water species at lower latitudes, and range expansion of warm-water and cool-water species into higher latitudes (Murdoch et al. 2000).

If California water temperatures rise significantly, the difficulty of managing the state's already threatened salmon and steelhead fisheries would increase. Higher atmospheric temperatures will make it more difficult to maintain rivers cold enough for cold-water fish, including anadromous fish. With reduced snowmelt, existing cold-water pools behind major foothill dams are likely to shrink. As a result, river water temperature could warm beyond a point that is tolerable for the salmon and steelhead that currently rely upon these rivers during the summer. Under this scenario, it is doubtful that the existing, cold-water temperature standards in the upper Sacramento River would be able to be maintained.

Nutrient loading generally increases with runoff, particularly in human-dominated landscapes (Alexander et al. 1996). Delivery of constituents like phosphorus, pesticides, or acids in pulses can have adverse consequences for fishes. Water-quality conditions that exceed ecological thresholds will limit the effectiveness of policies designed for average conditions (Murdoch et al. 2000).

Peak flows occurring much earlier in the season (Leong and Wigmosta 1999, Hay et al. 2000) could result in "washout" of early life-history stages of autumn-spawning salmonids. Changes in sediment loading and channel morphology in an altered climate can impact processes regulating nutrient cycling and community composition (Ward et al. 1992).

Although California, like many other regions, has lost most of its original wetlands, some vital remnants remain. Burkett and Kusler (2000) reviewed likely climate change impacts on wetlands. They concluded that expected changes in temperature and precipitation would alter hydrologic dynamics, biogeochemistry, plant species composition, and sediment flows. Because of land fragmentation resulting from past human activities, wetland plants often cannot migrate in response to temperature and water-level changes and, hence, are vulnerable to complete elimination. Wetland plant response to increased CO_2 could also lead to shifts in community structure with impacts to other species that are habituated to specific plant distributions. Small changes in the balance between precipitation and evapotranspiration can alter groundwater level by a few centimeters, which can significantly reduce the size of wetlands and shift wetland types. Burkett and Kusler (2000) note that practical options for protecting wetlands as a whole from expected climate changes are few and poorly understood. Some management measures could be applied to specific places to increase ecosystem resilience or to partially compensate for negative impacts, but there is often no explicit economic or institutional support for doing so. Among the options for mitigation are restricting development near existing coastal and estuarine wetlands, linking fragmented ecosystems to provide plant and animal migration routes, using infrastructures to enhance ecosystem function, and explicitly protecting and allocating of water needed for ecosystem health. Some research has been done on these issues, but far more is needed, including modeling, experimental work, and practical experiments (Power et al. 1995, Carpenter et al 2000).

Climate Change and Impacts on Managed Water-Resource Systems

There is a rapidly growing body of literature about how climate changes may affect U.S. water resources systems (see http://www.pacinst.org/resources and the Water Briefs section of this book for a searchable on-line bibliography). Research has been conducted on a wide range of water-system characteristics, including reservoir operations, water quality, hydroelectric generation, and others. But significant gaps in our information remain.

We do know that precipitation, temperature, and carbon dioxide levels affect both the supply of, and demand for, renewable water resources. Agricultural, urban, industrial, and environmental needs vary over the year. For example, irrigation is particularly sensitive to climatic conditions during the growing season. Also, while indoor domestic water use is not very sensitive to temperature and precipitation, outdoor uses for gardens and parks are very climate dependent. Higher water temperatures would reduce the efficiency of cooling systems and increase the demand for cooling water. Thus, climate will affect overall water use directly and indirectly.

The California Central Valley Project and the State Water Project (the federal- and state-operated water-supply and distribution systems in California) are each operated under strict guidelines, with constraints that determine the total amount of water that can be exported to urban and agricultural users. Flood-control storage space must be reserved in reservoirs during certain seasons. Water rights in the upper Sacramento and San Joaquin rivers restrict water availability downstream. Minimum flow requirements and dissolved oxygen concentrations in rivers and the Delta have been set for some fisheries. And wildlife and habitat restoration and salinity standards in the Delta are all considered in pumping operations. Even under existing supply and demand patterns, water requirements are barely met under dry and critical water years. As climate changes continue to materialize, water managers must begin, at a minimum, evaluating these constraints and ways of optimizing operations of the system for different conditions than those for which they were designed.

Water Supply Infrastructure

A major challenge facing hydrologists and water managers is to evaluate how changes in system reliability resulting from climate changes may differ from those anticipated from natural variability and, in theory, those already anticipated in original project designs. Both surface and groundwater systems are known to be sensitive to the kinds of changes in inflows and demands described earlier. Many regional studies have shown that large changes in the reliability of water yields from reservoirs could result from small changes in inflows (e.g., Lettenmaier and Sheer 1991, McMahon et al. 1989, Cole et al. 1991, Mimikou et al. 1991, Nash and Gleick 1991b, 1993). Lettenmaier and Sheer (1991), for example, noted the sensitivity of the California State Water Project to climate change under current operating rules. They concluded that changes in operating rules might improve the ability of the system to meet delivery requirements, but only at the expense of an increased risk of flooding. This kind of trade off is now being seen in a broader set of analyses.

Lettenmaier et al. (1999) identified changes in runoff as among the most important factors determining the climate sensitivity of system performance, even when they evaluated the direct effects of climate change on water demands. These sensitivities depended on the purposes for which water was needed and the priority given to those uses. Higher temperatures increased system use in many basins, but these increases tended to be modest, as were the direct effects of higher temperatures on system reliability.

Hydropower and Thermal Power Generation

California produces hydropower at a rate second only to the Pacific Northwest in the United States. The amount of hydropower production for a given facility is a function of amount of water available, the head over which the water falls, and the time of operation. Changes in precipitation amount or pattern will have a direct impact on hydropower generation (Lettenmaier et al. 1999). If snowpack decreases, hydropower generation during these months would be reduced. However, wetter winters might enable additional hydropower generation during winter and spring if adequate flood control can be provided, though the economic value of hydropower is higher in the summer period when peak electricity demand occurs.

Variability in climate already causes variations in hydroelectric generation. During a multi-year drought in California in the late 1980s and early 1990s, decreased hydropower generation led to increases in fossil-fuel combustion and higher costs to consumers. Between 1987 and 1991, these changes cost ratepayers more than US$3 billion and increased greenhouse gas emissions (Gleick and Nash 1991). Because of conflicts between flood-control functions and hydropower objectives, human-induced climate changes in California may require more water to be released from California reservoirs in spring to avoid flooding. This would result in a reduction in hydropower generation and the economic value of that generation. At the same time, production of power by fossil fuels would have to increase to meet the same energy demands in California at a cost of hundreds of millions of dollars and an increase in emissions of greenhouse gases (Hanemann and McCann 1993).

Agriculture

There are likely to be changes in water use, as well as in water supply, driven by climate change, and these will be imposed on top of any changes in water demand that result from changes in population, industrial structure, and technology. In general, agricultural water demands increase with temperature. Higher carbon dioxide levels, however, can reduce water consumption (at least in laboratory tests, for some plants), and increase yields in some crops (Korner 2000, Shaw et al. 2002). The net result for agriculture will depend on many factors that will vary from region to region and crop to crop, but relatively small changes in water availability could lead to relatively large impacts in the agricultural sector. In view of the vital importance of agriculture for California, and the pressures on irrigation imposed by growing urban uses, change in water demand due to climate change are likely to have major social and economic significance.

In California, the vast majority of agricultural production requires irrigation water from both surface and groundwater sources. Increases in water availability due to climate changes could help reduce the pressures faced by growers. Conversely, decreases in water availability are likely to affect growers more than other users for two reasons: urban and industrial users can pay more for water; and proportional reductions in water availability would lead to larger overall reductions to farmers. If irrigators holding senior water rights are allowed to sell or transfer those rights, some could actually benefit from decreases in water availability (Gleick and others 2000).

Agricultural losses are already apparent in periods of extremes, and farmers can, and do, take actions to cope. During California's 6-year drought from 1987 to 1992, agricultural losses were reduced by temporarily fallowing some land, pumping more groundwater, concentrating water supplies on the most productive soils and higher value crops, and purchasing water in spot markets to prevent the loss of tree crops. Direct economic losses to California's irrigated agriculture in 1991 were estimated at only US$250 million, less than 2 percent of the state's total agricultural revenues (Nash 1993, United States Army Corps of Engineers 1994).

Brumbelow and Georgakakos (2000) assessed changes in irrigation demands and crop yields using physiologically-based crop models, and reached several important conclusions for regional agricultural changes, though their results are dependent upon a single climate scenario and hence should be considered speculative. Durum wheat irrigation needs decreased significantly in California (82 percent decrease), though this is a relatively minor crop for the state. Corn irrigation demands also significantly decreased west of the 104th meridian (40 percent to 75 percent decrease) and were otherwise only slightly changed. In all regions, the length of the overall growing season increased. Little information is available on the economic implications of these kinds of changes because of the difficulty of integrating regional changes into global market models.

Moving From Climate Science to Water Policy

For over a decade, scientists working in the area of climate and water have been producing formal, peer-reviewed recommendations for integrating their work into policy. Box 7.1 summarizes these reports. Many of their suggestions for coping and adaptation are synthesized here, separated into specific categories of challenges. A general theme in the recommendations is the separate identification of impacts with both high and low degrees of likelihood, and suggestions for "no-regrets" strategies focused on high-risk concerns. No-regrets strategies are defined by the IPCC as policies that would have net social benefits independent of the scope and severity of anthropogenic climate change (McCarthy et al. 2001).

Some of the recommendations have been acted on, at least in part, and California policymakers are currently discussing additional responses.

Managing Existing Infrastructure

During the twentieth century dams, reservoirs, and other water infrastructures were designed with a focus on extreme events such as the critical drought period or the probable maximum flood. This approach provided a cushion to deal with uncertainties such as climate variability (Matalas and Fiering 1977).

Managing current infrastructure under conditions of climate change could, in principle, prove no different than managing for historical climate variability. But there are several reasons why we must pay special attention to our existing infrastructure and its ability to address climate change:

- Climate change could produce hydrologic conditions and extremes of a different nature than current systems were designed to manage.

- Climate change may produce similar kinds of variability, but outside of the range for which current infrastructure was designed.

- We must not assume that sufficient time and information will be available before the onset of large or irreversible climate impacts to permit managers to respond appropriately.

- The risk of surprises or uncertainties may require special planning and new tools (Gleick and others 2000).

There are two critical issues associated with using existing facilities to address future climate change: can they handle the kinds of changes that will occur and at what economic and ecological cost? There have been few detailed analyses of either of these questions, in part because of the remaining uncertainties about how the climate may actually change. The work of Lettenmaier et al. (1999), Georgakakos and Yao (2000a,b), and Carpenter and Georgakakos (2001) reinforces the conclusion that effective operation of complex systems can reduce impacts of climate change, but only if implemented in a timely and dynamic manner. Lettenmaier et al. (1999) addressed this question of

---

Box 7.1 Reports Recommending the Integration of Climate Science and Water Policy

- Waggoner 1990 —The American Association for the Advancement of Science published this book after a two-year assesment of climate and United States water resources. It was the most indepth, interdisciplinary, and scientifically sophisticated report on the subject until the Water Sector Report of the United States National Assessment (Gleick and others 2000).

- California Energy Commission (CEC) 1991—The first report on climate change produced by a California State agency was mandated by California Assembly Bill 4420 in 1988. The CEC report is specific to California. Its recommendations were based on the assumption that snowmelt timing will be the primary hydrologic variable altered by climate change, and precipitation was held constant in its scenarios. California water policymakers cited it repeatedly as an influential early document.

- American Water Works Association 1997—The Public Advisory Forum of the American Water Works Association issued a succinct set of recommendations to water managers. As the largest United States professional organization of water utilities and providers, this set of recommendations carries more weight with water managers than some of the more academic assessments.

- Gleick and others 2000—The report of the Water Sector of the National Assessment on the Potential Consequences of Climate Variability and Change for the United States provides a regional and national overview of the impacts of climate change on water resources.

- Wilkinson and others 2002—The report of the California Regional Assessment Group of the National Assessment provides an overview of impacts for the state's ecosystems, economy, society, human health, and other areas. It includes a major chapter on water resources. In its section on recommendations for adaptation, it quotes in full the Water Sector (Gleick and others 2000) and the AWWA reports (American Water Works Association 1997). In addition, it offers other recommendations specific to California.

response to climate change for a series of water systems around the United States. They noted that reservoir systems buffer modest hydrologic changes through operational adaptations. As a result, the effects of climate change on the systems they studied tend to be smaller than the underlying changes in hydrologic variables. They concluded that significant changes in design or scale of water-management systems might not be warranted to accommodate climate changes alone, although this obviously depends on the ultimate size of the changes. They urged a concerted effort to adjust current operating rules or demand patterns to better balance the conflicting roles that reservoirs must play. This requires planning and participation by water managers.

Other water-management options include conducting a systematic review and evaluation of all major multi-purpose reservoirs for water supply and flood control and for their ability to adapt to plausible scenarios of climate change with current operating rules. Then, alternative options for water management should include evaluation of measures to improve water supply and quality, reduce demands throughout the state, maintain and restore ecosystems, re-operate reservoirs, and adapt to sea level rise in the delta. The work will emphasize increased flexibility in both physical systems and institutional mechanisms in order to permit a greater range of response. Supply and quality measures will be particularly important in regions dependent on imported supplies.

Due to the many uncertainties in predicting peak flows under climate change scenarios, a closer look at the design practices of hydraulic infrastructure should be considered. Related to flood risk are the rainfall depth-duration-frequency data widely used for designing local stormwater control and drainage facilities. It has been suggested that these statistics be updated frequently, at least every 20 years or so. In this way, climate changes will be gradually incorporated into the record and in the rainfall statistics.

Another reason why management approaches may be of special importance is the high costs and environmental concerns that now make it difficult to get new supply projects approved. These costs also make it likely that the projects that are undertaken will have less redundancy built into their water supply and control facilities than existing projects (Frederick 1991). For these reasons, a large number of formal policy recommendations about how water planners and managers should deal with climate change focus on how infrastructure is managed.

Recommendations

- Governments and agencies should reevaluate legal, technical, and economic procedures for managing water resources in the light of the climate changes that are highly likely (Waggoner 1990, American Water Works Association 1997, Gleick and others 2000, Wilkinson and others 2002).

- Scientists and engineers should reexamine engineering designs, operating rules, contingency plans, and water allocation policies under a wider range of climate scenarios (American Water Works Association 1997, Gleick and others 2000).

- California should assess delta levees' strength with respect to increasing sea level rise (California Energy Commission 1991).

- Existing dams should have temperature controls added for fish species that require cold water downstream (California Energy Commission 1991).

Water Planning and Assessment

Decisions about long-term water planning depend on climatic conditions and what humans do to respond and adapt to those conditions. In the past, these decisions relied on the assumption that future climatic conditions would have the same characteristics and variability as past conditions. Dams are sized and built using available information on existing flows in rivers and the size and frequency of expected floods and droughts. Reservoirs are operated for multiple purposes using the past hydrologic record to guide decisions. Irrigation systems are designed using historical information on temperature, water availability, and soil water requirements.

This reliance on the past record now may lead us to make incorrect—and potentially dangerous or expensive—decisions. Given that risk, one of the most important coping strategies must be to try to understand what the consequences of climate change will be for water resources and to begin planning for those changes. This means more than planning new infrastructure: it must now include an emphasis on incorporating demand management and efficiency into planning. It may also mean accepting a different level or nature of risk—something water planners have trouble doing.

Recommendations

- Planning should occur over appropriate regions, which may or may not correspond to current boundaries (Waggoner 1990). This would elevate the importance of hydrologic boundaries over political boundaries.

- Information on the relative costs and benefits of non-structural managements options, like demand management or decreased floodplain development, should be produced (Gleick and others 2000).

- Climate change scientists should focus on the timeframes and spatial scales relevant to water managers, who are concerned with watershed-level predication and decadal time scales (Waggoner 1990).

- Timely flows of information between scientific community, public, and water management should be facilitated (American Water Works Association 1997; Gleick and others 2000).[3]

- Planners should reassess water transfer plans for the Sacramento-San Joaquin Delta, particularly in light of predicted sea-level rise (California Energy Commission 1991).

New "Supply" Options

Traditional water-supply options, such as dams, reservoirs, and aqueducts may still have an important role to play in meeting water needs in parts of the United States. Because new infrastructure often has a long lifetime, it is vital that the issue of climate change be factored into decisions about design and operation. At present, the California Department of Water Resources, in collaboration with United States Bureau of Reclamation (USBR) and local agencies, is looking into enlarging in-stream storages in major reser-

---

3. Several recent conferences illustrate that this is currently happening. For example, at a recent CALFED meeting detailing modeling projects, several local stakeholder groups were represented along with larger environmental groups and many branches of government.

voirs, off-stream storage options, and flooding four delta "islands"—actually low-lying lands protected by levees. These projects may increase supply reliability, improve water quality, and improve wildlife habitats. Because new infrastructure often has a long life-time, it is vital that the issue of climate change be factored into decisions about designs and operations.

In addition to more traditional new "supply," there are a number of new options. Among these are wastewater reclamation and reuse, water marketing and transfers, and even limited desalination where less costly alternatives are not available and where water prices are high. These alternatives have, in the past, been either expensive relative to tra-ditional water costs or their potential contributions to supplies too limited to make a sig-nificant impact in the long term. This is no longer the case in many regions. Each of these methods can now be considered as a possible supplement to traditional supply planning. In California, for example, more than 500 mcm per year of wastewater are now used for a wide range of purposes, ranging from restoring environmental flows to industrial cooling to indirect recharge of groundwater basins. Wastewater is a far more reliable source of water than many traditional sources of supply that can be affected adversely by drought, making it a particularly useful resource in the context of possible climate changes.

Recommendations

- New supply should come from both traditional and alternative places, such as wastewater reclamation and reuse, water marketing and transfers, and possibly desalination (Gleick and others 2000).

- Moving away from exclusive reliance on surface water by integrating surface and groundwater management can reduce vulnerability to climate fluctuations (O'Conner et al. 1999).

- Agencies should explore the vulnerability of both structural and nonstruc-tural water systems (American Water Works Association 1997).

- Private enterprises should decrease vulnerabilities to the hydrologic effects of climate change through water transfers or construction of new infrastruc-ture (Waggoner 1990).

- Water managers should carefully consider increased storage in new surface or underground storage facilities (California Energy Commission 1991; Gleick and others 2000).

- Site-dependent application of climate change science to stormwater man-agement strategies should be used, including approaches like increasing per-meable surfaces in urban areas (Wilkinson and others 2002).

Demand Management, Conservation, and Efficiency

As the economic and environmental costs of new water-supply options have risen, so has interest in exploring ways of improving the efficiency of both allocation and use of water resources. Demand management, especially in face of population increase, is crit-ical to mitigate loss of water supply or pressures to tap new supplies to meet growing needs. More water-efficient methods in agricultural, industrial, and urban water have been effective in the past in this capacity (Owens-Viani et al. 1999, Gleick et al. 2003) and should be further developed and implemented.

Improvements in the efficiency of end uses and sophisticated management of water demands are increasingly being considered as major tools for meeting future water needs, particularly in water-scarce regions where extensive infrastructure already exists (Vickers 1991, Postel 1997, Gleick 1998a, Dziegielewski 1999, Vickers 1999). Evidence is accumulating that such improvements can be made more quickly and economically, with fewer environmental and ecological impacts, than further investments in new supplies (Gleick et al. 1995, Owens-Viani et al. 1999, Gleick et al. 2003).

A number of water-system studies have begun to look at the effectiveness of reducing system demands to decrease the overall stresses on water supplies, both with and without climate changes. Wood et al. (1997) and Lettenmaier et al. (1999) noted that long-term demand growth estimates had a greater impact on system performance than climate changes when long-term withdrawals are projected to grow substantially. Actions to reduce demands or to moderate the rate of increase in demand growth can, therefore, play a major role in reducing the impacts of climate changes. Far more work is needed to evaluate the relative costs and benefits of demand management and water-use efficiency options in the context of a changing climate, but relevant recommendations for policymakers have already been crafted.

Recommendations

- Opportunities for water conservation, demand management, and efficiency should be explored and encouraged (Waggoner 1990, California Energy Commission 1991, Gleick and others 2000, Wilkinson and others 2002).

- Managers should plan and invest for multiple benefits (e.g., water supply, energy, wastewater, and environmental benefits) resulting from water-use efficiency increases (Wilkinson and others 2002).

Economics, Pricing, and Markets

Prices and markets are increasingly important tools for balancing supply and demand for water. They will also be valuable for coping with climate-induced changes. Because new construction and new concrete projects are increasingly expensive, environmentally damaging, and socially controversial, new tools such as the reduction or elimination of subsidies, sophisticated pricing mechanisms, and smart markets provide incentives to use less water, produce more with existing resources, and reallocate water among different users. As conditions change, markets can help resources move from lower- to higher-value uses (NRC 1992, Western Water Policy Review Advisory Commission 1998). However, when water is transferred, third parties are likely to be affected. A challenge for developing more effective water markets is to develop institutions that can expeditiously and efficiently take third-party impacts into account (Loh and Gomez 1996, Gomez and Steding 1998, Dellapenna 1999). As a result, despite their potential advantages, prices and markets have been slow to develop as tools for adapting to changing supply and demand conditions.

California's emergency Drought Water Banks in the early 1990s helped mitigate the impacts of a prolonged drought by facilitating water transfers among willing buyers and sellers. Dellapenna (1999) and others have noted, however, that the California Water Bank was not a true market, but rather a state-managed effort that reallocated and moved water from small users to large users at a price set by the state. More recent efforts to develop functioning markets on smaller scale have had some success (see California Department of Water Resources, http://rubicon.water.ca.gov/b16098/v2txt/ch6e. html, for a discussion of types of markets, water transfers, and institutional efforts underway or being considered in California).

Such transfers, both temporary and permanent, may be particularly useful for adapting to the consequences of climate variability. They may also be effective in dealing with long-term imbalances that might result from changing demographic and economic factors or social preferences. As the historical allocation of water becomes sufficiently out of balance with more modern needs and demands, a shift from temporary to permanent transfers may be warranted.

Recommendations

- Governments should encourage flexible institutions for water allocation including water markets (Waggoner 1990).

- Prices and markets should be adjusted to balance supply and demand (Gleick and others 2000).

- Economic and market tools should be explored, but not in the context of privatization (Wilkinson and others 2002).

- Economists should investigate economic effects of climate change and of adaptations to climate change (Gleick and others 2000).

New Institutions or Institutional Behaviors

Ultimately, climate changes may be sufficiently severe to lead to conditions different enough from current ones to warrant the development of new institutions, or changes in "behaviors" of existing institutions. As a related example, water management in California has changed significantly in the past few decades in response to increased awareness of environmental degradation. The creation of a new state agency—the California Bay-Delta Authority—to address these issues has spurred fundamental changes in the way Federal and state agencies and districts communicate with each other and work to define common goals. Increasing manifestations of climate change may prompt similar changes in water management among local, regional, and national water institutions.

Because climatic changes may also be slow to manifest themselves, dramatic institutional change is unlikely in the short term. Awareness of the issue of climate change within institutions, however, may have the effect of increasing collaboration between agencies, as short-term hydrologic changes impact water management at all levels.

In the long term, the unusual nature of some kinds of climate change may require more than simply modifying existing institutions. Non-linear or threshold effects may lead to abrupt changes in conditions. At a minimum, planning should be done to explore alternatives available to water managers in the event of these kinds of impacts.

Recommendations

- Flexible decisions should be encouraged, particularly in the design and construction of new projects (Waggoner 1990, Gleick and others 2000, Wilkinson and others 2002).

- Water laws should be updated and improved in the context of climate change, including review of the legal allocation of water rights (American Water Works Association 1997, Gleick and others 2000).

- Changing land use patterns should be examined as coping mechanisms (Gleick and others 2000).

- Local governments should consider creating more full-time water manager positions to attract top professionals capable of considering long-term issues and concerns in planning (O'Conner et al. 1999).

- The state should reevaluate risks to flood zones at intervals of 20 to 30 years (California Energy Commission 1991)—shorter than normal planning intervals.

Reducing Uncertainty, Developing New Knowledge, and Gathering Information

Water-system managers do not dismiss the issue of climate change, but they have been reluctant to consider it in their planning horizons until they perceive a greater degree of scientific certainty about regional impacts (O'Conner et al. 1999). Given the complexity of these issues, the uncertainties described earlier in this chapter, and remaining gaps in information and research, some specific research needs and priorities are identified in various recommendations. Certain actions by state policymakers or agencies can facilitate improvements in understanding climate change, the risks of climate impacts, and the likely effectiveness of alternative policy approaches.

Interestingly, most managers admit that they expect disruptions in daily operations caused by changes in climate variability. Experienced and full-time water managers were more likely to consider future climate scenarios in planning than inexperienced or part-time managers, but all wanted uncertainty to be reduced—a goal shared by scientists. Recommendations for reducing uncertainty are common.

Recommendations

- Increased funding is necessary for interdisciplinary research necessary to address the broad-based impacts and effects of climate change (Waggoner 1990).

- Water organizations should communicate regularly with scientists, with the dual goals of communicating scientific advances to managers, and communicating what knowledge is necessary from scientists for effective management (Waggoner 1990, American Water Works Association 1997, Gleick and others 2000).

- California should improve both weather and flood forecasting (California Energy Commission 1991).

- "Those reporting about climate change bear a special responsibility for accuracy, conveying the real complexities and uncertainties, and not oversimplifying. Scientists must make extra effort to explain clearly in conservative and understandable terms." (Waggoner 1990).

- Improve GCMs to more accurately represent hydrologic impacts, water resource availability, overall hydrologic impacts, and regional impacts (Waggoner 1990, Gleick and others 2000).

- Improve downscaling, or regional forecasting capabilities, of GCMs[4] (Gleick and others 2000).

Hydrologic and Environmental Monitoring

Better data on hydrology and land use are critical to California's successful adaptation to expected climate change. Changes in hydrology are among the most certain of climate change impacts and good hydro-meteorological data are the starting point for evaluating the capabilities of the current water-supply and flood-protection systems. Hydrological data are used in the design and operation of water supply systems and flood control works, the provision of environmental needs, and in design of other infrastructure. Several state agencies have ongoing climate, water, and land use/land cover monitoring programs. But there are important gaps, particularly in areas where greater changes are anticipated.

At a minimum, data must be collected in several important categories, which should include enhancement of measurements of precipitation and related climate data, streamflow, snowpack, and ocean and delta water levels. This sampling network should be sensitive enough to reveal changes expected from climate change. It is important to continue to collect, maintain, and evaluate records from existing California stations, incorporating data from recent years. Efforts should be made to prevent cuts in monitoring and data collection due to budget constraints. More systematic sea-level measurements are needed in the San Francisco Bay and delta region, and elsewhere along California's coast. On the whole, such efforts are as important to inform managers, as they are critical for progress in climate science.

Recommendations

- The state should improve hydrologic monitoring overall, but special focus should include variables known or thought to be sensitive to climate change (California Energy Commission 1991; Gleick and others 2000).

- Water-quality monitoring should be increased (California Energy Commission 1991) in the context of the risks of climate change.

Conclusions

As the science of global climate change continues to solidify, climate concerns are playing an increasing role in local and regional debates over effective water management. It is increasingly apparent, and accepted, that water planners and managers can no longer ignore climate in long-term planning, infrastructure design, and operations. At the same time, however, some serious challenges face regional water managers—even those willing to both try to understand the state-of-the-science and to take account of possible future climate changes in planning and management of a complex water system.

---

4. This is one area that continues to see significant advances (e.g., Knowles and Cayan 2002, Snyder et al. 2002). Interestingly, Knowles and Cayan (2002) acknowledge water managers at California's Department of Water Resources for providing motivation for their work.

California water planners and managers have been among the first in the United States to consider these issues, though most efforts in this field have been modest and informal. Initial research and analysis on climate risks facing California water resources began in the early 1980s and, in the last two decades, at least some work has been done by agencies such as the California Energy Commission, Department of Water Resources, floodplain management groups, and even local water districts. Some water agencies around the state have begun to consider the implications of climate change for the reliability and safety of water systems, and professional water organizations have begun urging managers and planners to integrate climate change into long-term planning. The failure to plan for climate change is likely to lead to costly and preventable surprises.

Many uncertainties remain. Responsible planning, however, requires that water experts and managers work with climate scientists and others to reduce those uncertainties and to begin to prepare for those impacts that are well understood, already appearing, or likely to appear. Experience in California may prove useful for other vulnerable regions around the world.

REFERENCES

Aguado, E., Cayan, D., Riddle, L., and Roos, M. 1992. *Climatic fluctuations and the timing of West Coast streamflow. Journal of Climate*, 5, no. 12:1468–1483.

Alexander, R. B., Murdoch, P. S., and Smith, R. A. 1996. Streamflow-induced variations in nitrate flux in tributaries to the Atlantic coastal zone. *Biogeochemistry*, 33, no. 3:149–177.

American Water Works Association. 1997. Climate change and water resources: Committee Report of the Public Advisory Forum. *Journal of the American Water Works Association*, 89, no. 11:107–110.

Auble, G. T., Friedman, J. M., and Scott, M. L. 1994. Relating riparian vegetation to present and future streamflows. *Ecological Applications*. 4:544–554.

Bates, G. T., Giorgi, F., and Hostetler, S. W. 1993. Toward the simulation of the effects of the Great Lakes on regional climate. *Monthly Weather Review*, 121, no. 5:1373–1387.

Benson, L., Kashgarian, M., Rye, R., Lund, S., Paillet, F., Smoot, J., Kester, C., Mensing, S., Meko, D., and Lindstrom, S. 2002. Holocene multidecadal and multicentennial droughts affecting Northern California and Nevada. *Quaternary Science Reviews*, 21, no. 4–6:659–682.

Boland, J. J. 1997. Assessing urban water use and the role of water conservation measures under climate uncertainty. *Climatic Change*, 37, no. 1:157–176.

Boland, J. J. 1998. Water supply and climate uncertainty. *Water Resources Update*, 112:55–63.

Booker, J. F. 1995. Hydrologic and economic impacts of drought under alternative policy responses. *Water Resources Bulletin*, 31, no. 5:889–906.

Brown, R. A., Rosenberg, N. J., and Izarraulde, R. C. 1999. Response of United States regional water resources to CO_2-fertilization and Hadley Centre climate model projections of greenhouse-forced climate change: A continental scale simulation using the HUMUS model, Pacific Northwest National Laboratory, 29.

Brumbelow, K., and Georgakakos, A. 2001. An assessment of irrigation needs and crop yield for the United States under potential climate changes. *Journal of Geophysical Research—Atmospheres*, 106, no. D21:27383–27405.

California Energy Commission. 1991. Planning for and adapting to climate change. Global climate change: Potential impacts and policy recommendations committee report. II:6.1–6.18. Sacramento, California.

Carpenter, T. M., and Georgakakos, K. P. 2001. Assessment of Folsom lake response to historical and potential future climate scenarios: 1. Forecasting. *Journal of Hydrology*, 249, no. 1–4:148–175.

Cayan, D. R. 1996. Interannual climate variability and snowpack in the Western United States. *Journal of Climate*, 9, no. 5:928–948.

Cayan, D. R., Redmond, K. T., and Riddle, L. G. 1999. ENSO and hydrologic extremes in the Western United States. *Journal of Climate*, 12, no. 9:2881–2893.

Changnon, S. A., and Demissie, M. 1996. Detection of changes in streamflow and floods resulting from climate fluctuations and land use-drainage changes. *Climatic Change*, 32, no. 4:411–421.

Cole, J. A., Slade, S., Jones, P. D., and Gregory, J. M. 1991. Reliable yield of reservoirs and possible effects of climate change. *Hydrological Sciences Journal*, 36, no. 6:579–598.

Covich, A. P., Fritz, S. C., Lamb, P. J., Marzolf, R. D., Matthews, W. J., Poiani, K. A., Prepas, E. E., Richman, M. B., and Winter, T. C. 1997. Potential effects of climate change on aquatic ecosystems of the Great Plains of North America. *Hydrological Processes*, 11, no. 8:993–1021.

Croley, T. E. 1990. Laurentian Great Lakes double-CO_2 climate change—Hydrological impacts. *Climatic Change*, 17, no. 1:27–47.

Dennis, A. S. 1991. Initial climate change scenario for the western United States. Global Climate Change Response Program. United States Department of the Interior, Bureau of Reclamation, Denver, Colorado.

Dettinger, M. D., and Cayan, D. R. 1995. Large-scale atmospheric forcing of recent trends toward early snowmelt runoff in California. *Journal of Climate*, 8, no. 3:606–623.

Earle, C. J. 1993. Asynchronous droughts in California streamflow as reconstructed from tree rings. *Quaternary Research*, 39:290–299.

Eddy, R. L. 1996. Variability of wet and dry periods in the Upper Colorado River Basin and the possible effects of climate change. Global Climate Change Response Program, United States Department of the Interior, Bureau of Reclamation, Denver, Colorado.

Eheart, J. W., Wildermuth, A. J., and Herricks, E. E. 1999. The effects of climate change and irrigation on criterion low streamflows used for determining Total Maximum Daily Loads. *Journal of the American Water Resources Association*, 35, no. 6:1365–1372.

Felzer, B., and Heard, P. 1999. Precipitation difference amongst GCMs used for the United States National Assessment. *Journal of the American Water Resources Association*, 35, no. 6:1327–1340.

Field, C. B., Jackson, R. B., and Mooney, H. A. 1995. Stomatal responses to increased CO_2-implications from the plant to the global scale. *Plant Cell and Environment*, 18:1214–1225.

Firth, P., and Fisher, S. G. (editors). 1992. *Global Climate Change and Freshwater Ecosystems*. New York: Springer-Verlag.

Frederick, K. D. 1993. Climate change impacts on water resources and possible responses in the MINK Region. *Climatic Change*, 24, no. 1–2, 83–115.

Frei, C., Schär, C., Lüthi, D., and Davies, H. C. 1998. Heavy precipitation processes in a warmer climate. *Geophysical Research Letters*, 25:1431–1434.

Gleick, P. H. 1986. Methods for evaluating the regional hydrologic impacts of global climatic changes. *Journal of Hydrology*, 88:97–116.

Gleick, P. H. 1987a. The development and testing of a water-balance model for climate impact assessment: Modeling the Sacramento Basin. *Water Resources Research*, 23, no. 6:1049–1061.

Gleick, P. H. 1987b. Regional hydrologic consequences of increases in atmospheric carbon dioxide and other trace gases. *Climatic Change*, 10, no. 2:137–161.

Gleick, P. H., 1998. Water planning and management under climate change. *Water Resources Update*, no. 112:25–32.

Gleick, P. H., and Chalecki, E. L. 1999. The impacts of climatic changes for water resources of the Colorado and Sacramento-San Joaquin River Basins. *Journal of the American Water Resources Association*, 35, no. 6:1429–1441.

Gleick, P. H., and Maurer, E. P. 1990. *Assessing the costs of adapting to sea-level rise: A case study of San Francisco Bay*. Pacific Institute for Studies in Development, Environment, and Security, Berkeley, California and the Stockholm Environment Institute, Stockholm.

Gleick, P. H., and Nash, L. 1991. *The societal and environmental costs of the continuing California drought*. Berkeley, California:Pacific Institute for Studies in Development, Environment, and Security. 66 pages.

Gleick, P. H., and others. 2000. Water: Potential consequences of climate variability and change for the water resources of the United States. Report of the water sector assessment team of the national assessment of the potential consequences of climate variability and change. Oakland, California:Pacific Institute for Studies on Development, Economics, and Security. 151 pages.

Grimm, N. B., Chacon, A., Dahm, C. N., Hostetler, S. W., Lind, O. T., Starkweather, P. L., and Wurtsbaugh, W. W. 1997. Sensitivity of aquatic ecosystems to climatic and anthropogenic changes: The Basin and Range, American Southwest and Mexico. *Hydrological Processes*, 11, no. 8:1023–1041.

Groisman, P. Y., Karl, T. R., and Knight, R. W. 1994. Observed impact of snow cover on the heat balance and the rise of continental spring temperatures. *Science*, 263:198–200.

Hamlet, A. F., and Lettenmaier, D. P. 1999. Effects of climate change on hydrology and water resources in the Columbia River basin. *Journal of the American Water Resources Association*, 35:1597–1623.

Haston, L., and Michaelsen, J. 1997. Spatial and temporal variability of southern California precipitation over the last 400 years and relationships to atmospheric circulation patterns. *Journal of Climate*, 10, no. 8:1836–1852.

Hauer, F. R., Baron, J. S., Campbell, D. H., Fausch, K. D., Hostetler, S. W., Leavesley, G. H., McKnight, D. M., Leavitt, P. R., and Stanford, J. A. 1997. Assessment of climate change and freshwater ecosystems of the Rocky Mountains, United States of America and Canada. *Hydrologic Processes,* 11:903-924.

Hay, L. E., Wilby, R. J. L., and Leavesley, G. H. 2000. A comparison of delta change and downscaled GCM scenarios for three mountainous basins in the United States. *Journal of the American Water Resources Association,* 36, no. 2:387–397.

Hayden, B. P. 1999. Climate change and extratropical storminess in the United States: An assessment. *Journal of the American Water Resources Association,* 35, no. 6:1387–1398.

Hennessy, K. J., Gregory, J. M., and Mitchell, J. B. F. 1997. Changes in daily precipitation under enhanced greenhouse conditions. *Climate Dynamics,* 13:667–680.

Hostetler, S. W., and Small, E. E. 1999. Response of North American freshwater lakes to simulated future climates. *Journal of the American Water Resources Association,* 35, no. 6:1625–1638.

Hughes, M. K., and Brown, P. M. 1992. Drought frequency in central California since 101 B.C. recorded in giant sequoia tree rings. *Climate Dynamics,* 6:161–167.

Intergovernmental Panel on Climate Change (IPCC). 2001. *The scientific basis: IPCC third assessment report.* Cambridge, United Kingdom:Cambridge University Press.

Jacoby, H. D. 1990. *Water quality. Climate change and U.S. water resources,* 307–328. Waggoner, P. E. New York:John Wiley & Sons.

Karl, T. R., and Knight, R. W. 1998. Secular trends of precipitation amount, frequency, and intensity in the United States. *Bulletin of the American Meteorological Society,* 79, no. 2:231–241.

Karl, T. R., and Trenberth, K. E. 2003. Modern global climate change. *Science,* 302:1719–1723 December 5.

Kim, J., Kim, T. K., Arritt, R. W., and Miller, N. L. 2002. Impacts of increased atmospheric CO_2 on the hydroclimate of the western United States. *Journal of Climate,* 15:1926–1943.

Knowles, N., and Cayan, D. R. 2002. Potential effects of global warming on the Sacramento/San Joaquin watershed and the San Francisco estuary. *Geophysical Research Letters,* 29, no. 18:1891.

Korner, C. 1996. The response of complex multispecies systems to elevated CO_2, 20–42. *Global Change and Terrestrial Ecosystems.* Walker, B. R., and Steffen, W. Cambridge, United Kingdom: Cambridge University Press.

Laird, K. R., Fritz, S. C., Grimm, E. C., and Mueller, P. G. 1996. Century-scale paleoclimatic reconstruction from Moon Lake, a closed-basin lake in the northern Great Plains. *Limnology and Oceanography,* 41:890–902.

Lambert, S. J. 1995. The effect of enhanced greenhouse warming on winter cyclone frequencies and strengths. *Journal of Climate,* 8, no. 5:1447–1452.

Lettenmaier, D. P., and Gan, T. Y. 1990. Hydrologic sensitivities of the Sacramento/San Joaquin river basin, California, to global warming. *Water Resources Research,* 26, no. 1:69–86.

Lettenmaier, D. P., and Sheer, D. P. 1991. Climatic sensitivity of California water resources. *Journal of Water Resources Planning and Management,* 117, no. 1:108–125.

Lettenmaier, D. P., Wood, E. F., and Wallis, J. R. 1994. Hydro-climatological trends in the continental United States, 1948–1988. *Journal of Climatology,* 7:586–607.

Lettenmaier, D. P., Wood, A. W., Palmer, R. N., Wood, E. F., and Stakhiv, E. Z. 1999. Water resources implications of global warming: A United States regional perspective. *Climatic Change,* 43, no. 3:537–579.

Lins, H. F., and Slack, J. R. 1999. Streamflow trends in the United States."*Geophysical Research Letters,* 26, no. 2:227–230.

Lord, W. B., Booker, J. F., Getches, D. H., Harding, B. L., Kenney, D. S., and Young, R. A. 1995. Managing the Colorado River in a severe sustained drought: An evaluation of institutional options. *Water Resources Bulletin,* 31, no. 5:939–944.

Major, D. C. 1998. Climate change and water resources: The role of risk management methods. *Water Resources Update,* no. 112:47–50.

Matalas, N. C., and Fiering, M. B. 1977. *Water-resource systems planning. Climate, climatic change, and water supply,* 99–110. National Academy of Sciences, Washington, D. C.:National Academy Press.

McCabe, G. J. 1996. Effects of winter atmospheric circulation on temporal and spatial variability in annual streamflow in the western United States. *Hydrological Sciences Journal,* 41:873–888.

McCabe, G. J., and Wolock, D. M. 1999. Future snowpack conditions in the western United States derived from general circulation model climate simulations. *Journal of the American Water Resources Association,* 35:1473–1484.

McCarthy, J. J., Canziani, O. F., Leary, N. A., Dokken, D. J., and White, K. S. (editors). 2001. *Climate change 2001: Impacts, adaptation & vulnerability. Contribution of working group II to the third assessment report of the Intergovernmental Panel on Climate Change (IPCC)*. Cambridge:Cambridge University Press.

McCormick, M. J. 1990. Potential changes in thermal structure and cycle of Lake Michigan due to global warming. *Transactions of the American Fisheries Society*, 119, no. 2:183–194.

Meehl, G. A., Zwiers, F., Evans, J., Knutson, T., Mearns, L., and Whetton, P. 2000. Trends in extreme weather and climate events: Issues related to modeling extremes in projections of future climate change. *Bulletin of the American Meteorological Society*, 81, no. 3:427–436.

Meko, D. M., Stockton, C. W., and Boggess, W. R. 1980. A tree-ring reconstruction of drought in southern California. *Water Resources Bulletin*, 16:594–600.

Meko, D. M., Therrell, M. D., Baisan, C. H., and Hughes, M. K. 2001. Sacramento River flow reconstructed to AD 869 from tree rings. *Journal of the American Water Resources Association*, 37, no. 4:1029–1039.

Melack, J. M., Dozier, J., Goldman, C. R., Greenland, D., Milner, A. M., and Naiman, R. J. 1997. Effects of climate change on inland waters of the Pacific Coastal Mountains and western Great Basin of North America. *Hydrological Processes*, 11, no. 8:971–992.

Meyer, J. L., Sale, M. J., Mulholland, P. J., and Poff, N. L. 1999. Impacts of climate change on aquatic ecosystem functioning and health. *Journal of the American Water Resources Association*, 35, no. 6:1373–1386.

Michaelsen, J., Haston, L., and Davis, F. W. 1987. 400 years of central California precipitation variability reconstructed from tree-rings. *Water Resources Bulletin*, 23:809–818.

Miller, N., Kim, J. W., and Dettinger, M. D. 1999. California streamflow evaluation based on a dynamically downscaled 8-year hindcast (1988–1995), observations, and physically based hydrologic models: Eos, American Geophysical Union Fall Meeting supplement, 80:F406.

Mimikou, M., Kouvopoulos, Y., Cavadias, G., and Vayianos, N. 1991a. Regional hydrological effects of climate change. *Journal of Hydrology*, 123, no. 1–2:119–146.

Mimikou, M. A., Hadjisavva, P. S., Kouvopoulos, Y. S., and Afrateos, H. 1991b. Regional climate change impacts. II. Impacts on water management works. *Hydrological Sciences Journal*, 36, no. 3:259–270.

Mimikou, M. A., and Kouvopoulos, Y. S. 1991. Regional climate change impacts: Impacts on water resources. *Hydrological Sciences Journal*, 36, no. 3:247–258.

Moore, M. V., Pace, M. L., Mather, J. R., Murdoch, P. S., Howarth, R. W., Folt, C. L., Chen, C. Y., Hemond, H. F., Flebbe, P. A., and Driscoll, C. T. 1997. Potential effects of climate change on freshwater ecosystems of the New England/Mid Atlantic Region. *Hydrological Processes*, 11, no. 8:925–947.

Mulholland, P. J., Best, G. R., Coutant, C. C., Hornberger, G. M., Meyer, J. L., Robinson, P. J., Stenberg, J. R., Turner, R. E., Vera-Herrera, F., and Wetzel, R. G. 1997. Effects of climate change on freshwater ecosystems of the Southeastern United States and the Gulf of Mexico. *Hydrological Processes*, 11, no. 8:949–970.

Murdoch, P. S., Burns, D. A., and Lawrence, G. B. 1998. Relation of climate change to the acidification of surface waters by nitrogen deposition. *Environmental Science & Technology*, 32:1642–1647.

Murdoch, P. S., Baron, J. S., and Miller, T. L. 2000. Potential effects of climate change on surface water quality in North America. *Journal of the American Water Resources Association*, 36:347–366.

Nash, L. L., and Gleick, P. H. 1991. The sensitivity of streamflow in the Colorado Basin to climatic changes. *Journal of Hydrology*, 125:221–241.

Nash, L. L., and Gleick, P. H. 1991. The implications of climate change for water resources in the Colorado River basin. First National Conference on Climate Change and Water Resources Management, Albuquerque, New Mexico, United States Army Corps of Engineers.

Nash, L. L., and Gleick, P. H. 1993. *The Colorado River basin and climatic change: The sensitivity of streamflow and water supply to variations in the temperature and precipitation*, 121. Washington, D.C.:United States Environmental Protection Agency.

Noda, A., and Tokioka, T. 1989. The effect of doubling the CO_2 concentration on convective and non-convective precipitation in a general circulation model coupled with a simple mixed layer ocean model. *Journal of the Meteorological Society of Japan*, 67:1057–1069.

O'Conner, R. E., Yarnal, B., Neff, R., Bord, R., Wiefek, N., Reenock, C., Shudak, R., Jocoy, C. L., Pacale, P., and Knight, C. G. 1999. Weather and climate extremes, climate change, and planning: Views of community water system managers in Pennsylvania's Susquehanna River basin. *Journal of the American Water Resources Association*, 35, no. 6:1411–1420.

Owens-Viani, L., Wong, A. K., and Gleick, P. H., editors. 1999. *Sustainable use of water: California success stories*. Oakland, California:Pacific Institute for Studies in Development, Environment, and Security.

Piechota, T. C., Dracup, J. A., and Fovell, R. G. 1997. Western United States streamflow and atmospheric circulation patterns during El Niño-Southern Oscillation. *Journal of Hydrology*, 201: 249–271.

Revelle, R. R., and Waggoner, P. E. 1983. Effects of a carbon dioxide-induced climatic change on water supplies in the western United States. In *Changing climate*, National Academy of Sciences, Washington, D. C.:National Academy Press.

Rind, D., Goldberg, R., Hansen, J., Rosenzweig, C., and Ruedy, R. 1990. Potential evapotranspiration and the likelihood of future drought. *Journal of Geophysical Research: Atmospheres*, 95, no. D7:9983–10004.

Robinson, P. J., Samel, A. N., and Madden, G. 1993. Comparisons of modelled and observed climate for impacts assessments. *Theoretical and Applied Climatology*, 48:75–87.

Roos, M. 1987. Possible changes in California snowmelt runoff patterns. Fourth Annual Pacific Climate (PACLIM) Workshop, Pacific Grove, California.

Rötter, R., and Geijn v.d., S. C. 1999. Climate change effects on plant growth, crop yield and livestock. *Climatic Change*, 43, no. 4:651–681.

Schaake, J. C. 1990. From climate to flow. In *Climate change and U.S. water resources*. Waggoner, P. E., editor. 177–206. New York:J. Wiley and Sons.

Schindler, D. W. 1997. Widespread effects of climatic warming on freshwater ecosystems in North America. *Hydrological Processes*, 11, no. 8:1043–1067.

Shaw, M. R., Zavaleta, E. S., Chiariello, N. S., Cleland, E. E., Mooney, H. A., and Field, C. B. 2002. Grassland responses to global environmental changes suppressed by elevated CO_2 *Science*, 298, no. 5600:1987–1990.

Snyder, M. A., Bell, J. L., Sloan, L. C., Duffy, P. B., and Govindasamy, B. 2002. Climate responses to a doubling of atmospheric carbon dioxide for a climatically vulnerable region. *Geophysical Research Letters*, 29, no. 11:10.1029/2001GL014431.

Solley, W. G., Pierce, R., and Perlman, H. A. 1998. Estimated use of water in the United States, 1995. United States Geological Survey Circular #1200. Denver.

Stine, S. 1994. Extreme and persistent drought in California and Patagonia during medieval time. *Nature*, 369, no. 6481:546–549.

Stockton, C. W., and Boggess, W. R. 1979. Geohydrological implications of climate change on water resource development. United States Army Coastal Engineering Research Center, Fort Belvoir, Virginia.

Tarlock, A. D. 1991. Western water law, global warming, and growth limitations. *Loyola Law Review*, 24, no. 4:979–1014.

Timmermann, A., Oberhuber, J., Bacher, A., Esch, M., Latif, M., and Roeckner, E. 1999. Increased El Niño frequency in a climate model forced by future greenhouse warming. *Nature*, 398:694–696.

Trelease, F. J. 1977. *Climatic change and water law. Climate, climatic change, and water supply.* National Academy of Sciences, Washington, D.C.:National Academy Press.

Vogel, R. M., Bell, C., and Fennessey, N. M. 1997. Climate, streamflow and water supply in the northeastern United States. *Journal of Hydrology*, 198, no. 1–4:42–68.

Waggoner, P. E., editor. 1990. *Climate change and U.S. water resources*. New York:John Wiley & Sons.

Ward, A. K., Ward, G. M., Harlin, J., and Donahoe, R. 1992. Geological mediation of stream flow and sediment and solute loading to stream ecosystems due to climate change. In *Global climate change and freshwater ecosystems*. Firth, P., and Fisher, S. G., editors 116–142. New York: Springer-Verlag.

Westman, W. E., and Malanson, G. P. 1992. Effects of climate change on Mediterranean-type ecosystems in California and Baja California. In *Global warming and biological diversity*. Peters, R. L., and Lovejoy, T. E., editors. 258–276. New Haven, Connecticut:Yale University Press.

Wetherald, R. T., and Manabe, S. 1999. Detectability of summer dryness caused by greenhouse warming. *Climatic Change*, 43, no. 3:495–511.

Wilkinson, R., and others. 2002. Potential consequences of climate variability and change for California. Draft final report of the California regional assessment group for the United States global change research program. Santa Barbara, California. September. http://www.ncgia.ucsb.edu/pubs/CA_Report.pdf

Wood, A. W., Lettermaier, D. P., and Palmer, R. N. 1997. Assessing climate change implications for water resources planning. *Climatic Change*, 37, no. 1:203–228.

Yao, H., and Georgakakos, A. 2001. Assessment of Folsom Lake response to historical and potential future climate scenarios 2. Reservoir management. *Journal of Hydrology*, 249, nos. 1–4:176–196.

3rd World Water Forum in Kyoto Disappointment and Possibility

Nicholas L. Cain

Introduction

The 3rd World Water Forum, held in March 2003, attracted over 24,000 people to the picturesque Kinki region of Japan.[1] The massive 8-day conference and exhibition—spread over the cities of Kyoto, Osaka, and Shiga—featured hundreds of sessions, exhibitors, and activities all related to water. But despite the diversity of voices and topics, the key question of the conference revolved around whether water should be thought of first as a human right or as an economic good.

The forum's stated focus was on meeting basic needs for the billions of people without access to clean water or adequate sanitation. Issues of water and governance, water and sanitation, water and food, protecting natural systems, financing dams and other larger infrastructure, water privatization, and water and conflict were all emphasized in the official agenda. There was broad agreement on the critical importance of doing more for people without adequate water or sanitation, and on the importance of empowering the world's poor and women to play a greater role in addressing water problems. But, in many ways, the question of whether water should be treated primarily as an economic or social good dominated the conference and caused a split between those favoring more private investment and those who want governments and communities to take the lead in providing basic water and sanitation services.

The sheer size and breadth of the conference, combined with effective efforts by the organizers to reach out to a wide range of groups, ensured that there were many informative sessions, much energetic debate, and some productive action. One of the forum's successes

---

1. The author and three other Pacific Institute staff attended this conference, giving at least six presentations to a wide range of sessions. A paper co-written by Gary H. Wolff, the Pacific Institute's Principal Economist and Engineer, was presented to the CEO roundtable, but Dr. Wolff did not attend.

was its active outreach to people marginalized in past international discussions over water. Another success was the consistent emphasis on developing nations and the need for action to reduce the terrible death toll from water-related disease. Logistically, both the World Water Council and the forum's Japanese hosts deserve praise.

Despite the impressive diversity of the conference and its admirable focus on important issues, many of the groups and individuals working to forge a new, more progressive global water policy were deeply disappointed by the short shrift given to efficiency and conservation, the skewing of the conference's agenda toward privatization and globalization, and the weak ministerial statement, which neglected to affirm a human right to water and failed to produce any significant commitments of money or other resources to solve global water problems.

The lack of specific new actions was also disappointing. Although hundreds of "water actions" were pledged prior to the festival, only 123 of these have actually been confirmed in the final report. Of these, only a small number are concrete, on-the-ground projects. Although there is some emphasis on the sustainable use of resources, projects that encourage efficiency, conservation, or community-scale infrastructure are few and far between.

For those working to map a more sustainable, efficient, and equitable path to solving the global water crisis, there were successes as well. Water experts, indigenous people, and public interest and environmental advocates from around the world were able to build alliances, learn from each other's work, and coordinate their activities. A coalition of non-governmental organizations (NGOs), led by the "Water is Life" network, drafted their own NGO Statement, which was presented to the forum and included in the final report. The NGO Statement contains valuable language on protecting water as a human right, using community-scale approaches to provide water, and ensuring that any large-scale developments meet the highest environmental and social standards.

Given the high cost of the forum (which we estimated to be at least US$50 million) and the dearth of real progress—not to mention the wealth of other water-related conferences that have been held of late—some participants question the value of future Water Forums. Perhaps, along with alternative solutions, we must explore alternative venues.

A forum as large and complex as the 3rd World Water Forum is impossible to summarize in just a few pages; thus we will look at the key themes and most critical documents while noting that there is comprehensive information about the forum online.[2] After a brief overview, this Water Brief will look in detail at the forum's main "products:" the Minister's Statement, Camdessus Report, Summary Forum Statement, and NGO Statement.

Background to the 3rd World Water Forum

Many major water conferences have been held in the past decade, at great expense in time, resources, and labor. While many different groups organize water meetings, from the United Nations to academic and industry organizations, the recent World Water Forums have become the major focus of effort for many. The Kyoto Conference in 2003 was officially called the 3rd World Water Forum by the World Water Council, which has been the lead organizer of these triennial conferences.[3] The 3rd World Water Forum was meant to

---

2. The 3rd World Water Forum's website has complete proceedings: http://www.world.water-forum3.com.

3. The 1st World Water Forum was held in Marrakech, Morocco in March 1997. The 2nd World Water Forum was held in The Hague, The Netherlands in 2000.

focus on the themes identified by the 2nd World Water Forum in the Hague in 2000: "meeting basic water needs, securing food supply, protecting ecosystems, sharing water resources, managing risks, valuing water, and governing water wisely."[4] The forum covered 33 different themes and 5 geographic areas, but questions of finance and privatization framed many of the key debates.

Mindful that large conferences are often thought of as little more than talkfests, participants were encouraged to commit to specific "water actions." Projects like the Virtual Water Forum and Water Voices attempted to put conference attendees in touch more viscerally with the problems of the billions of people without access to clean water or adequate sanitation.

Although events were held in several venues in Kyoto, Shiga, and Osaka, the Forum was centered at the Kyoto International Conference Hall, which has played host to many notable conferences including the meeting that led to the 1997 Kyoto Protocol on Climate Change. Dozens of booths of every kind—selling filter technology, singing the praises of development banks, and celebrating indigenous people—lined one large area off the hall's entryway. In rooms ranging from tiny to cathedral-like, over 350 sessions took place with seminars, panel discussions, plenary sessions, and other meetings.

There was also a set of minister's meetings, which were closed to most attendees. To bridge the gap between the closed and open meetings, the forum held two "dialogue" sessions, in which several hundred ministers and other officials met with several hundred forum goers.

Although the forum should be commended for its focus on the water problems of the world's poorest people and on its outreach to voices traditionally marginalized, the remedies put forth by the World Water Council and in the Minister's Statement were disappointingly retrograde.

Mahmoud Abu-Zeid, president of the World Water Council, called for a doubling of the number of large dams (New Scientist 2003). Three of the most important documents created for or by the forum—the Ministerial Statement, the Summary Forum Statement, and the Camdessus Report—all mapped out a congruent vision for meeting the UN Millennium Development Goals (MDG)[5] based on building more large dams and centralized infrastructure, greatly increasing international investment and aid to US$180 billion a year, opening markets to privatization, and speeding the flow of private capital.

In part because of the high costs of this approach, privatization and private capital are now seen as essential. And because of the growing controversy over privatization (see Gleick et al. 2002), this issue was vigorously debated at the forum.

One trend that was apparent at Kyoto was the growing call for public-private partnerships. Many corporations and development banks see them as a good way to minimize risk and assuage the fears of residents and governments about lack of local control and accountability. Although these types of partnerships can help protect the most vulnerable people, they can also unfairly shift risks to the public or leave taxpayers footing the bill for market fluctuations.

According to research by the Pacific Institute presented at the Kyoto conference (Palaniappan et al. 2003), and other research published beforehand (Gleick et al. 2002), privat-

---

4. Final Report, Secretariat of 3rd World Water Forum, Tokyo, Japan, page 58, no date. http://www. world.water-forum3.com/en/finalreport.html

5. These are human development goals set forth by the United Nations Millennium Declaration (2003), which seek, among many other things, to halve the proportion of people without access to clean water by 2015 (see Chapter 1).

ization can play a valuable role, especially as part of a public-private partnership. But as a tool for helping the world's poorest people, it has limited value unless explicit and clear principles to protect the public good are used. According to the Pacific Institute working paper released at the conference, Principles and Practices for the New Economy of Water, the authors could not find "any single case where all of these principles were applied or met" (see Chapter 3).

Efficiency, conservation, and community-scale projects can, in many cases, be more helpful to poor people without basic water or sanitation service than large, centralized projects. Not surprisingly, by reducing the amount that must be spent on expensive centralized infrastructure, the needs for sophisticated technology, privatization, and corporate equity are also reduced.

The Ministerial Statement

As the forum neared its end, representatives from over 130 different nations produced a Ministerial Statement. A skeptical reader might wonder what the value of another piece of paper might be to those struggling to find clean water; and indeed some forum-goers wondered whether it might be better to write a statement for the ministers rather than expect a useful document from them.

That said, as the top policy document produced by the forum, the Ministerial Declaration has value as a spur and guide to national water development efforts, as a set of priorities for development banks and international development agencies, and as a vision for people at all levels to strive toward. In this case, the vision seems heavily tilted towards privatization and large-scale, centralized infrastructure with little attention to efficiency and community-scale projects, and no mention of water as a human right (see Water Brief 2).

The declaration adopted in Kyoto begins by focusing on the role water plays "as a driving force for sustainable development, including environmental integrity and the eradication of poverty and hunger," calling it "indispensable for human health and welfare…" and notes that "prioritizing water issues is an urgent global requirement." These kinds of statement are common to conference reports and are fine as far as they go, but the declaration omits stronger language that acknowledges water as a human right. As such, the statement misses a key opportunity to ensure that greater importance is given at the national level to ensuring that basic human needs are met. And considering that the United Nations adopted a human right to water in November 2002 (United Nations 2002), this omission makes the conference statement seem timid and regressive (see Water Brief 2). Lane (2003) describes the failure of the Ministerial Statement to acknowledge the UN General Comment as "hardly an edifying example of political leadership on water."

The Ministerial Statement affirms the importance of good governance on the part of nations, the need for stakeholder participation, for "transparency and accountability," and for "technical assistance from the international community"—language very similar to that found in the 2000 Declaration adopted in the Hague for the 2nd World Water Forum (The Hague 2000).[6]

Where the Kyoto Declaration becomes controversial is in its call, in paragraph 5, for nations to move to more aggressive use of privatization to extend water service and fund large-

---

6. Both Declarations are can be found online; the 2nd World Water Forum's Ministerial Declarations is online at http://www.worldwaterforum.net/index2.html and the 3rd World Water Forum's is online at http://www.world.water-forum3.com/en/finalreport.html

scale water infrastructure projects. Paragraph 6 builds on this by calling for the development of new kinds of "public-private partnerships."

This portion of the statement attracted the most criticism from NGOs like Public Citizen and labor organizations like Public Services International (Hall 2003, Public Citizen 2003a,b). They criticized the document's adoption of the Camdessus Report and the statement's strong endorsement of increased privatization as the best solution for bringing water to people in the developing world. The ministers then affirm their desire to support regional and sub-regional water efforts and also "reaffirm the necessity for countries to better coordinate monitoring and assessment systems at local, basin, and national levels." Non-structural approaches to managing future water systems are not mentioned until paragraph 13, well into the first subsection of policy recommendations. Efficiency and conservation, for the purposes of this document, seem to consist mainly of "reducing losses from distribution systems and other [unspecified] water demand management measures." This is unfortunate given research showing that even developing nations can benefit from cost-effective efficiency and conservation efforts (e.g., IUCN 2000). There is only one other nod to efficiency, in paragraph 19, calling for "every effort to reduce unsustainable water management and improve the efficiency of agriculture water use."

The Kyoto Declaration goes on to support the development of "innovative and environmentally sound technologies, such as the desalination of seawater, water recycling, and water harvesting." With the exception of this reference to water harvesting, the document gives scant attention to appropriate technology and community-scale projects, which have been used successfully in the developing world for years. Other points reaffirm the non-controversial issues of preventing water pollution, protecting ecosystems, combating deforestation, and using "structural" and "non-structural" methods to mitigate disaster.

Although there is much salutatory guidance and many worthwhile recommendations in the Ministerial Declaration, the lack of strong language affirming a human right to water, the emphasis on privatization, the uncritical acceptance of the Camdessus Report, and the short shrift given to water efficiency and community-scale infrastructure undercut the potential of the document and make for a weak statement.

Camdessus Report: In Brief

In preparation for the Kyoto Conference, a panel was created to find the "financial means for achieving" the Millennium Development Goals.[7] Headed by Michael Camdessus, former managing director of the International Monetary Fund (IMF), the final report of this committee, Financing Water for All (and often referred to as the Camdessus Report), has become the most significant, and controversial, product of the entire Kyoto Conference. Although it is beyond the scope of this paper to look in depth at the Camdessus Report, we summarize it here given its role in setting guidance for global water policy, and its importance as a planning tool for international development organizations.

The Camdessus Report calls for "at least double" the current amount of money to be spent on water financing. In its survey of water infrastructure, the report draws attention to the vast differences in dam and reservoir storage around the world noting that in "hydraulic infrastructure" the United States exceeds Ethiopia by a factor of 100.

Like the Summary Statement and the Ministerial Statement, the Camdessus Report has laudable language on the importance of providing basic human needs and on the im-

---

7. The title of the committee was "World Panel on Financing Water Infrastructure," which already indicates a bias toward "hard path" solutions.

portance of empowering women, poor people, and those traditionally marginalized. What has aroused the ire of a broad group of water advocates is that the report seems to lay the groundwork for a return to large-scale, centralized projects and it calls for private capital and corporations to be given a much larger role in providing water to people across the globe.

As the Camdessus panel officially launched its report in a session in Kyoto, union officials, water activists, and representatives from NGOs unfurled a banner, reading "Water for People, Not for Profits," and marched through the venue. In a press conference and in statements released afterwards, representatives of the Water for Life coalition ripped the Camdessus Report as a blatantly pro-privatization prescription for "big money" financing of large-scale dams and infrastructure.

Although more dams may, in fact, be needed in certain developing nations, it is not clear that this is the best way to help the world's poorest people. Water experts from many different disciplines have noted that dams and other large-scale projects often cost more and help less than their proponents claim and in the meantime devastate indigenous communities, destroy pristine environments, and forever alter the flow of rivers and lakes. This was essentially the finding of the World Commission On Dams, which recommended a set of strict criteria be applied to all large-scale water infrastructure projects (WCD 2002).

The Camdessus Report notes that private investment in water systems globally, and in the developing world in particular, is quite low and has actually been declining in recent years due to concerns over various kinds of risk. The report calls for a "devaluation liquidity backstopping facility," which protects international corporations from currency fluctuations, and for other financial instruments and procedures to minimize risk.

But water experts from different organizations questioned whether corporations, even with the risk reduction tools recommended by the report, would ever be able to bring water to the world's poorest people. Richard Jolly, chair of the Water Supply and Sanitation Collaborative Council (WSSCC), in a press conference given after the release of the Camdessus Report, noted that private companies fund only 5 percent of current investments globally and that reaching people in the poorest places usually isn't profitable (Jolly 2003). An analysis from Public Services International questioned whether money needed to reduce corporate risk would channel funds away from on-the-ground projects (Hall 2003).

Others at the conference called attention to the paltry amounts industrialized nations spend on global water issues; Agnes van Ardenne, Dutch minister of a development cooperation said in an interview, "Rich countries donate US$50 billion annually [on aid in general; OECD estimates direct aid to water projects at only US$3 billion a year] but they spend US$300 billion to subsidize farm prices (Kongrut 2003)." Chapter 1, Table 1.1, summarizes current overseas development assistance given to water projects and more data are provided in the Data Tables section at the end of this volume.

Summary Forum Statement and Water "Actions"

The Summary Forum Statement (Kyoto 2003) included in the Water Forum's Final Report boils down various recommendations from over 30 different thematic and regional statements into just 9 pages and provides critical top-level guidance for policymakers and development officials, researchers, and advocates. Although the Summary Statement does give a nod to demand management and more efficient use of water, it quickly goes on to note the importance of using "all options to augment the available water supply" (Kyoto 2003, p. 58).

Unlike the Ministerial Statement, the Summary Statement sets out the recommendations from the World Commission on Dams as a reference and urges the use of good practices "in order to avoid the environmental and social costs and risks of the past."

The Summary Statement then moves to touch on key themes: governance, capacity building, finance, and participation. It urges nations to make water a priority and reiterates language regarding the importance of good governance; the need "for capacity building, education, and access to information;" and benefits from approaching water issues in a multi-stakeholder fashion.

The Summary Statement, under the Financing heading, echoes the figures of the Camdessus Report, calling for US$180 billion annually to "produce global water security over the next 25 years." The statement notes that this huge amount of capital will require "greater efficiency and better financial management" and goes on to say that several different mixes of public and private funding have been tried to generate more funding "and the results have been mixed." The "debate concerning public-private partnerships," the authors point out, "has not been resolved." An understatement, to say the least.

Under the heading Creating Global Awareness and Political Support the conference Summary noted "the human right to water was defined in a General Comment by the United Nations Committee on Economic, Social and Cultural Rights (United Nations 2002)." It then goes on to list other actions that have taken place since the 2nd World Water Forum to bolster water governance and build bridges between different agencies, organizations, disciplines, and themes. The authors note the "awareness raising, information sharing" and partnership activities between a wide range of organizations that have taken place or are in progress.

The statement closes with a summary of recommendations. The first involves various suggestions to improve "alliances, partnerships, networking, participation and dialogue" in water governance. The second recommendation involves nature and ecosystems and sets down a strong emphasis on sustainable development, although efficiency and demand management are once again at the end of the list of applicable tools. Under recommendation three, the statement, like the Camdessus Report, suggests that, "governments and local authorities should take adequate measures to reduce risk and improve cost recovery" in order to encourage private investment.

The conference as a whole gave much better attention to efficiency, conservation, and community-scale technology than is reflected in the final statements and the Camdessus Report. Several dozen sessions touched on efficiency and conservation with special attention to the role of demand management in sustainable development, water and climate, and water and sanitation. But here again, considering the overall size of the conference, support was not as strong as it should have been and the forum lacked a dedicated efficiency theme.

One innovative feature of the 3rd World Water Forum was its focus on specific "water actions." Each nation, development organization, and NGO was encouraged to pledge a specific action that they were going to take to improve water access. Although hundreds of actions were pledged by the end of the forum, only 123 of these actions were confirmed in the final report. Forum organizers do, however, expect this to increase over the coming months.

A brief analysis of these action pledges[8] shows that of the 123 confirmed actions only a modest fraction" are on-the-ground projects. The majority is focused on research, data sharing, communications, or organizational actions—all-important, but not the same as concrete water actions.

---

8. A full list is in the Water Forum's Final Report.

NGO Statement

The Kyoto Conference made an early and strong effort to reach NGOs, especially small NGOs and NGOs from the developing world. And from the perspective of those working to craft a more progressive vision of the world's water future, the progress made by non-governmental organizations, labor organizers, water experts, and advocates was a silver lining to the tepid ministerial and forum statements. In scores of sessions, pointed questions and well-informed protesters expanded the debate on key issues, although even public advocates complained that in some cases there was too much protest and not enough debate (Lane 2004). Representatives from various groups also crafted an NGO Statement that is included in the final report (Kyoto 2003).

This NGO Statement begins with an unequivocal assertion that "water is a public good and access to safe, affordable water is a human right." The assembled NGOs opposed "the renewed push for large-scale infrastructure projects that undermine local, participatory, decentralized actions" and instead called for local actions coordinated "into policies and budgets at all levels." Financing, according to the NGO Statement, should be directed "toward low cost technologies and community developed systems." The statement also explicitly rejects the pro-privatization conclusions and call for large-scale infrastructure in the Camdessus Report.

In other regards, however, the NGO statement shares much in common with the Forum Summary and the Ministerial Statement—all call for participatory Integrated Water Resource Management, for building the capacity of service providers, and for partnerships between development agencies, NGOs, and national and local governments. But despite the similarities of rhetoric, there are two distinctly different visions for the future of global water policy presented. And with the lives of tens of millions of people at stake, the question of the best way forward is not trivial.

After a decade or more of efforts at internal reform by organizations like The World Bank, and after the publication of the World Commission on Dams report, it seemed for a time that global water policy was finally beginning to move in a more sustainable and effective direction. Instead of embracing and accelerating this reform, the 3rd World Water Forum rejected past advice and advocated for increased privatization, greater flows of private capital, and a major increase in the number of large-scale projects. Water experts from a broad range of organizations and many people who work and live in the developing world believe that this is the wrong way forward, and so a split is developing in the water community. Until this schism is resolved and an effective, equitable, and practical plan is developed, billions of people will continue to suffer.

REFERENCES

Camdessus, M., (chair) and Winpenny, J. (author). March 2003. *Financing water for all: Report of the world panel on financing water infrastructure.* World Water Council, 3rd World Water Forum, Global Water Partnership, February 2004.
http://www.gwpforum.org/gwp/library/FinPanRep.MainRep.pdf
Gleick, P. H. 1999. The human right to water. *Water Policy* 1, no. 5:487–503.
Gleick, P. H. 2002. *Dirty water: Estimated deaths from water-related diseases by 2020.* Oakland, California: A Report of the Pacific Institute for Studies in Development, Environment, and Security.
Gleick, P. H., Wolff, G., Chalecki, E. L., and Reyes, R. 2002. *The new economy of water: The risks and benefits of globalization and privatization of fresh water.* Oakland, California: A Report of the Pacific Institute for Studies in Development, Environment, and Security. Oakland, California.

Hall, D. 2003. *Financing water for the world—An alternative to guaranteed profits.* Public Services International Research Unit. http://www.psiru.org/reports/2003-03-W-finance.doc

Hague. 2000. The Ministerial Declaration of the 2nd World Water Forum. Hague, Netherlands. http://www.worldwaterforum.net/Ministerial/declaration.html

IUCN. 2001. Water demand management programme for Southern Africa: Phase II. Pretoria, South Africa:The World Conservation Union.

Jolly, R. 2003. Transcript of press conference remarks, Kyoto, Japan, March 21. http://www.wsscc.org/download/Camdessus%20document.doc

Kongrut, A. 2003. Water Forum Report draws controversy in Kyoto. Bangkok Post, March 23.

Kyoto. 2003. Summary statement: Final report. 3rd World Water Forum. Kyoto, Japan. http://www.world.water-forum3.com/. (This document includes the NGO statement.)

Lane, J. Invited editorial: The 3rd World Water Forum, Kyoto, March 2003. *Water Policy*, 5, no. 4:381–382.

New Scientist. 2003. Dismay over call to build more dams. *New Scientist*, March 22.

Palaniappan, M., Gleick, P. H., Srinivasan, V., and Hunt, C. 2003. *Principles and practices for the new economy of water: Protecting the public interest in water privatization.* Oakland, California: A Report of the Pacific Institute for Studies in Development, Environment, and Security, Working Paper.

Public Citizen. 2003a. The myth of private sector financing: Global water corporations seek new public hand-outs. *Public Citizen*, March. http://www.citizen.org/documents/mythofprivatefinancing.pdf

Public Citizen. 2003b. Daily report: News updates from Kyoto for March 21. *Public Citizen*, March. http://www.citizen.org/cmep/Water/cmep_Water/articles.cfm?ID=9166

United Nations. 2003. Millennium Development Goals. United Nations Development Programme. http://www.undp.org/mdg/

World Commission on Dams (WCD). 2000. *Dams and development: A new framework for decision making.* The World Commission on Dams, November. http://www.dams.org

Ministerial Declaration of the 3rd World Water Forum: Message from Lake Biwa & Yodo River Basin March 23, 2003

We, the Ministers and Heads of Delegation, assembled in Kyoto, Japan on 22–23 March 2003, on the occasion of the 3rd World Water Forum. Building upon the outcomes of the Monterrey Conference on Financing for Development, the World Summit on Sustainable Development (WSSD), and the United Nations Secretary General's Water, Energy, Health, Agriculture and Biodiversity (WEHAB) initiative as well as other water-related events, we assert our common resolve to implement the appropriate recommendations in order to achieve the internationally agreed targets and goals including the United Nations Millennium Development Goals (MDGs).

Taking note of the thematic and regional statements and recommendations from the 3rd World Water Forum, we declare the following:

General Policy

1. Water is a driving force for sustainable development including environmental integrity, and the eradication of poverty and hunger, indispensable for human health and welfare. Prioritizing water issues is an urgent global requirement. Each country has the primary responsibility to act. The international community as well as international and regional organizations should support this. Empowerment of local authorities and communities should be promoted by governments with due regard to the poor and gender.

2. Whilst efforts being undertaken so far on water resources development and management should be continued and strengthened, we recognize that good governance, capacity building and financing are of the utmost importance to succeed in our efforts. In this context, we will promote integrated water resources management.

3. In managing water, we should ensure good governance with a stronger focus on household and neighborhood community-based approaches by addressing equity in sharing benefits, with due regard to pro-poor and gender perspectives in water policies. We should further promote the participation of all stakeholders, and ensure transparency and accountability in all actions.

4. We are committed, in the long term, to fortify the capacity of the people and institutions with technical and other assistance from the international community. This must include, among others, their ability to measure and monitor performance, to share innovative approaches, best practices, information, knowledge and experiences relevant to local conditions.

5. Addressing the financial needs is a task for all of us. We must act to create an environment conducive to facilitating investment. We should identify priorities on water issues and reflect them accordingly in our national development plans/sustainable development strategies including Poverty Reduction Strategy Papers (PRSPs). Funds should be raised by adopting cost recovery approaches which suit local, climatic, environmental and social conditions

and the "polluter-pays" principle, with due consideration to the poor. All sources of financing, both public and private, national and international, must be mobilized and used in the most efficient and effective way. We take note of the report of the World Panel on Financing Water Infrastructure.

6. We should explore the full range of financing arrangements, including private sector participation in line with our national policies and priorities. We will identify and develop new mechanisms of public-private partnerships for the different actors involved, while ensuring the necessary public control and legal frameworks to protect the public interests, with a particular emphasis on protecting the interests of the poor.

7. As water situations differ from region to region, we will support established regional and sub-regional efforts such as the vision of the African Ministerial Conference on Water (AMCOW) to facilitate the New Partnership for Africa's Development (NEPAD) and the Central American Integration System (SICA), and the implementation of the program of action in favor of Least Developed Countries (LDCs). Recognizing the uniquely fragile nature of water resources in small island developing states, we support specific programs of collaboration such as the Caribbean Pacific Joint Program for Action on Water and Climate in Small Island Countries.

8. We reaffirm the necessity for countries to better coordinate monitoring and assessment systems at local, basin, and national levels, with development of relevant national indicators where appropriate. We call upon the United Nations, inter alia through the Commission on Sustainable Development, to take a leading role and cooperate with other organizations involved in the water sector to work in a transparent and cooperative way. We welcome the willingness of the Organization for Economic Cooperation and Development and other organizations to periodically inform the international community of aid activities in water-related areas. Ways to track progress on water issues may be usefully explored on the basis of existing facilities and relying upon information from countries and relevant United Nations agencies, regional development banks and other stakeholders, including civil society organizations.

9. We welcome the proposal to establish a new network of websites to follow up the Portfolio of Water Actions that will publicize actions planned and taken on water-related issues by countries and international organizations in order to share information and promote cooperation.

Water Resources Management and Benefit Sharing

10. As we aim to develop integrated water resources management and water efficiency plans by 2005, we will assist developing countries, particularly the least developed countries, and countries with economies in transition, by providing tools and further required assistance. In this context, among others, we encourage regional development banks to take a facilitating role. To this end, we invite all stakeholders, including private donors and civil society organizations, concerned to participate in this process.

11. Recognizing that cooperation between riparian states on transboundary and/or boundary watercourses contributes to sustainable water management and mutual benefits, we encourage all those states to promote such cooperation.

12. We will further encourage scientific research on predicting and monitoring the global water cycle, including the effect of climate change, and develop information systems that will enable the sharing of such valuable data worldwide.

13. We will promote measures for reducing losses from distribution systems and other water demand management measures as a cost-effective way of meeting demand.

14. We will endeavor to develop and deploy non-conventional water resources by promoting innovative and environmentally sound technologies, such as the desalination of seawater, water recycling, and water harvesting.

15. We recognize the role of hydropower as one of the renewable and clean energy sources, and that its potential should be realized in an environmentally sustainable and socially equitable manner.

Safe Drinking Water and Sanitation

16. Achieving the target established in the MDGs to halve the proportion of people without access to safe drinking water by 2015 and that established in the Plan of Implementation of the WSSD to halve the proportion of people without access to basic sanitation by 2015 requires an enormous amount of investment in water supply and sanitation. We call on each country to develop strategies to achieve these objectives. We will redouble our collective efforts to mobilize financial and technical resources, both public and private.

17. We will address water supply and sanitation in urban and rural areas in ways suitable for the respective local conditions and management capacities, with a view to achieving short-term improvement of water and sanitation services as well as cost-effective infrastructure investments and sound management and maintenance over time. In so doing, we will enhance poor people's access to safe drinking water and sanitation.

18. While basic hygiene practices starting from hand washing at the household level should be encouraged, intensified efforts should also be launched to promote technical breakthroughs, especially the development and practical applications of efficient and low-cost technologies tailored to daily life for the provision of safe drinking water and basic sanitation. We encourage studies for innovative technologies to be locally owned.

Water for Food and Rural Development

19. Water is essential for broad-based agricultural production and rural development in order to improve food security and eradicate poverty. It should continuously contribute to a variety of roles including food production, economic growth, and environmental sustainability. We are concerned with increasing pressure on the limited fresh water resources and on the environment. Noting that a diverse array of agricultural practices and agricultural economies has evolved in the world, we should make every effort to reduce unsustainable water management and improve the efficiency of agricultural water use.

20. Through effective and equitable water use and management, and extending irrigation in areas of need, we will promote neighborhood community based development, which should result in income-generating activities and opportunities and contribute to poverty eradication in rural areas.

21. We encourage innovative and strategic investment, research, and development and international cooperation for the progressive improvement of agricultural water management, by such means as demand-driven management including participatory irrigation management, rehabilitation and modernization of existing water facilities, water-harvesting, water-saving/drought-resistant crop varieties, water storage, and dissemination of agricultural best practices.

22. Inland fisheries being a major source of food, freshwater fish production should be addressed through intensified efforts to improve water quality and quantity in rivers and protection or restoration of breeding areas.

Water Pollution Prevention and Ecosystem Conservation

23. We recognize the need to intensify water pollution prevention in order to reduce hazards to health and the environment and to protect ecosystems, including control of invasive species. We recognize traditional water knowledge and will promote the awareness of positive and negative impacts of human activities on watersheds for the entire water cycle through public information and education, including for children, in order to avoid pollution and unsustainable use of water resources.

24. To ensure a sustainable water supply of good quality, we should protect and use in a sustainable manner the ecosystems that naturally capture, filter, store, and release water, such as rivers, wetlands, forests, and soils.

25. We urge countries to review and, when necessary, to establish appropriate legislative frameworks for the protection and sustainable use of water resources and for water pollution prevention.

26. In view of the rapid degradation of watersheds and forests, we will concentrate our efforts to combat deforestation, desertification and land degradation through programs to promote greening, sustainable forest management, the restoration of degraded lands and wetlands, and the conservation of biodiversity.

Disaster Mitigation and Risk Management

27. The growing severity of the impacts of floods and droughts highlights the need for a comprehensive approach that includes strengthened structural measures such as reservoirs and dikes and also non-structural measures such as land-use regulation and guidance, disaster forecasting and warning systems, and national risk management systems, in harmony with the environment and different water uses, including inland waterway navigation.

28. We will cooperate to minimize damage caused by disasters through enhancing the sharing and exchange, where appropriate, of data, information, knowledge, and experiences at the international level. We encourage the continuation of collaboration between scientists, water managers, and relevant stakeholders to reduce vulnerability and make the best prediction and forecasting tools available to water managers.

29. Finally, we thank the Government and people of Japan for hosting this Ministerial Conference and the Forum.

NGO Statement of the 3rd World Water Forum

As taken from page 15 of the World Water Council's Final Report (http://www.world.water-forum3.com/en/finalreport_pdf/fr160_175.pdf)

We as a diverse group of experienced NGOs and community organisations working in the water sector convey our recommendations and concerns to the Kyoto Ministerial Conference, which are the result of deliberations in the NGO Panel Debate.

We assert that water is a public good and access to safe, affordable water is a human right. We welcome that this has now been recognised by the United Nations Committee on Social, Economic and Cultural Rights.

We oppose the promotion of the development paradigm exemplified by the commodification of water and the renewed push for large-scale infrastructure projects that undermine local, participatory, decentralised actions.

Despite ample and credible evidence of the value of local actions within river basins, they continue to be marginalized and trivialised. It's time to mainstream these locally rooted strategies, by incorporating them into policies and budgets at all levels.

Recommendations

Governments must reaffirm that access to water and sanitation is a basic human right in the Kyoto Declaration and must always respect and protect human rights in all water policy and water resource management decisions.

Governments must recognise the legitimacy of NGOs and community organisations as contributors to sustainable development. Community based approaches require more formal political and financial support so they can be replicated and scaled up. Build the capacity of public service providers, community cooperatives and small-scale entrepreneurs to provide water. Base Integrated Water Resource Management strategies on people's participation and provide adequate mechanisms for dialogue and conflict resolution within river basins and across national boundaries. NGOs are ready to work with governments to set up these mechanisms.

Governments, International Financial Institutions and the private sector should cease to promote water mega-projects without reference to international agreements and must always incorporate the recommendations of the World Commission on Dams into water and energy planning processes, including reparations.

Governments have to close the financing gap, increase financial efficiency and eliminate corruption and Donors must meet their commitments to debt relief.

Financing should be targeted towards low cost technologies and community developed systems.

Governments must keep water resources and services out of the World Trade Organization and all other regional and international trade negotiations and agreements and in particular the General Agreement on Trade in Services.

NGOs will continue to work with governments to develop water policies that prioritise ecosystem security and environment health of river basins to provide sufficient flow, quantity and quality of water for human livelihoods, development and biodiversity.

As enablers of participation in planning, implementing and monitoring we require full and free transparent access to information held by governments, international agencies and corporations.

We call for a rejection of the Camdessus Report as the product of an unaccountable, unrepresentative, inaccessible process no longer suitable for this day and age.

The Human Right to Water: Two Steps Forward, One Step Back

Peter H. Gleick

Introduction

An earlier volume of *The World's Water* (Gleick 2000) addressed the issue of the human right to water. That piece and an earlier journal article (Gleick 1999) argued that access to basic water services is a fundamental human right implicitly and explicitly supported by international law, declaration, and customary law derived from state practice. It also argued that governments, international aid agencies, non-governmental organizations, and local communities should work to provide all humans with a basic water requirement and to guarantee the right to water as a human right. In November 2002, the United Nations Economic and Social Council passed General Comment No.15, explicitly acknowledging the "human right to water" as "indispensable for leading a life in human dignity" and "a prerequisite for the realization of other human rights." It also concludes, "States parties have to adopt effective measures to realize, without discrimination, the right to water, as set out in this General Comment" (United Nations 2002). This United Nations General Comment is the strongest and clearest official statement yet of the legal basis for a human right to water.

At the same time, most governments continue to do little to formally address the issue. With few notable exceptions, politicians have shown themselves to be reluctant to acknowledge explicit human rights that may impose additional requirements or obligations on them. Similarly, the international water community, after issuing several breakthrough statements on water and human rights at international meetings in the 1970s and 1980s, has more recently found itself unable to move forward on this issue. At the most recent major water conferences—the 2nd World Water Conference in the Hague in 2000, the International Conference on Freshwater in Bonn in 2001, and the 3rd World Water Conference in Kyoto in 2003 (see Water Brief 1)—the issue is regularly put onto the agenda by academics, non-governmental organizations, and community groups and regularly taken off again by government delegations. Because these major water conferences are increasingly organized and funded by governments or by a combination of government, industry, and self-appointed water councils rather than the broader international water community, more political restrictions are placed on the content of conference statements. As a result, if a single influential diplomatic delegation or conference organizer is opposed to inclusion of an issue in a conference statement, it is too easily discarded without discussion. The United States, for example, is both highly influential and highly reluctant to discuss major issues in "human rights" terms, and for at least one of the major water conferences held in the past few years, they explicitly removed the phrase "water is a human right" from the final conference statement (see, the United States edits on the draft Bonn statement, Bonn 2001).

Billions of people still lack basic water services. The World Health Organization (WHO) estimates, as regularly noted in this biennial report, that more than a billion people lack clean drinking water and 2.4 billion people lack adequate sanitation services (WHO 2000). This is considered such a fundamental development failure that two separate Millennium Development Goals were set to address it (see Chapter 1). The need to provide basic water services is also the driving force behind much of the discussion about making the human right to water an explicit objective.

Two Steps Forward—Progress Toward Acknowledging a Right to Water

The term "right" to water is used here in the sense of genuine rights under international law, where States have a duty to protect and promote those rights for an individual. There is an extensive body of agreements, covenants, international statements, and state custom formally identifying and declaring a range of human rights. These previously acknowledged instruments both implicitly and explicitly support the right to water.

Implicit support comes from the Universal Declaration on Human Rights (UDHR) and the associated binding Covenants, the International Covenant on Economic, Social and Cultural Rights (ICESCR), and the International Covenant on Civil and Political Rights (ICCPR), which call for a right to food, health, human well-being, and life. Water is a "component element" of these explicit rights, and indeed, is considered more fundamental than many other rights formally acknowledged.

The 2002 United Nations General Comment also makes clear that the human right to water is implicit in, and indispensable for, the rights acknowledged in the binding rights Covenants.

> "The right to water clearly falls within the category of guarantees essential for securing an adequate standard of living, particularly since it is one of the most fundamental conditions for survival...The right to water is also inextricably related to the right to the highest attainable standard of health and the rights to adequate housing and adequate food" (United Nations 2002, Paragraph 3).

There is also support for the human right to water in international statements, agreements, and state practice. Beginning in the 1970s, a series of international environmental or water conferences have taken on the issue of access to basic resources such as water. Statements and conclusions from these sources are relevant to this analysis, and such statements offer strong evidence of international intent and "state practice," which is another mechanism of enshrining international rights policy. South Africa and France are exploring mechanisms and wording for incorporating an explicit human right to water in national law, and others are moving in a similar direction (see Box WB 2.1). The South African case has been highly publicized and discussed. In France, discussion is now underway about how to advance this concept legally (Académie de l'Eau 2003).

Box WB 2.1 International Documents, Treaties, Declarations, and Standards Recognizing the Right to Water and Related Forms of Health and Human Development

Articles 20, 26, 29, 46, Geneva Convention III (1949)

Articles 85, 89, 127, Geneva Convention IV (1949)

Articles 54, 55, Geneva Convention Additional Protocol I (1977)

Articles 5, 14, Geneva Convention Additional Protocol II (1977)

Preamble, Mar del Plata Declaration of the United Nations Water Conference (1977)

Article 14(2)(h), Convention on the Elimination of All Forms of Discrimination Against Women (1979)

Article 8 of the United Nations Declaration on the Right to Development (1986)

Article 11, American Convention on Human Rights in the Area of Economic, Social, and Cultural Rights (1988)

Article 24(2)(c), Convention on the Rights of the Child (1989)

Paragraph 18:47, Agenda 21 Report of the United Nations Conference on Environment and Development (1992)

Principle No. 3, The Dublin Statement on Water and Sustainable Development, International Conference on Water and the Environment (1992)

Article I(10), Vienna Declaration, World Conference on Human Rights (1993)

Principles 2 and 3, Cairo Programme of Action, United Nations International Conference on Population and Development (ICPD) (1994)

Section 27(1)(b), Bill of Rights of the Constitution of South Africa (1994)

Articles 93 and 107 of the Beijing Platform of Action (1995)

Commitment 1(n) of the Copenhagen Declaration on Social Development (2000)

Paragraphs 5 and 19, Recommendation 14 of the Committee of Ministers to Member States on the European Charter on Water Resources (2001)

Promotion of the Realization of the Right to Drinking Water, Sub-Commission on Human Rights Resolution 2002/6.

General Comment 15 of the United Nations General Assembly (2002)

Other Examples

The African Consultative Forum on Water Supply and Sanitation lists access to safe drinking water as a basic right…and therefore a responsibility for all governments in its "Africa Statement" of November 1998 (WSSCC 1998).

Ethiopia—Constitution, 1995, Article 90.
To the extent the country's resources permit, policies shall aim to provide all Ethiopians access to public health and education, clean water, housing, food, and social security (Ethiopian Constitution 1995).

Gambia—Constitution, 1996. Article 216(4).
The State shall endeavour to facilitate equal access to clean and safe water.

South Africa—Constitution, 1996. Section 27.
(1) Everyone has the right to have access to:
    (a) health care services, including reproductive health care;
    (b) sufficient food and water; and
    (c) social security, including, if they are unable to support themselves and their dependents, appropriate social assistance.

(2) The state must take reasonable legislative and other measures, within its available resources, to achieve the progressive realization of each of these rights…"

Uganda—Constitution, 1995. Article 14 of the National Objectives.
The State shall endeavour to fulfill the fundamental rights of all Ugandans to social justice and economic development and shall, in particular, ensure that …all Ugandans enjoy rights and opportunities and access to education, health services, clean and safe water, decent shelter, adequate clothing, food, security, and pension and retirements benefits.
http://www.government.go.ug/constitution/index.php

Zambia—Constitution, as amended in 1996. Article 112.
The State shall endeavour to provide clean and safe water.

In Burkina Faso, Article 2 of the Law of 8 February 2001 concerning water management states: "The law recognizes to each person the right to water in a quantity necessary to its needs and to the fulfillment of fundamental requirements of life and dignity."

The Supreme Court of India considers that the right to water is a fundamental right (originating in Article 21 of the Indian Constitution on the right to life).

Sources: Gleick 1999, United Nations 2002, footnote 5. Smets 2003, footnote 32.

Box WB 2.1 lists some of the many international documents, treaties, declarations, and other sources recognizing the right to water and closely related forms of health and development for which basic water is a fundamental requirement.

Among the most important of these statements and declarations is the 1977 Mar del Plata conference, which stated that "all peoples...have the right to have access to drinking water in quantities and of a quality equal to their basic needs." In 1986, the United Nations General Assembly adopted the Declaration on the Right to Development, which calls for the right to access to basic resources. The United Nations explicitly includes water as a basic resource and has previously noted that persistent conditions of underdevelopment in which millions of humans are "denied access to such essentials as food, water, clothing, housing, and medicine in adequate measure" represent a clear and fragrant "mass violation of human rights" (United Nations 1995).

Recognition of water is also explicit in the 1989 Convention of the Rights of the Child (CRC). Article 24 of the CRC states that a child has the right to enjoy the highest attainable standard of health and that states must secure this right "through, inter alia,...the provision of adequate nutritious foods and clean drinking water" (United Nations 1989). The 2002 United Nations General Comments greatly expand upon these explicit indications (United Nations 2002).

In just the past two years, several other efforts have acknowledged the right to water. In 2003, for example, the Holy See published a message stating: "There is a growing movement to formally adopt a human right to water. The dignity of the human person mandates its acknowledgment. ...The right to water is thus an inalienable right."

During the opening ceremony of the 3rd World Water Forum in Kyoto, French President J. Chirac in a video presentation proposed "access to water be recognized as a fundamental right."

Similarly, in September 2003, the European Parliament discussing the Commission Communication on water management in developing countries and priorities for European Union (EU) development cooperation, declared "access to drinking water is a basic human right" (European Council on Environmental Law [ECEL] 2004). And in early 2004, the European Council on Environmental Law passed the "Resolution on the Recognition of the Right to Drinking Water in the Member States of the European Union" on January 17. Among other things, this resolution states:

1. "Access to drinking water and sanitation is a fundamental right of the individual. The implementation of this right shall be ensured by law which shall specify the conditions for its exercise. Everyone is obliged to act so as to protect the sustainability of this resource with respect to quantity and quality.

2. Everyone in urban areas and in areas with water supply and sanitation networks has the right to be provided with sufficient drinking water for his or her fundamental needs and is also entitled to benefit from the provision of sanitation services. Everyone is obliged to contribute to the recovery of the cost of water supply and sanitation. Where networks for supplying water are not available, residents shall be able to obtain drinking water from a source within their local authority area at an affordable price (ECEL 2004)."

One Step Back—Barriers

As the above examples indicate, the progress in declaring, defining, and implementing a human right to water has been substantial. Nevertheless, there are still governmental and institutional barriers and roadblocks to putting this right into practical use. The issue of a human right to water is now raised in every international water forum, including, for example, the 1st, 2nd, and 3rd World Water Forums in, respectively, Marrakech (1997), The Hague (2000) and Kyoto (2003), with the objective of getting an acknowledgement of this right included in the formal statements to be issued at the close of each meeting.

Despite President Chirac's opening support for the concept in Kyoto, more than 100 ministers and high-ranking civil servants could not ultimately agree whether to include a reference to United Nations General Comment 15 in their Ministerial Declaration. The United Kingdom proposed that ministers agree to discuss the implications of this Comment at the domestic level. There was extensive support for this among the conference participants, including supportive statements by conference hosts and various dignitaries, but, ultimately, the final text adopted by Ministers fails to make reference to a right to water (see Water Brief 1. Thus, in all three World Water Forums, language on the human right to water was included and then removed from the final conference statements, at the request of a small number of influential representatives.[1] This represents a failure of political will.

Such conference statements are not binding, which makes the reluctance on the part of some countries to include a right to water in them seem puzzling. But, in fact, such reluctance is part of an overall effort to limit human rights debates to a subset of political issues and because some see a conflict between treating water as an economic good and a human right simultaneously (Smets 2003). There is also a fear that inclusion of language acknowledging this right in conference statements will contribute to making it part of "customary international law." Customary law and conventional state law can be primary sources of international law and can contribute to strengthening human rights standards beyond those established in international agreements.

What is "customary international law?" Customary international law develops from the practice and behavior of states. To international lawyers, the practice of states includes official government acts, statements at international conferences and in diplomatic exchanges, formal instructions to diplomats, national court decisions, legislation, and other actions taken by governments in the international arena (Buergenthal et al. 2002).

Customary international law is recognized when states follow certain practices out of a sense of legal obligation, when they are intended for adherence generally, and when they are widely accepted (LII 2004). Customary law and law made through international agreement have equal authority as international law. General principles common to systems of national law also can be sources of international law. In this case, a general principle may be invoked as a rule of international law because it is common to major legal systems of the world. Indeed, the human right to water has been acknowledged so widely in international water meetings, state practice, and now the General Comment 15 of the United Nations and the 2004 resolution of the European Council on Environmental Law that some argue that it should already be considered customary law (Gleick 1999, United Nations 2002).

---

1. In the "Marrakech Declaration" the word "right" was replaced with "needs" at the last moment, making the sentence grammatically awkward ("In particular the Forum recommends actions to recognize the basic human needs to have access to clean water and sanitation...") (Marrakech 1997). In 2000 at The Hague, I served as a non-aligned "Science" delegate to the Ministerial meetings and was present when rights language was struck. In Bonn 2001, I served as a member of the United States delegation, where I argued unsuccessfully for the United States to acknowledge the human right to water, and the United States edits of that document show that they asked that it be eliminated (Bonn 2001). At Kyoto in 2003, I and others again met with the United States delegation formally and informally to discuss this issue and was told that the delegation would again not only refuse to support any official declaration of a human right to water but would work to remove any such language.

"As the right to adequate food and the right to water are also recognized in a large number of other binding and non-binding legal instruments they are arguably part of customary international law as well" (UNFAO 2003).

Others are not yet convinced that a general human right to water rises to the level of customary law, though Professor Joseph Dellapenna has said:

"…it seems fairly clear, if you turn to the law of human rights and the rights of an indigenous people to maintain their traditional lifestyle, that at the very least governments are under an obligation not to deprive people of the water they need to maintain that lifestyle" (Fuller, no date).

Others incorrectly interpret acknowledging a human right as requiring that it be provided free (see, the arguments against this in Gleick 1999 and Smets 2003). In a direct challenge to this interpretation, the 2004 Resolution passed by the European Council on Environmental Law explicitly declares a human right to water, but also states "Everyone is obliged to contribute to the recovery of the cost of water supply and sanitation" (ECEL 2004). As of early 2004, the United States position in this area continues to be limiting and regressive, though consistent with its positions on other human rights laws. The United States, for example, is a signatory to the 1966 Covenant on Social, Economic and Cultural Rights, but has not ratified that Covenant—one of an increasingly long series of international agreements related to human rights or the environment that the United States has failed to ratify (see Box WB 2.2, as of this writing.

Box WB 2.2 Selected International Agreements Related to Human Rights or the Environment Not Ratified by the United States, 2004

- 1966 Covenant on Social, Economic and Cultural Rights (signed, not ratified)
- Convention on the Rights of the Child

First and Second Optional Protocol to the International Covenant on Civil and Political Rights

- Convention on the Elimination of All Forms of Discrimination Against Women (CEDAW)
- Rome Statute of the International Criminal Court
- Comprehensive Nuclear Test Ban Treaty
- Kyoto Protocol on Climate Change
- Convention on Biological Diversity
- Stockholm Convention on Persistent Organic Pollutants (POPs).

Source: United Nations 2004.

Why Bother?

In the past two years, the legal and institutional basis for a human right to water has solidified, with clear declarations from the United Nations, the European Community, and others. The debate is now shifting to how to implement this right, how to define the responsibilities of states in meeting that right, and the consequences that should flow from the failure to do so. And while the legal questions are increasingly being resolved, even many advocates of more aggressive action to address basic water problems do not know why getting governments to acknowledge a human right to water would be useful. Does it impose an obligation on governments to provide water to its citizens? Would it open the door to legal action for countries that have failed to do so? I don't believe so, but such legal rights do obligate states to provide the institutional, economic, and social environment necessary to help individuals to progressively realize those rights, and it prohibits states from denying access to water. Meeting basic needs for water should also take precedence over allocations of spending for other development needs, and it will require a redirection of current economic spending priorities. There is also a moral imperative to meeting basic water rights, however, and governments must step forward and both acknowledge their responsibilities and then work toward meeting them.

REFERENCES

Académie de l'Eau. 2003. Solidarity for drinking water: Economic aspects. Paris:Water Academy of France, February. (Adopted by the French Water Academy, December 2002.)

Bonn. 2001. United States comments and edits on 2nd draft recommendations for action. Bonn International Freshwater Conference.
http://www.water-2001.de/co_doc/proposals/usa.pdf

Buergenthal, T., Shelton, D., and Stewart, D. 2002. *International human rights in a nutshell*, 3rd edition. Eagan, Minnesota:West Law.

Ethiopian Constitution. 1995. Article 90 of the Constitution of The Federal Democratic Republic of Ethiopia. http://www.ethiopar.net/English/cnstiotn/conchp10.htm

European Council on Environmental Law. 2004. Resolution on the Recognition of the Right to Drinking Water in the Member States of the European Union. (Adopted January 17.)

Fuller, J. no date. Legal expert describes Iraqi treatment of marsh Arabs as genocide. United States Embassy, Japan. January 2004.
http://usembassy.state.gov/tokyo/wwwhsec20021119a3.html

Gleick, P. H. 1999. The human right to water. *Water Policy*. 1, no. 5:487–503.

Gleick, P. H. 2000. The human right to water. *The world's water 2000–2001*, 1–17, Washington, D.C.: Island Press.

Holy See. 2003. Water, an essential element for life. The Vatican.
http://www.zenit.org/english/visualizza.phtml?sid=33276

Legal Information Institute. 2004. International law: An overview. Cornell University.
http://www.law.cornell.edu/topics/international.html

Marrakech. 1997. *The Marrakech Declaration. The 1st World Water Forum*. February 2, 2004.
http://www.cmo.nl/pe/pe7/pe-772.html, and
http://www.unesco.org.uy/phi/gwpsamtac/uruguay/articulo2.htm

Pacific Institute. 2003. Ministers ignore human right to water despite consensus, press release, March 19. http://www.pacinst.org/kyoto/

Phiri, Z. 2000. Water law, water rights and water supply in Zambia—Issues and perspectives. Presented at the 1st WARFSA/WaterNet Symposium: Sustainable Use of Water Resources, Maputo, November 1–2.

Smets, H. 2003. The Right to Water at the Time of the 3rd World Water Forum. Paris:French Water Academy. October draft.

United Nations. 1989. Convention on the Rights of the Child. http://www.unicef.org/crc/crc.htm

United Nations. 2002. Substantive issues arising in the implementation of the International Covenant on Economic, Social and Cultural Rights: The right to water. Economic and Social Council, Committee on Economic, Social, and Cultural Rights, E/C.12/2002/11. November 26. New York.

United Nations. 2004. Status of ratifications of the principal international human rights treaties. Office of the United Nations High Commissioner for Human Right.
http://www.unhchr.ch/pdf/report.pdf

United Nations Food and Agricultural Organization. 2003. *Agriculture, food and water.* Annex 1: The right to adequate food and the right to water.
http://www.fao.org/DOCREP/006/Y4683E/y4683e13.htm

Water Supply and Sanitation Collaborative Council. 1998. Africa statement—Water and sanitation Africa initiative, January 4, 2004. http://www.wsscc.org/load.cfm?edit_id=164

World Health Organization. 2000. Global water supply and sanitation assessment 2000 report.
http://www.who.int/docstore/water_sanitation_health/Globassessment/GlobalTOC.htm

Substantive Issues Arising in the Implementation of International Covenent on Economic, Social, & Cultural Rights

United Nations General comment no. 15 (2002). The Right to Water (sections 11 & 12 of the International Covenant on Economic, Social, & Cultural Rights)

I. Introduction

1. Water is a limited natural resource and a public good fundamental for life and health. The human right to water is indispensable for leading a life in human dignity. It is a prerequisite for the realization of other human rights. The Committee has been confronted continually with the widespread denial of the right to water in developing as well as developed countries. Over one billion persons lack access to a basic water supply, while several billion do not have access to adequate sanitation, which is the primary cause of water contamination and diseases linked to water.[1] The continuing contamination, depletion and unequal distribution of water is exacerbating existing poverty. States parties have to adopt effective measures to realize, without discrimination, the right to water, as set out in this general comment.

The Legal Bases of the Right to Water

2. The human right to water entitles everyone to sufficient, safe, acceptable, physically accessible and affordable water for personal and domestic uses. An adequate amount of safe water is necessary to prevent death from dehydration, to reduce the risk of water-related disease and to provide for consumption, cooking, personal and domestic hygienic requirements.

3. Article 11, paragraph 1, of the Covenant specifies a number of rights emanating from, and indispensable for, the realization of the right to an adequate standard of living "including adequate food, clothing and housing." The use of the word "including" indicates that this catalogue of rights was not intended to be exhaustive. The right to water clearly falls within the category of guarantees essential for securing an adequate standard of living, particularly since it is one of the most fundamental conditions for survival. Moreover, the Committee has previously recognized that water is a human right contained in article 11, paragraph 1, (see General Comment No. 6, 1995).[2] The right to water is also inextricably related to the right to the highest attainable standard of health (article 12, paragraph 1)[3] and the rights to ade-

---

1. In 2000, the World Health Organization estimated that 1.1 billion persons did not have access to an improved water supply (80 percent of them rural dwellers) able to provide at least 20 litres of safe water per person a day; 2.4 billion persons were estimated to be without sanitation. (See WHO, The Global Water Supply and Sanitation Assessment 2000, Geneva, 2000, page 1.) Further 2.3 billion persons each year suffer from diseases linked to water: see United Nations, Commission on Sustainable Development, Comprehensive Assessment of the Freshwater Resources of the World, New York, 1997, p. 39.

2. See paragraphs 5 and 32 of the Committee's General Comment No. 6 (1995) on the economic, social and cultural rights of older persons.

3. See General Comment No. 14 (2000) on the right to the highest attainable standard of health, paragraphs 11, 12(a), 12(b), 12(d), 15, 34, 36, 40, 43, and 51.

quate housing and adequate food (article 11, paragraph 1).[4] The right should also be seen in conjunction with other rights enshrined in the International Bill of Human Rights, foremost amongst them the right to life and human dignity.

4. The right to water has been recognized in a wide range of international documents, including treaties, declarations and other standards.[5] For instance, Article 14, paragraph 2, of the Convention on the Elimination of All Forms of Discrimination Against Women stipulates that States parties shall ensure to women the right to "enjoy adequate living conditions, particularly in relation to…water supply." Article 24, paragraph 2, of the Convention on the Rights of the Child requires States parties to combat disease and malnutrition "through the provision of adequate nutritious foods and clean drinking-water."

5. The right to water has been consistently addressed by the Committee during its consideration of States parties' reports, in accordance with its revised general guidelines regarding the form and content of reports to be submitted by States parties under articles 16 and 17 of the International Covenant on Economic, Social and Cultural Rights, and its general comments.

6. Water is required for a range of different purposes, besides personal and domestic uses, to realize many of the Covenant rights. For instance, water is necessary to produce food (right to adequate food) and ensure environmental hygiene (right to health). Water is essential for securing livelihoods (right to gain a living by work) and enjoying certain cultural practices (right to take part in cultural life). Nevertheless, priority in the allocation of water must be given to the right to water for personal and domestic uses. Priority should also be given to the water resources required to prevent starvation and disease, as well as water required to meet the core obligations of each of the Covenant rights.[6]

---

4. See paragraph 8(b) of General Comment No. 4 (1991). See also the report by Commission on Human Rights' Special Rapporteur on adequate housing as a component of the right to an adequate standard of living, Mr. Miloon Kothari (E.CN.4/2002/59), submitted in accordance with Commission on Human Rights Resolution 2001/28 of 20 April 2001, E.CN.4/2002/59. In relation to the right to adequate food, see the report by the Special Rapporteur of the Commission on the Right to Food, Mr. Jean Ziegler (E/CN.4/2002/58), submitted in accordance with Commission on Human Rights Resolution 2001/25 of 20 April 2001, E/CN.4/2002/58.

5. See Article 14, paragraph 2(h), Convention on the Elimination of All Forms of Discrimination Against Women (1979); Article 24, paragraph 2(c), Convention on the Rights of the Child (1989); Articles 20, 26, 29 and 46, of the Geneva Convention III relative to the Treatment of Prisoners of War, of 1949; Articles 85, 89, and 127, of the Geneva Convention IV (relative to the Treatment of Civilian Persons in Time of War, of 1949; Articles 54 and 55 of Additional Protocol I thereto of 1977; Articles 5 and 14 Additional Protocol II (of 1977); Preamble, Mar Del Plata Declaration Action Plan of the United Nations Water Conference (1977); see Paragraph 18.47 of Agenda 21, Report of the United Nations Conference on Environment and Development (1992), Vol. 1, Rio de Janeiro, June 3–14 1992 (A/CONF.151/ 26); Rev.1 (Vol. I and Vol. I/Corr.1, Vol. II, Vol. III, and Vol. III/Corr.1) (United Nations publication, Sales No. E.93.I.8); Resolutions adopted by the Conference, resolution 1, annex II; Principle No. 3, The Dublin Statement on Water and Sustainable Development, International Conference on Water and the Environment (1992) (A/CONF.151/PC/112); Principle No. 2, Programme of Action, of *Report of the United Nations International Conference on Population and Development*, Cairo, September 5–13, 1994 (United Nations publication, Sales No. E.95.XIII.18), chap. I, resolution 1, annex; Paragraphs 5 and 19, Recommendation 14 (2001) of the Committee of Ministers to Member States on the European Charter on Water Resources; resolution 2002/6 of the United Nations Sub-Commission on the Promotion and Protection of Human Rights and on the promotion of the realization of the right to drinking water. See also the report on the relationship between the enjoyment of economic, social and cultural rights and the promotion of the realization of the right to drinking water supply and sanitation (E/CN.4/Sub.2/2002/10) submitted by the Special Rapporteur of the Sub-Commission on the right to drinking water supply and sanitation, Mr. El Hadji Guissé.

6. See World Summit on Sustainable Development, Plan of Implementation 2002, paragraph 25(c).

Water and Covenant Rights

7. The Committee notes the importance of ensuring sustainable access to water resources for agriculture to realize the right to adequate food (see General Comment No.12, 1999).[7] Attention should be given to ensuring that disadvantaged and marginalized farmers, including women farmers, have equitable access to water and water management systems, including sustainable rain harvesting and irrigation technology. Taking note of the duty in article 1, paragraph 2, of the Covenant, which provides that a people may not "be deprived of its means of subsistence," states parties should ensure that there is adequate access to water for subsistence farming and for securing the livelihoods of indigenous peoples.[8]

8. Environmental hygiene, as an aspect of the right to health under article 12, paragraph 2 (b), of the Covenant, encompasses taking steps on a non-discriminatory basis to prevent threats to health from unsafe and toxic water conditions.[9] For example, States parties should ensure that natural water resources are protected from contamination by harmful substances and pathogenic microbes. Likewise, States parties should monitor and combat situations where aquatic eco-systems serve as a habitat for vectors of diseases wherever they pose a risk to human living environments.[10]

9. With a view to assisting States parties' implementation of the Covenant and the fulfilment of their reporting obligations, this General Comment focuses in Part II on the normative content of the right to water in articles 11, paragraph 1, and 12, on States parties' obligations (Part III), on violations (Part IV) and on implementation at the national level (Part V), while the obligations of actors other than States parties are addressed in Part VI.

II. Normative Content of the Right to Water

10. The right to water contains both freedoms and entitlements. The freedoms include the right to maintain access to existing water supplies necessary for the right to water, and the right to be free from interference, such as the right to be free from arbitrary disconnections or contamination of water supplies. By contrast, the entitlements include the right to a system of water supply and management that provides equality of opportunity for people to enjoy the right to water.

11. The elements of the right to water must be *adequate* for human dignity, life and health, in accordance with articles 11, paragraph 1, and 12. The adequacy of water should not be interpreted narrowly, by mere reference to volumetric quantities and technologies. Water should be treated as a social and cultural good, and not primarily as an economic good. The man-

---

7. This relates to both availability and to accessibility of the right to adequate food (See General Comment No.12, 1999, paragraphs 12 and 13).

8. See Statement of Understanding accompanying the United Nations Convention on the Law of Non-Navigational Uses of Watercourses (A/51/869 of April 11 1997), which declared that, in determining vital human needs in the event of conflicts over the use of watercourses "special attention is to be paid to providing sufficient water to sustain human life, including both drinking water and water required for production of food in order to prevent starvation." United Nations General Assembly Doc. A/51/869 (April 11, 1997).

9. See paragraph 15, General Comment No. 14.

10. According to the WHO definition, vector-borne diseases include diseases transmitted by insects, (malaria, filariasis, dengue, Japanese encephalitis and yellow fever), diseases for which aquatic snails serve as intermediate hosts (schistosomiasis), and zoonoses with vertebrates as reservoir hosts.

ner of the realization of the right to water must also be sustainable, ensuring that the right can be realized for present and future generations.[11]

12. While the adequacy of water required for the right to water may vary according to different conditions, the following factors apply in all circumstances:

(a) *Availability.* The water supply for each person must be sufficient and continuous for personal and domestic uses.[12] These uses ordinarily include drinking, personal sanitation, washing of clothes, food preparation, personal and household hygiene.[13] The quantity of water available for each person should correspond to World Health Organization (WHO) guidelines.[14] Some individuals and groups may also require additional water due to health, climate, and work conditions;

(b) *Quality.* The water required for each personal or domestic use must be safe, therefore free from micro-organisms, chemical substances and radiological *hazards* that constitute a threat to a person's health.[15] Furthermore, water should be of an acceptable colour, odour and taste for each personal or domestic use.

(c) *Accessibility.* Water and water facilities and services have to be accessible to *everyone* without discrimination, within the jurisdiction of the State party. Accessibility has four overlapping dimensions:

(i) *Physical accessibility:* water, and adequate water facilities and services, must be within safe physical reach for all sections of the population. Sufficient, safe and acceptable water must be accessible within, or in the immediate vicinity, of each household, educational institution and workplace.[16] All water facilities and services must be of sufficient quality, culturally appropriate and sensitive to gender, life-cycle and privacy requirements. Physical security should not be threatened during access to water facilities and services;

(ii) *Economic accessibility:* Water, and water facilities and services, must be affordable for all. The direct and indirect costs and charges associated with securing water must be affordable, and must not compromise or threaten the realization of other Covenant rights;

---

11. For a definition of sustainability, see the *Report of the United Nations Conference on Environment and Development*, Rio de Janeiro, 3–14, 1992, Declaration on Environment and Development, principles 1, 8, 9, 10, 12, and 15; and the Agenda 21, in particular principles 5.3, 7.27, 7.28, 7.35, 7.39, 7.41, 18.3, 18.8, 18.35, 18.40, 18.48, 18.50, 18.59, and 18.68.

12. "Continuous" means that the regularity of the water supply is sufficient for personal and domestic uses.

13. In this context, "drinking" means water for consumption through beverages and foodstuffs. "Personal sanitation" means disposal of human excreta. Water is necessary for personal sanitation where water-based means are adopted. "Food preparation" includes food hygiene and preparation of foodstuffs, whether water is incorporated into, or comes into contact with, food. "Personal and household hygiene" means personal cleanliness and hygiene of the household environment.

14. See J. Bartram and G. Howard, "Domestic water quantity, service level and health: what should be the goal for water and health sectors", WHO, 2002. See also P .H. Gleick, 1996, Basic water requirements for human activities: meeting basic needs, *Water International*, 21; pp. 83–92.

15. The Committee refers States parties to WHO, *Guidelines for drinking-water quality*, 2nd edition, vols. 1–3 (Geneva, 1993) that are "intended to be used as a basis for the development of national standards that, if properly implemented, will ensure the safety of drinking water supplies through the elimination of, or reduction to a minimum concentration, of constituents of water that are known to be hazardous to health."

16. See also General Comment No. 4 (1991), paragraph 8(b), General Comment No. 13 (1999) paragraph 6(a), and General Comment No. 14 (2000) paragraphs 8(a) and 8(b). Household includes a permanent or semi-permanent dwelling, or a temporary halting site.

(iii) *Non-discrimination:* Water and water facilities and services must be accessible to all, including the most vulnerable or marginalized sections of the population, in law and in fact, without discrimination on any of the prohibited grounds; and

(iv) *Information accessibility:* accessibility includes the right to seek, receive and impart information concerning water issues.[17]

Special Topics of Broad Application

Non-discrimination and Equality

13. The obligation of States parties to guarantee that the right to water is enjoyed without discrimination (article 2, paragraph 2), and equally between men and women (article 3), pervades all of the Covenant obligations. The Covenant thus proscribes any discrimination on the grounds of race, colour, sex, age, language, religion, political or other opinion, national or social origin, property, birth, physical or mental disability, health status (including HIV/AIDS), sexual orientation and civil, political, social or other status, which has the intention or effect of nullifying or impairing the equal enjoyment or exercise of the right to water. The Committee recalls paragraph 12 of General Comment No. 3 (1990), which states that even in times of severe resource constraints, the vulnerable members of society must be protected by the adoption of relatively low-cost targeted programmes.

14. States parties should take steps to remove de facto discrimination on prohibited grounds, where individuals and groups are deprived of the means or entitlements necessary for achieving the right to water. States parties should ensure that the allocation of water resources, and investments in water, facilitate access to water for all members of society. Inappropriate resource allocation can lead to discrimination that may not be overt. For example, investments should not disproportionately favour expensive water supply services and facilities that are often accessible only to a small, privileged fraction of the population, rather than investing in services and facilities that benefit a far larger part of the population.

15. With respect to the right to water, States parties have a special obligation to provide those who do not have sufficient means with the necessary water and water facilities and to prevent any discrimination on internationally prohibited grounds in the provision of water and water services.

16. Whereas the right to water applies to everyone, States parties should give special attention to those individuals and groups who have traditionally faced difficulties in exercising this right, including women, children, minority groups, indigenous peoples, refugees, asylum seekers, internally displaced persons, migrant workers, prisoners and detainees. In particular, States parties should take steps to ensure that:

(a) Women are not excluded from decision-making processes concerning water resources and entitlements. The disproportionate burden women bear in the collection of water should be alleviated;

---

17. See paragraph 48 of this General Comment.

(b) Children are not prevented from enjoying their human rights due to the lack of adequate water in educational institutions and households or through the burden of collecting water. Provision of adequate water to educational institutions currently without adequate drinking water should be addressed as a matter of urgency;

(c) Rural and deprived urban areas have access to properly maintained water facilities. Access to traditional water sources in rural areas should be protected from unlawful encroachment and pollution. Deprived urban areas, including informal human settlements, and homeless persons, should have access to properly maintained water facilities. No household should be denied the right to water on the grounds of their housing or land status;

(d) Indigenous peoples' access to water resources on their ancestral lands is protected from encroachment and unlawful pollution. States should provide resources for indigenous peoples to design, deliver and control their access to water;

(e) Nomadic and traveller communities have access to adequate water at traditional and designated halting sites;

(f) Refugees, asylum-seekers, internally displaced persons and returnees have access to adequate water whether they stay in camps or in urban and rural areas. Refugees and asylum-seekers should be granted the right to water on the same conditions as granted to nationals;

(g) Prisoners and detainees are provided with sufficient and safe water for their daily individual requirements, taking note of the requirements of international humanitarian law and the United Nations Standard Minimum Rules for the Treatment of Prisoners;[18]

(h) Groups facing difficulties with physical access to water, such as older persons, persons with disabilities, victims of natural disasters, persons living in disaster-prone areas, and those living in arid and semi-arid areas, or on small islands are provided with safe and sufficient water.

III. States Parties' Obligations

General legal obligations

17. While the Covenant provides for progressive realization and acknowledges the constraints due to the limits of available resources, it also imposes on States parties various obligations which are of immediate effect. States parties have immediate obligations in relation to the right to water, such as the guarantee that the right will be exercised without discrimination of any kind (article 2, paragraph 2) and the obligation to take steps (article 2, paragraph 1) towards the full realization of articles 11, paragraph 1, and 12. Such steps must be deliberate, concrete and targeted towards the full realization of the right to water.

---

18. See Articles 20, 26, 29, and 46, of the third Geneva Convention III (of August 12, 1949); Articles 85, 89, and 127 of the Geneva Convention IV (of August 12, 1949); Articles 15 and 20, paragraph 2, United Nations Standard Minimum Rules for the Treatment of Prisoners (1955), in *Human Rights: A Compilation of International Instruments* (United Nations publication, Sales No. E.88.XIV.1).

18. States parties have a constant and continuing duty under the Covenant to move as expeditiously and effectively as possible towards the full realization of the right to water. Realization of the right should be feasible and practicable, since all States parties exercise control over a broad range of resources, including water, technology, financial resources and international assistance, as with all other rights in the Covenant.

19. There is a strong presumption that retrogressive measures taken in relation to the right to water are prohibited under the Covenant.[19] If any deliberately retrogressive measures are taken, the State party has the burden of proving that they have been introduced after the most careful consideration of all alternatives and that they are duly justified by reference to the totality of the rights provided for in the Covenant in the context of the full use of the State party's maximum available resources.

Specific Legal Obligations

20. The right to water, like any human right, imposes three types of obligations on States parties: obligations to *respect*, obligations to *protect* and obligations to *fulfil*.

(a) Obligations to Respect

21. The obligation to *respect* requires that States parties refrain from interfering directly or indirectly with the enjoyment of the right to water. The obligation includes, inter alia, refraining from engaging in any practice or activity that denies or limits equal access to adequate water; arbitrarily interfering with customary or traditional arrangements for water allocation; unlawfully diminishing or polluting water, for example through waste from State-owned facilities or through use and testing of weapons; and limiting access to, or destroying, water services and infrastructure as a punitive measure, for example, during armed conflicts in violation of international humanitarian law.

22. The Committee notes that during armed conflicts, emergency situations and natural disasters, the right to water embraces those obligations by which States parties are bound under international humanitarian law.[20] This includes protection of objects indispensable for survival of the civilian population, including drinking water installations and supplies and irrigation works, protection of the natural environment against widespread, long-term and severe damage and ensuring that civilians, internees and prisoners have access to adequate water.[21]

(b) Obligations to Protect

23. The obligation to *protect* requires State parties to prevent third parties from interfering in any way with the enjoyment of the right to water. Third parties include individuals, groups, corporations and other entities as well as agents acting under their authority. The obligation includes, inter alia, adopting the necessary and effective legislative and other measures to restrain, for example, third parties from denying equal access to adequate

---

19. See General Comment No. 3 (1990), paragraph 9.

20. For the interrelationship of human rights law and humanitarian law, the Committee notes the conclusions of the International Court of Justice in *Legality of the Threat or Use of Nuclear Weapons (Request by the General Assembly), ICJ Reports* (1996) p. 226, paragraph 25.

21. See Articles 54 and 56, Additional Protocol I to the Geneva Conventions (1977), article 54, Additional Protocol II (1977), Articles 20 and 46 of the third Geneva Conventions III (of August 12, 1949), and common article 3 of the Geneva Conventions (of August 12,1949).

water; and polluting and inequitably extracting from water resources, including natural sources, wells and other water distribution systems.

24. Where water services (such as piped water networks, water tankers, access to rivers and wells) are operated or controlled by third parties, States parties must prevent them from compromising equal, affordable, and physical access to sufficient, safe and acceptable water. To prevent such abuses an effective regulatory system must be established, in conformity with the Covenant and this General Comment, which includes independent monitoring, genuine public participation and imposition of penalties for non-compliance.

(c) Obligations to Fulfil

25. The obligation to *fulfil* can be disaggregated into the obligations to facilitate, promote and provide. The obligation to facilitate requires the State to take positive measures to assist individuals and communities to enjoy the right. The obligation to promote obliges the State party to take steps to ensure that there is appropriate education concerning the hygienic use of water, protection of water sources and methods to minimize water wastage. States parties are also obliged to fulfil (provide) the right when individuals or a group are unable, for reasons beyond their control, to realize that right themselves by the means at their disposal.

26. The obligation to fulfil requires States parties to adopt the necessary measures directed towards the full realization of the right to water. The obligation includes, inter alia, according sufficient recognition of this right within the national political and legal systems, preferably by way of legislative implementation; adopting a national water strategy and plan of action to realize this right; ensuring that water is affordable for everyone; and facilitating improved and sustainable access to water, particularly in rural and deprived urban areas.

27. To ensure that water is affordable, States parties must adopt the necessary measures that may include, inter alia: (a) use of a range of appropriate low-cost techniques and technologies; (b) appropriate pricing policies such as free or low-cost water; and (c) income supplements. Any payment for water services has to be based on the principle of equity, ensuring that these services, whether privately or publicly provided, are affordable for all, including socially disadvantaged groups. Equity demands that poorer households should not be disproportionately burdened with water expenses as compared to richer households.

28. States parties should adopt comprehensive and integrated strategies and programmes to ensure that there is sufficient and safe water for present and future generations.[22] Such strategies and programmes may include: *(a)* reducing depletion of water resources through unsustainable extraction, diversion and damming; *(b)* reducing and eliminating contamination of watersheds and water-related eco-systems by substances such as radiation, harmful chemicals and human excreta; *(c)* monitoring water reserves; *(d)* ensuring that proposed developments do not interfere with access to adequate water; *(e)* assessing the impacts of actions that may impinge upon water availability and natural-ecosystems watersheds, such as climate changes, desertification and increased soil salinity, deforestation and loss of biodiversity;[23] *(f)* increasing the efficient use of water by end-users; *(g)* reducing water

---

22. See footnote 5, the United Nations Conference on Environment and Development (UNCED), Rio de Janeiro, 1992, Agenda 21, Chapters 5, 7, and 18; and the World Summit on Sustainable Development, Plan of Implementation (2002), paragraphs 6(a), (l), and (m), 7, 36, and 38.

23. See the Convention on Biological Diversity, the Convention to Combat Desertification, and the United Nations Framework Convention on Climate Change, and subsequent protocols.

wastage in its distribution; *(h)* response mechanisms for emergency situations; *(i)* and establishing competent institutions and appropriate institutional arrangements to carry out the strategies and programmes.

29. Ensuring that everyone has access to adequate sanitation is not only fundamental for human dignity and privacy, but is one of the principal mechanisms for protecting the quality of drinking water supplies and resources.[24] In accordance with the rights to health and adequate housing (see General Comments No. 4 (1991) and 14 (2000)) States parties have an obligation to progressively extend safe sanitation services, particularly to rural and deprived urban areas, taking into account the needs of women and children.

International Obligations

30. Article 2, paragraph 1, and articles 11, paragraph 1, and 23 of the Covenant require that States parties recognize the essential role of international cooperation and assistance and take joint and separate action to achieve the full realization of the right to water.

31. To comply with their international obligations in relation to the right to water, States parties have to respect the enjoyment of the right in other countries. International cooperation requires States parties to refrain from actions that interfere, directly or indirectly, with the enjoyment of the right to water in other countries. Any activities undertaken within the State party's jurisdiction should not deprive another country of the ability to realize the right to water for persons in its jurisdiction.[25]

32. States parties should refrain at all times from imposing embargoes or similar measures, that prevent the supply of water, as well as goods and services essential for securing the right to water.[26] Water should never be used as an instrument of political and economic pressure. In this regard, the Committee recalls its position, stated in its General Comment No. 8 (1997), on the relationship between economic sanctions and respect for economic, social and cultural rights.

33. Steps should be taken by States parties to prevent their own citizens and companies from violating the right to water of individuals and communities in other countries. Where States parties can take steps to influence other third parties to respect the right, through legal or political means, such steps should be taken in accordance with the Charter of the United Nations and applicable international law.

34. Depending on the availability of resources, States should facilitate realization of the right to water in other countries, for example through provision of water resources, finan-

---

24. Article 14(2), paragraph 2, of the Convention on the Elimination of All Forms of Discrimination Against Women (1979) stipulates States parties shall ensure to women the right to "adequate living conditions, particularly in relation to...sanitation". Article 24(2), paragraph 2, of the Convention on the Rights of the Child (1989) requires States parties to "To ensure that all segments of society...have access to education and are supported in the use of basic knowledge of...the advantages of...hygiene and environmental sanitation."

25. The Committee notes that the United Nations Convention on the Law of Non-Navigational Uses of Watercourses (1997) requires that social and human needs be taken into account in determining the equitable utilization of watercourses, that States parties take measures to prevent significant harm being caused, and, in the event of conflict, special regard must be given to the requirements of vital human needs: see Articles 5, 7, and 10 of the Convention.

26. In General Comment No. 8 (1997), the Committee noted the disruptive effect of sanctions upon sanitation supplies and clean drinking water, and that sanctions regimes should provide for repairs to infrastructure essential to provide clean water.

cial and technical assistance, and provide the necessary aid when required. In disaster relief and emergency assistance, including assistance to refugees and displaced persons, priority should be given to Covenant rights, including the provision of adequate water. International assistance should be provided in a manner that is consistent with the Covenant and other human rights standards, and sustainable and culturally appropriate. The economically developed States parties have a special responsibility and interest to assist the poorer developing States in this regard.

35. States parties should ensure that the right to water is given due attention in international agreements and, to that end, should consider the development of further legal instruments. With regard to the conclusion and implementation of other international and regional agreements, States parties should take steps to ensure that these instruments do not adversely impact upon the right to water. Agreements concerning trade liberalization should not curtail or inhibit a country's capacity to ensure the full realization of the right to water.

36. States parties should ensure that their actions as members of international organizations take due account of the right to water. Accordingly, States parties that are members of international financial institutions, notably the International Monetary Fund, The World Bank, and regional development banks, should take steps to ensure that the right to water is taken into account in their lending policies, credit agreements and other international measures.

Core Obligations

37. In General Comment No. 3 (1990), the Committee confirms that States parties have a core obligation to ensure the satisfaction of, at the very least, minimum essential levels of each of the rights enunciated in the Covenant. In the Committee's view, at least a number of core obligations in relation to the right to water can be identified, which are of immediate effect:

(a) To ensure access to the minimum essential amount of water, that is sufficient and safe for personal and domestic uses to prevent disease;

(b) To ensure the right of access to water and water facilities and services on a non-discriminatory basis, especially for disadvantaged or marginalized groups;

(c) To ensure physical access to water facilities or services that provide sufficient, safe and regular water; that have a sufficient number of water outlets to avoid prohibitive waiting times; and that are at a reasonable distance from the household;

(d) To ensure personal security is not threatened when having to physically access to water;

(e) To ensure equitable distribution of all available water facilities and services;

(f) To adopt and implement a national water strategy and plan of action addressing the whole population; the strategy and plan of action should be devised, and periodically reviewed, on the basis of a participatory and transparent process; it should include methods, such as right to water indicators and benchmarks, by which progress can be closely monitored; the process by which the strategy and plan of action are devised, as well as their content, shall give particular attention to all disadvantaged or marginalized groups;

(g) To monitor the extent of the realization, or the non-realization, of the right to water;

(h) To adopt relatively low-cost targeted water programmes to protect vulnerable and marginalized groups;

(i) To take measures to prevent, treat and control diseases linked to water, in particular ensuring access to adequate sanitation;

38. For the avoidance of any doubt, the Committee wishes to emphasize that it is particularly incumbent on States parties, and other actors in a position to assist, to provide international assistance and cooperation, especially economic and technical which enables developing countries to fulfil their core obligations indicated in paragraph 37 above.

IV. Violations

39. When the normative content of the right to water (see Part II) is applied to the obligations of States parties (Part III), a process is set in motion, which facilitates identification of violations of the right to water. The following paragraphs provide illustrations of violations of the right to water.

40. To demonstrate compliance with their general and specific obligations, States parties must establish that they have taken the necessary and feasible steps towards the realization of the right to water. In accordance with international law, a failure to act in good faith to take such steps amounts to a violation of the right. It should be stressed that a State party cannot justify its non-compliance with the core obligations set out in paragraph 37 above, which are non-derogable.

41. In determining which actions or omissions amount to a violation of the right to water, it is important to distinguish the inability from the unwillingness of a State party to comply with its obligations in relation to the right to water. This follows from articles 11, paragraph 1, and 12, which speak of the right to an adequate standard of living and the right to health, as well as from article 2, paragraph 1, of the Covenant, which obliges each State party to take the necessary steps to the maximum of its available resources. A State which is unwilling to use the maximum of its available resources for the realization of the right to water is in violation of its obligations under the Covenant. If resource constraints render it impossible for a State party to comply fully with its Covenant obligations, it has the burden of justifying that every effort has nevertheless been made to use all available resources at its disposal in order to satisfy, as a matter of priority, the obligations outlined above.

42. Violations of the right to water can occur through *acts of commission*, the direct actions of States parties or other entities insufficiently regulated by States. Violations include, for example, the adoption of retrogressive measures incompatible with the core obligations (outlined in paragraph 37 above), the formal repeal or suspension of legislation necessary for the continued enjoyment of the right to water, or the adoption of legislation or policies which are manifestly incompatible with pre-existing domestic or international legal obligations in relation to the right to water.

43. Violations through *acts of omission* include the failure to take appropriate steps towards the full realization of everyone's right to water, the failure to have a national policy on water, and the failure to enforce relevant laws.

44. While it is not possible to specify a complete list of violations in advance, a number of typical examples relating to the levels of obligations, emanating from the Committee's work, may be identified:

(a) Violations of the obligation to respect follow from the State party's interference with the right to water. This includes, inter alia: (i) arbitrary or unjustified disconnection or exclusion from water services or facilities; (ii) discriminatory or unaffordable increases in the price of water; and (iii) pollution and diminution of water resources affecting human health;

(b) Violations of the obligation to protect follow from the failure of a State to take all necessary measures to safeguard persons within their jurisdiction from infringements of the right to water by third parties.[27] This includes, inter alia: (i) failure to enact or enforce laws to prevent the contamination and inequitable extraction of water; (ii) failure to effectively regulate and control water services providers; (iv) failure to protect water distribution systems (e.g., piped networks and wells) from interference, damage and destruction; and

(c) Violations of the obligation to fulfil occur through the failure of States parties to take all necessary steps to ensure the realization of the right to water. Examples includes, inter alia: (i) failure to adopt or implement a national water policy designed to ensure the right to water for everyone; (ii) insufficient expenditure or misallocation of public resources which results in the non-enjoyment of the right to water by individuals or groups, particularly the vulnerable or marginalized; (iii) failure to monitor the realization of the right to water at the national level, for example by identifying right-to-water indicators and benchmarks; (iv) failure to take measures to reduce the inequitable distribution of water facilities and services; (v) failure to adopt mechanisms for emergency relief; (vi) failure to ensure that the minimum essential level of the right is enjoyed by everyone (vii) failure of a State to take into account its international legal obligations regarding the right to water when entering into agreements with other States or with international organizations.

V. Implementation at the National Level

45. In accordance with article 2, paragraph 1, of the Covenant, States parties are required to utilize "all appropriate means, including particularly the adoption of legislative measures" in the implementation of their Covenant obligations. Every State party has a margin of discretion in assessing which measures are most suitable to meet its specific circumstances. The Covenant, however, clearly imposes a duty on each State party to take whatever steps are necessary to ensure that everyone enjoys the right to water, as soon as possible. Any national measures designed to realize the right to water should not interfere with the enjoyment of other human rights.

Legislation, Strategies and Policies

46. Existing legislation, strategies, and policies should be reviewed to ensure that they are compatible with obligations arising from the right to water, and should be repealed, amended or changed if inconsistent with Covenant requirements.

---

27. See paragraph 23 for a definition of "third parties".

47. The duty to take steps clearly imposes on States parties an obligation to adopt a national strategy or plan of action to realize the right to water. The strategy must: *(a)* be based upon human rights law and principles; *(b)* cover all aspects of the right to water and the corresponding obligations of States parties; *(c)* define clear objectives; *(d)* set targets or goals to be achieved and the time—frame for their achievement; *(e)* formulate adequate policies and corresponding benchmarks and indicators. The strategy should also establish institutional responsibility for the process; identify resources available to attain the objectives, targets and goals; allocate resources appropriately according to institutional responsibility; and establish accountability mechanisms to ensure the implementation of the strategy. When formulating and implementing their right to water national strategies, States parties should avail themselves of technical assistance and cooperation of the United Nations specialized agencies (see Part VI).

48. The formulation and implementation of national water strategies and plans of action should respect, inter alia, the principles of non-discrimination and people's participation. The right of individuals and groups to participate in decision-making processes that may affect their exercise of the right to water must be an integral part of any policy, programme or strategy concerning water. Individuals and groups should be given full and equal access to information concerning water, water services and the environment, held by public authorities or third parties.

49. The national water strategy and plan of action should also be based on the principles of accountability, transparency and independence of the judiciary, since good governance is essential to the effective implementation of all human rights, including the realization of the right to water. In order to create a favourable climate for the realization of the right, States parties should take appropriate steps to ensure that the private business sector and civil society are aware of, and consider the importance of, the right to water in pursuing their activities.

50. States parties may find it advantageous to adopt framework legislation to operationalize their right to water strategy. Such legislation should include: *(a)* targets or goals to be attained and the time-frame for their achievement; *(b)* the means by which the purpose could be achieved; *(c)* the intended collaboration with civil society, private sector and international organizations; *(d)* institutional responsibility for the process; *(e)* national mechanisms for its monitoring; and *(f)* remedies and recourse procedures.

51. Steps should be taken to ensure there is sufficient coordination between the national ministries, regional and local authorities in order to reconcile water-related policies. Where implementation of the right to water has been delegated to regional or local authorities, the State party still retains the responsibility to comply with its Covenant obligations, and therefore should ensure that these authorities have at their disposal sufficient resources to maintain and extend the necessary water services and facilities. The States parties must further ensure that such authorities do not deny access to services on a discriminatory basis.

52. States parties are obliged to monitor effectively the realization of the right to water. In monitoring progress towards the realization of the right to water, States parties should identify the factors and difficulties affecting implementation of their obligations.

Indicators and Benchmarks

53. To assist the monitoring process, right to water indicators should be identified in the national water strategies or plans of action. The indicators should be designed to monitor, at the national and international levels, the State party's obligations under articles 11 (paragraph 1) and 12. Indicators should address the different components of adequate water (such as sufficiency, safety and acceptability, affordability and physical accessibility), be disaggregated by the prohibited grounds of discrimination, and cover all persons residing in the State party's territorial jurisdiction or under their control. States parties may obtain guidance on appropriate indicators from the ongoing work of WHO, the Food and Agriculture Organization of the United Nations (FAO), the United Nations Centre for Human Settlements (Habitat), the International Labour Organization (ILO), the United Nations Children's Fund (UNICEF), the United Nations Environment Programme (UNEP), the United Nations Development Programme (UNDP) and the United Nations Commission on Human Rights.

54. Having identified appropriate right to water indicators, States parties are invited to set appropriate national benchmarks in relation to each indicator.[28] During the periodic reporting procedure, the Committee will engage in a process of "scoping" with the State party. Scoping involves the joint consideration by the State party and the Committee of the indicators and national benchmarks which will then provide the targets to be achieved during the next reporting period. In the following five years, the State party will use these national benchmarks to help monitor its implementation of the right to water. Thereafter, in the subsequent reporting process, the State party and the Committee will consider whether or not the benchmarks have been achieved, and the reasons for any difficulties that may have been encountered (see General Comment No.14, 2000, paragraph 58). Further, when setting benchmarks and preparing their reports, States parties should utilize the extensive information and advisory services of specialized agencies with regard to data collection and disaggregation.

Remedies and Accountability

55. Any persons or groups who have been denied their right to water should have access to effective judicial or other appropriate remedies at both national and international levels (see General Comment No. 9, 1998, paragraph 4, and Principle 10 of the Rio Declaration on Environment and Development).[29] The Committee notes that the right has been constitutionally entrenched by a number of States and has been subject to litigation before national courts. All victims of violations of the right to water should be entitled to adequate reparation, including restitution, compensation, satisfaction or guarantees of non-repetition. National ombudsmen, human rights commissions, and similar institutions should be permitted to address violations of the right.

---

28. See E. Riedel, "New bearings to the state reporting procedure: practical ways to operationalize economic, social and cultural rights—The example of the right to health", in S. von Schorlemer (ed.); *Praxishandbuch UNO*, 2002, pp. 345–358. The Committee notes, for example, the commitment in the 2002 World Summit on Sustainable Development Plan of Implementation "to halve, by the year 2015, the proportion of people who are unable to reach or to afford safe drinking water (as outlined in the Millennium Declaration) and the proportion of people who do not have access to basic sanitation."

29. Principle 10 of the Rio Declaration on Environment and Development, (*Report of the United Nations Conference on Environment and Development*, 1992, see footnote 5), states with respect to environmental issues that "effective access to judicial and administrative proceedings, including remedy and redress, shall be provided."

56. Before any action that interferes with an individual's right to water is carried out by the State party, or by any other third party, the relevant authorities must ensure that such actions are performed in a manner warranted by law, compatible with the Covenant, and that comprises: *(a)* opportunity for genuine consultation with those affected; *(b)* timely and full disclosure of information on the proposed measures; *(c)* reasonable notice of proposed actions; *(d)* legal recourse and remedies for those affected; and *(e)* legal assistance for obtaining legal remedies (see also General Comments No. 4, 1991, and No. 7, 1997). Where such action is based on a person's failure to pay for water their capacity to pay must be taken into account. Under no circumstances shall an individual be deprived of the minimum essential level of water.

57. The incorporation in the domestic legal order of international instruments recognizing the right to water can significantly enhance the scope and effectiveness of remedial measures and should be encouraged in all cases. Incorporation enables courts to adjudicate violations of the right to water, or at least the core obligations, by direct reference to the Covenant.

58. Judges, adjudicators and members of the legal profession should be encouraged by States parties to pay greater attention to violations of the right to water in the exercise of their functions.

59. States parties should respect, protect, facilitate and promote the work of human rights advocates and other members of civil society with a view to assisting vulnerable or marginalized groups in the realization of their right to water.

VI. Obligations of Actors Other Than States

60. United Nations agencies and other international organizations concerned with water, such as WHO, FAO, UNICEF, UNEP, UN-Habitat, ILO, UNDP, the International Fund for Agricultural Development (IFAD), as well as international organizations concerned with trade such as the World Trade Organization (WTO), should cooperate effectively with States parties, building on their respective expertise, in relation to the implementation of the right to water at the national level. The international financial institutions, notably the International Monetary Fund and The World Bank, should take into account the right to water in their lending policies, credit agreements, structural adjustment programmes and other development projects (see General Comment No. 2, 1990), so that the enjoyment of the right to water is promoted. When examining the reports of States parties and their ability to meet the obligations to realize the right to water, the Committee will consider the effects of the assistance provided by all other actors. The incorporation of human rights law and principles in the programmes and policies by international organizations will greatly facilitate implementation of the right to water. The role of the International Federation of the Red Cross and Red Crescent Societies, International Committee of the Red Cross, the Office of the United Nations High Commissioner for Refugees (UNHCR), WHO and UNICEF, as well as non-governmental organizations and other associations, is of particular importance in relation to disaster relief and humanitarian assistance in times of emergencies. Priority in the provision of aid, distribution and management of water and water facilities should be given to the most vulnerable or marginalized groups of the population.

The Water & Climate Bibliography

Peter H. Gleick and Michael Kiparsky

Introduction and Background

International scientific consensus states that climate change is expected to alter hydrologic regimes throughout the world, with impacts on water supply, water quality, and water management (Gleick and others 2000, Houghton et al. 2001). The effects of climate change on water resources will vary regionally because of differences in climate impacts, vulnerability, and adaptive capacity, requiring widely different societal and governmental responses.

With the support of the Government of the Netherlands through the Dialogue on Water and Climate, the California Energy Commission, and the California Department of Water Resources, the Pacific Institute compiled a database of global scientific references for climate and water (Kiparsky et al. 2004). This database was expanded from a comprehensive United States-centered bibliography (Chalecki and Gleick 1999), compiled as part of the National Assessment called for by the 1990 United States Global Change Research Act (Public Law 101-606). The complete database can be accessed free at www.pacinst.org/resources. The Pacific Institute intends to update the database on an ongoing basis and to keep it freely available as funding permits. As of early 2004, the database contained around 3,600 references, including journal articles, book chapters, conference proceedings, and gray literature.

References were gathered by reviewing the contents of selected journals through the end of 2002, adding articles related to both climate and water. References in relevant papers were also evaluated for further relevant titles and related research, online contributions, and suggestions by colleagues added further to the dataset (Kiparsky et al. 2004). Subject keywords were assigned to all references using a list developed at the Institute and summarized in Table WB 3.1. Regional keywords were assigned on a continental scale, based on the locale of the subject of study (see Table WB 3.2). While study sites are most often described as specific locations in titles and abstracts (e.g., Sacramento River Basin), climatic changes operate over larger spatial scales (e.g., Western United States). These keywords enable aggregation

TABLE WB 3.1 Keywords Comprising the Thesaurus Used to Analyze the Bibliography and the Higher-Order Categories to Which They Are Assigned

Climate and Hydrology	Natural Systems	Human Systems
Modeling	Agriculture	Adaptation
Remote sensing	Biodiversity	Climate change assessment
Climate feedbacks	Coastal systems	Economics
Climate variability	Desertification	Human health
Drought	Ecology/ecosystems	Hydropower
ENSO	Estuaries	Irrigation
Evapotranspiration	Fisheries	Land use
Extreme/nonlinear events	Forests	Legal issues
Flood	Groundwater	Navigation/transportation
Glaciers	Lakes	Policy
Historical climate	Permafrost	Population
Hydrogeological cycle	Plant ecology	Recreation/tourism
Hydrology	Snowmelt	Reservoirs/storage
Ice cover	Snowpack	Security/defense
Lake ice cover	Water supply	Social issues
Lake level fluctuation	Wetlands	Urban issues
Paleoclimate		Water conservation
Precipitation		Water infrastructure
Rainfall		Water management
Rivers/streams/runoff/streamflow		Water use
Sea level rise		
Snowfall		
Soil moisture		
Storms		

Note: Assignment of ambiguous keywords is discussed in the text.

TABLE WB 3.2 Regional Keywords Grouped as Proxies for Analysis of Developing and Developed Countries

Developing Regions	Developed Regions	Other Regions
Africa	Europe	Global
Asia	Latin America	Islands
China	North America	unspecified/NA
India	Canada	
Former Soviet Union	Great Lakes	
Middle East	Mexico	
	United States	
	United States East	
	United States West	
	California	
	United States-Mexico border	
	Oceania	
	Australia	

of the literature by subject, and allow for analysis of patterns on scales relevant to climatic changes and geopolitical changes.

The database contains almost entirely references in English, introducing the obvious possibility of linguistic bias accounting for regional trends. However, even internationally, English is widely recognized as the *lingua franca* of science (Garfield 1989). For example, although its authors are of diverse geographical origin, almost all of the citations in the comprehensive work of the Intergovernmental Panel on Climate Change (IPCC) are in English (Houghton et al. 2001, McCarthy et al. 2001, Metz et al. 2001).

Analysis of the Database

This database provides a resource for researchers and policymakers. It allows for analysis of a broadly defined, but specific, set of research. For example, it is possible to study the regional availability of information on how climate change may affect water resources, or the temporal trend in research by subject or locale. This kind of information may be of value to policymakers trying to set or evaluate legislation or water planning, or researchers looking for gaps and holes in important results.

The Pacific Institute's Water and Climate Bibliography reveals regional patterns in the distribution of scientific references. The most striking pattern is the relative paucity of research in developing areas of the world in general, and in specific topics such as natural and human systems. For example, the human systems category includes the keywords "adaptation" and "climate change assessment," as well as many terms that relate to "coping" with projected climatic changes. Thus, this finding may have implications for issues of equity if it reflects a real pattern of differential scientific preparation.

An explanation for these patterns may be the later start on the part of research institutions in developing countries to address these issues. Such a factor is probably worsened by the disparities in financial resources available to developing country research institutions and the higher priority given to other more pressing scientific and social issues.

Figure WB 3.1 from Kiparsky et al. (2004) shows the total regional distribution of references in the bibliography. North America has been the subject of more work in this bibliography than the sum total of all the other continental regions combined. After the many studies of global scope, Europe has the next highest number of references, followed by Asia, Africa, and Oceania. These data reflect the unsurprising conclusion that more scientific effort has been put into research into climate change and water in developed countries. Figure WB 3.2 shows the temporal distribution of references for each regional grouping. In absolute terms, more work on these issues has been carried out in developed than in developing regions, but research on these topics has been increasing rapidly over time throughout the world.

Discussion

Scientific study may be a prerequisite for preparedness for climatic change. If so, then the results of analyzing the Bibliography indicate that developing countries are behind not only in their economic ability to cope with the potential effects of climatic change on water resources, as has been pointed out elsewhere (Gleick 1989, McCarthy et al. 2001), but also in the body of knowledge that would allow the them to adapt to and prepare for changes.

In addition to its influence on adaptive capacity in the sense of preparedness, scientific research may be an important component of the information leading to local perception of vulnerability to climate change (Wallner et al. 2003). As this perception is a critical driver of local actions (Bullock and Connor 2003), the lack of science in the developing world may hinder local responses.

It is important to acknowledge that there may be "shortcuts," through which developing nations can learn from and apply principles explored in other areas. However, water management in the developing world is often fundamentally different from that in developed areas (e.g., Briscoe 1996, Briscoe 1999, Kahn and Siddique 2000, World Commission on Dams 2000) and so the transferable lessons may be limited.

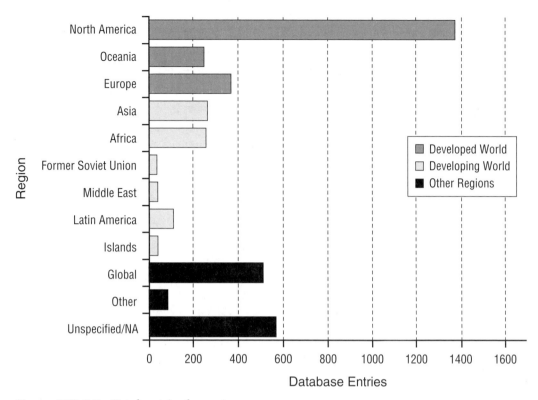

FIGURE WB 3.1 Total entries by region.

Note: Regional distribution of all references in the bibliography. The "other" category includes references to regions such as "northern hemisphere" and "tropical."

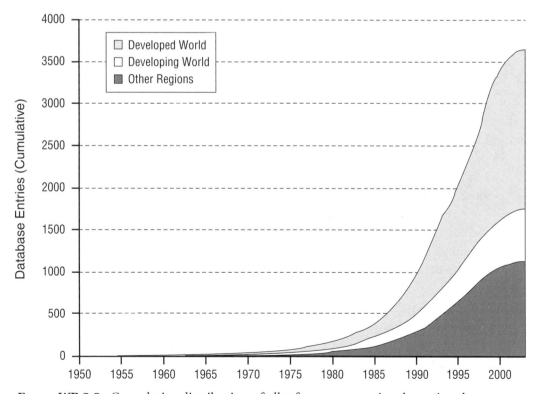

FIGURE WB 3.2 Cumulative distribution of all references over time by regional category.

Note: This figure may reflect the development of the field of climate and water regionally, and the disparity in the total bodies of knowledge on the subject between developing and developed regions of the world. "Other" regions include categories such as "tropical" or "northern hemisphere."

The patterns described above relate to larger issues of equity in the regional distribution of the impacts of climate change on societies. Regional analysis of the impacts of climate change acknowledges that adaptive capacity is unevenly distributed globally. Researchers have long observed that developing countries are expected to suffer more adverse impacts from, and to be more vulnerable to, climate change than developed countries (Gleick 1989, McCarthy et al. 2001, Baer 2002). It is important to acknowledge that scientific information and adaptive capacity are each unevenly distributed *within* individual countries (Kates 2000), although the information that can be culled from analysis of this Bibliography does not allow addressing these disparities.

The work of the IPCC emphasizes that human and natural capital are primary resources for adaptive capacity in a society. While it notes that information is a component of adaptive capacity, it does not explicitly address science as a factor. Kiparsky et al. (2004) argue that not only is science an important ingredient in integrating climate change to water management, but that, in some cases, lack of sufficient knowledge may potentially limit the ability to adapt. In addition to the limited capacity for adaptive response by developing countries, lack of science could hinder work by international development agencies. McCarthy et al. (2001) note that inclusion of climatic risks in international development initiatives could help mitigate disproportionate climatic impacts. However, without region-specific projections of impacts on institutions targeted for such initiatives, it will be difficult to motivate such efforts, and they may not be as effective. Lack of science to drive either legislative or development initiatives could skew adaptation options in developing countries toward cultural and behavioral ones and away from legislative or technical ones.

Conclusion

The development and ongoing maintenance of the Water and Climate Bibliography may prove a useful tool for both researchers and policymakers. By collecting these citations in one place, it is possible to see more clearly the large and growing efforts in this area, to lay to rest arguments that too little research is available to begin identifying and implementing policies, and to begin analyzing the scope and state of our knowledge. We expect further improvements in this Bibliography as time and resources permit.

An initial meta-analysis of the data in the Bibliography (Kiparsky et al. 2004) suggests that scientific effort on this subject has not been evenly distributed between developed and developing regions of the world or among the types of research done. More work has been carried out in all categories of research in the developed world. Overall, more analysis has been done on climate and hydrology than on natural and human systems, which are likely to be the subjects of most relevance to adaptation of human society to climatic changes. The disparity between these types of work is even greater in developing regions of the world.

IPCC and others have defined adaptive capacity as primarily related to the degree of economic resources available in a particular country. However, we argue that scientific study of the potential effects of climatic changes may be an important and underemphasized aspect of preparedness for adaptation. In addition, the disparity in the areas of closer direct relevance to human society may imply that the difference in knowledge needed to adapt to future changes in climate and water is more skewed than has previously been noted.

If the regional disparity in scientific effort reflected in the Pacific Institute database reflects real trends, then there may be significant implications for developing countries.

Efforts to increase basic scientific research on the effects of climate change on water resources in developing countries could represent significant steps toward reducing the disparity in the ultimate consequences we all face.

REFERENCES

Baer, P. 2002. Equity, greenhouse gas emissions and global common resources. In *Climate change policy: A survey*. Niles, J., editor. Washington, D.C.:Island Press.

Briscoe, J. 1996. Financing water and sanitation services: The old and new challenges. *Water Supply*. 14, no. 3/4:1–17.

Briscoe, J. 1999. The financing of hydropower, irrigation and water supply infrastructure in developing countries. *Water Resources Development*. 15, no. 5:459–491.

Bullock, A., and Connor, R. 2003. Coping options. In *Climate changes the water rules: How water managers can cope with today's climate variability and tomorrow's climate change*, Appleton, B., editor. 60–83. Dialogue on Water and Climate. Delft, Netherlands.

Chalecki, E. L., and Gleick, P. H. 1999. A framework of ordered climate effects on water resources: A comprehensive bibliography. *Journal of the American Water Resources Association*. Vol. 35, no. 6:1657–1665.

Garfield, E. 1989. The English language: The lingua franca of international science. *The Scientist*. Vol. 3, no. 10:12–13.

Gleick, P. H. 1989. Climate change and international politics: Problems facing developing countries. *Ambio*. 18, no. 6:333–339.

Gleick, P. H., and others. 2000. *Water: The Potential Consequences of Climate Variability and Change*. A Report of the National Water Assessment Group, United States Global Change Research Program, United States Geological Survey, United States Department of the Interior and the Pacific Institute for Studies in Development, Environment, and Security. Oakland, California.

Houghton, J. T., Ding, Y., Griggs, D. J., Noguer, M., van der Linden, P. J., and Xiaosu, D., editors. 2001. *Climate change 2001: The scientific basis*. Contribution of Working Group I to the Third Assessment Report of the Intergovernmental Panel on Climate Change (IPCC). Cambridge, United Kingdom:Cambridge University Press.

Kahn, H. R., and Siddique, Q. I. 2000. Urban water management problems in developing countries with particular reference to Bangladesh. *Water Resources Development*. 16, no. 1:21–33.

Kates, R. W. 2000. Cautionary tales: Adaptations and the global poor. *Climatic Change*. 45, no. 1:5–17.

Kiparsky, M., Brooks, C., and Gleick, P. H. 2004. Adaptive capacity and regional disparities in climate and water research. Submitted to *Climatic Change*.

McCarthy, J. J., Canziani, O. F., Leary, N. A., Dokken, D. J., and White, K. S., editors. 2001. *Climate change 2001: Impacts, adaptation & vulnerability*. Contribution of Working Group II to the Third Assessment Report of the Intergovernmental Panel on Climate Change (IPCC). Cambridge, United Kingdom:Cambridge University Press.

Metz, B., Davidson, O., Swart, R., and Pan, J., editors. 2001. *Climate change 2001: Mitigation*. Contribution of Working Group III to the Third Assessment Report of the Intergovernmental Panel on Climate Change (IPCC). Cambridge, United Kingdom:Cambridge University Press.

Wallner, A., Hunziker, M., and Kienast, F. 2003. Do natural science experiments influence public attitudes towards environmental problems? *Global Environmental Change*. 13, no. 3:185–194.

World Commission on Dams. 2000. *Dams and development: A new framework for decisionmaking*. London:Earthscan Publications.

Environment & Security: Water Conflict Chronology Version 2004-2005

Peter H. Gleick

Beginning with the first volume of *The World's Water*, the Environment and Security: Water Conflict Chronology has been a regular and popular feature. It is available online at www.worldwater.org, and new additions are often sent by readers. Increasingly, it is being used in both the popular literature and in research into water and conflict trends and types.

As these pages regularly note, water resources have rarely been the sole source of violent conflict or war, and indeed, water is far more often a source of cooperation. But this fact has led some international security "experts" to ignore or belittle the complex and real relationships between water and security. It is easy to draw a narrow definition of "security" in a way that excludes water (or other resources) from the debate over international security, or to require that security threats be narrow, single-issue factors. Such an approach both misunderstands the connections between water and security and misleads policymakers and the public seeking ways of reducing tensions and violence. In fact, there is a long and highly informative history of conflicts and tensions over water resources, the use of water systems as weapons during war, and the targeting of water systems during conflicts caused by other factors.

The Pacific Institute has been doing research into the connections between water resources, water systems, and international security and conflict since its inception in 1987. In October 2001, the Institute held a workshop together with Oregon State University and the Sandia National Laboratories to evaluate lessons from both arms control negotiations and international water negotiations. A report on this meeting, Fire and Water: An Examination of the Technologies, Institutions, and Social Issues in Arms Control and Transboundary Water-Resources Agreements was published in 2002 by the Pacific Institute (http://www. pacinst.org/reports/fire_and_water.pdf) and in Volume 8 of the Woodrow Wilson Center in its *Environmental Change and Security Project* Report, Volume 8.

One significant change to this volume's chronology is the addition of a set of events from the myths, legends, and history of the Middle East beginning 5,000 years before the present day. This ancient Middle East list was originally developed and published in Hatami and Gleick (1994) but was not previously included in the larger chronology. Also added to this list is a series of disputes over early canals and aqueducts.

Unfortunately, world events continue to expand the modern list in places like Iraq, Afghanistan, Nepal, India, and even the United States, with entries representing both home-grown and foreign terrorism. Al-Qaida even issued a threat in 2003 against United States drinking water systems, according to a Saudi Arabian magazine, and terrorists in Baghdad launched direct attacks on the city's water supply pipelines after the United States-led war. Whether or not the United States-led invasion of Iraq itself involved water as a tool or weapon or target of war is, as always, a matter of opinion and perspective. I have included it here because of specific events reported by the United Nations.

The updated chronology is presented here, with new entries and a range of corrections and modifications to the older ones. The current categories, or types of conflict, include:

- Control of Water Resources (state and non-state actors): where water supplies or access to water is at the root of tensions.

- Military Tool (state actors): where water resources, or water systems themselves, are used by a nation or state as a weapon during a military action.

- Political Tool (state and non-state actors): where water resources, or water systems themselves, are used by a nation, state, or non-state actor for a political goal.

- Terrorism (non-state actors): where water resources, or water systems, are either targets or tools of violence or coercion by non-state actors. This is a new category.

- Military Target (state actors): where water resource systems are targets of military actions by nations or states.

- Development Disputes (state and non-state actors): where water resources or water systems are a major source of contention and dispute in the context of economic and social development.

These definitions remain imprecise and single events can fall into more than one category, depending on perception and definition. For example, intentional military attacks on water-supply systems can fall into both the Targets and Tools categories, depending on one's point of view. Disputes over control of water resources may reflect either political power disputes or disagreements over approaches to economic development, or both. I believe these differences in perception and definition are inevitable and even desirable—international security is not a clean, precise field of study and analysis. No doubt there is a dissertation to be written here. The field of environment and security continues to change as international and regional politics evolve and as new factors become increasingly, or decreasingly, important in the affairs of humanity. In all this, however, one factor remains constant: the importance of water to life means that providing for water needs and demands will never be free of politics. As social and political systems change and evolve, this chronology and the kinds of entries and categories will change and evolve. I continue to look forward to contributions and comments from readers. Please email contributions with full citations and supporting information to pgleick@pipeline.com.

WATER BRIEF 4

WATER CONFLICT CHRONOLOGY

Peter H. Gleick

Pacific Institute for Studies in Development, Environment, and Security

Date	Parties Involved	Basis of Conflict	Violent Conflict or in the Context of Violence?	Description	Source
3000 BC	Ea, Noah	Religious account	Yes	Ancient Sumerian legend recounts the deeds of the deity Ea, who punished humanity for its sins by inflicting the Earth with a six-day storm. The Sumerian myth parallels the Biblical account of Noah and the deluge, although some details differ.	Hatami and Gleick 1994
2500 BC	Lagash, Umma	Military tool	Yes	Lagash-Umma Border Dispute—The dispute over the "Gu'edena" (edge of paradise) region begins. Urlama, King of Lagash from 2450 to 2400 BC, diverts water from this region to boundary canals, drying up boundary ditches to deprive Umma of water. His son Il cuts off the water supply to Girsu, a city in Umma.	Hatami and Gleick 1994
1790 BC	Hammurabi	Political tool	No	Code of Hammurabi for the State of Sumer—Hammurabi lists several laws pertaining to irrigation that provide for possible negligence of irrigation systems and water theft.	Hatami and Gleick 1994
1720–1684 BC	Abi-Eshuh, Iluma-Ilum	Military tool	Yes	Abi-Eshuh v. Iluma-Ilum—A grandson of Hammurabi, Abish or Abi-Eshuh, dams the Tigris to prevent the retreat of rebels lead by Iluma-Ilum, who declared the independence of Babylon. This failed attempt marks the decline of the Sumerians who had reached their apex under Hammurabi.	Hatami and Gleick 1994
circa 1300BC	Sisra, Barak, God	Religious account, Military tool	Yes	This is an Old Testament account of the defeat of Sisera and his "nine hundred chariots of iron" by the unmounted army of Barak on the fabled Plains of Esdraelon. God sends heavy rainfall in the mountains, and the Kishon River overflows the plain and immobilizes or destroys Sisera's technologically superior forces ("...the earth trembled, and the heavens dropped, and the clouds also dropped water," Judges 5:4; "...The river of Kishon swept them away, that ancient river, the river Kishon," Judges 5:21).	New Scofield Reference Bible, KJV; Judges 4:7–15 and Judges 5:4–22
1200 BC	Moses, Egypt	Military tool, Religious account	Yes	Parting of the Red Sea—When Moses and the retreating Jews find themselves trapped between the Pharoah's army and the Red Sea, Moses miraculously parts the waters of the Red Sea, allowing his followers to escape. The waters close behind them and cut off the Egyptians.	Hatami and Gleick 1994

Date	Parties	Use type	Conflict	Description	Source
720–705 BC	Assyria, Armenia	Military tool	Yes	After a successful campaign against the Halidians of Armenia, Sargon II of Assyria destroys their intricate irrigation network and floods their land.	Hatami and Gleick 1994
705–682 BC	Sennacherib, Babylon	Military weapon /target	Yes	In quelling rebellious Assyrians in 695 BC, Sennacherib razes Babylon and diverts one of the principal irrigation canals so that its waters wash over the ruins.	Hatami and Gleick 1994
Unknown	Sennacherib, Jerusalem	Military tool	Yes	As recounted in Chronicles 32.3, Hezekiah digs into a well outside the walls of Jerusalem and uses a conduit to bring in water. Preparing for a possible siege by Sennacherib, he cuts off water supplies outside of the city walls, and Jerusalem survives the attack.	Hatami and Gleick 1994
681–699 BC	Assyria, Tyre	Military tool, Religious account	Yes	Esarhaddon, an Assyrian, refers to an earlier period when gods, angered by insolent mortals, created destructive floods. According to inscriptions recorded during his reign, Esarhaddon besieges Tyre, cutting off food and water.	Hatami and Gleick 1994
669–626 BC	Assyria, Arabia, Elam	Military tool and target	Yes	Assurbanipal's inscriptions also refer to a siege against Tyre, although scholars attribute it to Esarhaddon. In campaigns against both Arabia and Elam in 645 BC, Assurbanipal, son of Esarhaddon, dries up wells to deprive Elamite troops. He also guards wells from Arabian fugitives in an earlier Arabian war. On his return from victorious battle against Elam, Assurbanipal floods the city of Sapibel, and ally of Elam. According to inscriptions, he dams the Ulai River with the bodies of dead Elamite soldiers and deprives dead Elamite kinds of their food and water offerings.	Hatami and Gleick 1994
612 BC	Egypt, Persia, Babylon, Assyria	Military tool	Yes	A coalition of Egyptian, Median (Persian), and Babylonian forces attacks and destroys Ninevah, the capital of Assyria. Nebuchadnezzar's father, Nebopolassar, leads the Babylonians. The converging armies divert the Khosr River to create a flood, which allows them to elevate their siege engines on rafts.	Hatami and Gleick 1994
605–562 BC	Babylon	Military tool	No	Nebuchadnezzar builds immense walls around Babylon, using the Euphrates and canals as defensive moats surrounding the inner castle.	Hatami and Gleick 1994
558–528 BC	Babylon	Military tool	Yes	On his way from Sardis to defeat Nabonidus at Babylon, Cyrus faces a powerful tributary of the Tigris, probably the Diyalah. According to Herodotus' account, the river drowns his royal white horse and presents a formidable obstacle to his march. Cyrus, angered by the "insolence" of the river, halts his army and orders them to cut 360 canals to divert the river's flow. Other historians argue that Cyrus	Hatami and Gleick 1994

WATER CONFLICT CHRONOLOGY continued

Date	Parties Involved	Basis of Conflict	Violent Conflict or in the Context of Violence?	Description	Source
				needed the water to maintain his troops on their southward journey, while another asserts that the construction was an attempt to win the confidence of the locals.	
539 BC	Babylon	Military tool	Yes	According to Herodotus, Cyrus invades Babylon by diverting the Euphrates above the city and marching troops along the dry riverbed. This popular account describes a midnight attack that coincided with a Babylonian feast.	Hatami and Gleick 1994
355–323 BC	Babylon	Military tool	Yes	Returning from the razing of Persepolis, Alexander proceeds to India. After the Indian campaigns, he heads back to Babylon via the Persian Gulf and the Tigris, where he tears down defensive weirs that the Persians had constructed along the river. Arrian describes Alexander's disdain for the Persians' attempt to block navigation, which he saw as "unbecoming to men who are victorious in battle."	Hatami and Gleick 1994
1503	Florence and Pisa warring states.	Military tool	Yes	Leonardo da Vinci and Machievelli plan to divert Arno River away from Pisa during conflict between Pisa and Florence.	Honan 1996
1573–1574	Holland and Spain	Military tool	Yes	In 1573 at the beginning of the 80-year war against Spain, the Dutch flooded the land to break the siege of Spanish troops on the town, Alkmaar. The same defense was used to protect Lieden in 1574. This strategy became known as the Dutch Water Line and was used frequently for defense in later years.	Dutch Water Line 2002
1642	China, Ming Dynasty	Military tool	Yes	The Huang He's dikes breached for military purposes. In 1642, "toward the end of the Ming dynasty (1368–1644), General Gao Mingheng used the tactic near Kaifeng in an attempt to suppress a peasant uprising."	Hillel 1991
1672	French, Dutch	Military tool	Yes	Louis XIV starts the third of the Dutch Wars in 1672, in which the French over-ran the Netherlands. In defense, the Dutch opened their dikes and flooded the country, creating a watery barrier that was virtually impenetrable.	Columbia 2000
1841	Canada	Development dispute, Terrorism	Yes	A reservoir in Ops Township, Upper Canada (now Ontario) was destroyed by neighbors who considered it a hazard to health.	Forkey 1998

Date	Parties	Basis of conflict	Violent conflict or in the context of violence	Description	Sources
1844	United States	Development dispute, Terrorism	Yes	A reservoir in Mercer County, Ohio, was destroyed by a mob that considered it a hazard to health.	Scheiber 1969
1850s	United States	Development dispute	Yes	Attack on a New Hampshire dam that impounded water for factories downstream by local residents unhappy over its effect on water levels.	Steinberg 1990
1853–1861	United States	Development dispute, Terrorism	Yes	Repeated destruction of the banks and reservoirs of the Wabash and Erie Canal in southern Indiana by mobs regarding it as a health hazard.	Fatout 1972, Fickle 1983
1870s	China	Development dispute	No	Local construction and government removal (twice) of an unauthorized dam in Hubei, China.	Rowe 1988
1870s–1881	United States	Development dispute	Yes	Recurrent friction and eventual violent conflict over water rights in the vicinity of Tularosa, New Mexico, involving villagers, ranchers, and farmers.	Rasch 1968
1887	United States	Development dispute, Terrorism	Yes	Dynamiting of a canal reservoir in Paulding County, Ohio, by a mob regarding it as a health hazard. State militia called out to restore order.	Walters 1948
1990	Canada	Development dispute	Yes	Partly successful attempt to destroy a lock on the Welland Canal in Ontario, Canada either by Fenians protesting English Policy in Ireland or by agents of Buffalo, New York grain handlers unhappy at the diversion of trade through the canal.	Styran and Taylor 2001
1908–1909	United States	Development dispute	Yes	Violence, including a murder, directed against agents of a land company that claimed title to Reelfoot Lake in northwestern Tennessee who attempted to levy charges for fish taken and threatened to drain the lake for agriculture.	Vanderwood 1969
1863	United States Civil War	Military tool	Yes	General U.S. Grant, during the Civil War campaign against Vicksburg, cut levees in the battle against the Confederates.	Grant 1885, Barry 1997
1898	Egypt, France, Britain	Military and political tool, Control of water resources	Military maneuvers	Military conflict nearly ensues between Britain and France in 1898 when a French expedition attempted to gain control of the headwaters of the White Nile. While the parties ultimately negotiated a settlement of the dispute, the incident has been characterized as having "dramatized Egypt's vulnerable dependence on the Nile, and fixed the attitude of Egyptian policy makers ever since."	Moorhead 1960
1907–1913	Owens Valley, Los Angeles, California	Political tool, Control of water resources, Terrorism, and	Yes	The Los Angeles Valley aqueduct/pipeline suffers repeated bombings in an effort to prevent diversions of water from the Owens Valley to Los Angeles.	Reisner 1986, 1993

239

Water Conflict Chronology *continued*

Date	Parties Involved	Basis of Conflict	Violent Conflict or in the Context of Violence?	Description	Source
		Development dispute			
1915	German Southwest Africa	Military tool	Yes	Union of South African troops capture Windhoek, capital of German Southwest Africa. (May). Retreating German troops poison wells—"a violation of the Hague convention."	Daniel 1995
1935	California, Arizona	Political tool, development dispute	Military maneuvers	Arizona calls out the National Guard and militia units to the border with California to protest the construction of Parker Dam and diversions from the Colorado River; dispute ultimately is settled in court.	Reisner 1986, 1993
1938	China and Japan	Military target and tool	Yes	Chiang Kai-shek orders the destruction of flood-control dikes of the Huayuankou section of the Huang He (Yellow) river to flood areas threatened by the Japanese army. West of Kaifeng, dikes are destroyed with dynamite, spilling water across the flat plain. The flood destroyed part of the invading army and its heavy equipment was mired in thick mud, though Wuhan, the headquarters of the Nationalist government, was taken in October. The waters flooded an area variously estimated as between 3,000 and 50,000 square kilometers, and killed Chinese estimated in numbers between "tens of thousands" and "one million."	Hillel 1991, Yang Lang 1989, 1994
1939–1942	Japan, China	Military target and tool	Yes	Japanese chemical and biological weapons activities reportedly include tests by "Unit 731" against military and civilian targets by lacing water wells and reservoirs with typhoid and other pathogens.	Harris 1994
1940–1945	Multiple parties	Military target	Yes	Hydroelectric dams routinely bombed as strategic targets during World War II.	Gleick 1993
1943	Britain, Germany	Military target	Yes	British Royal Air Force bombed dams on the Möhne, Sorpe, and Eder Rivers, Germany (May 16, 17). Möhne Dam breech killed 1,200, destroyed all downstream dams for 50 km. The flood that occurred after breaking the Eder dam reached a peak discharge of 8500 m³/s, which is 9 times higher than the highest flood observed. Many houses and bridges were destroyed and 68 were killed.	Kirschner 1949, Semann 1950

Date	Parties	Basis	Violent	Description	Sources
1944	Germany, Italy, Britain, United States	Military tool	Yes	German forces used waters from the Isoletta Dam (Liri River) in January and February to successfully destroy British assault forces crossing the Garigliano River (downstream of Liri River). The German Army then dammed the Rapido River, flooding a valley occupied by the American Army.	Corps of Engineers 1953
1944	Germany, Italy, Britain, United States	Military tool	Yes	German Army flooded the Pontine Marches by destroying drainage pumps to contain the Anzio beachhead established by the Allied landings in 1944. Over 40 square miles of land were flooded; a 30-mile stretch of landing beaches was rendered unusable for amphibious support forces.	Corps of Engineers 1953
1944	Germany, Allied forces	Military tool	Yes	Germans flooded the Ay River, France (July) creating a lake two meters deep and several kilometers wide, slowing an advance on Saint Lo, a German communications center in Normandy.	Corps of Engineers 1953
1944	Germany, Allied forces	Military tool	Yes	Germans flooded the Ill River Valley during the Battle of the Bulge (winter 1944–1945) creating a lake 16 kilometers long, 3 to 6 kilometers wide, and 1 to 2 meters deep, greatly delaying the American Army's advance toward the Rhine.	Corps of Engineers 1953
1947 onwards	Bangladesh, India	Development disputes, Control of water resources	No	Partition divides the Ganges River between Bangladesh and India; construction of the Farakka barrage by India, beginning in 1962, increases tension; short-term agreements settle dispute in 1977–1982, 1982–1984, and 1985–1988, and 30-year treaty is signed in 1996.	Butts 1997, Samson & Charrier 1997
1947–1960s	India, Pakistan	Development disputes, Control of water resources, Political tool	No	Partition leaves Indus basin divided between India and Pakistan; disputes over irrigation water ensue, during which India stems flow of water into irrigation canals in Pakistan; Indus Waters Agreement reached in 1960 after 12 years of World Bank-led negotiations.	Bingham et al. 1994, Wolf 1997
1948	Arabs, Israelis	Military tool	Yes	Arab forces cut off West Jerusalem's water supply in first Arab-Israeli war.	Wolf 1995, 1997
1950s	Korea, United States, others	Military target	Yes	Centralized dams on the Yalu River serving North Korea and China are attacked during Korean War.	Gleick 1993
1951	Korea, United Nations	Military tool and target	Yes	North Korea released flood waves from the Hwachon Dam damaging floating bridges operated by UN troops in the Pukhan Valley. U.S. Navy plans were then sent to destroy spillway crest gates.	Corps of Engineers 1953
1951	Israel, Jordan, Syria	Political tool, Military tool, Development disputes	Yes	Jordan makes public its plans to irrigate the Jordan Valley by tapping the Yarmouk River; Israel responds by commencing drainage of the Huleh swamps located in the demilitarized zone between Israel and Syria; border skirmishes ensue between Israel and Syria.	Wolf 1997, Samson & Charrier 1997

WATER CONFLICT CHRONOLOGY *continued*

Date	Parties Involved	Basis of Conflict	Violent Conflict or in the Context of Violence?	Description	Source
1953	Israel, Jordan, Syria	Development dispute, Military target, Political tool	Yes	Israel begins construction of its National Water Carrier to transfer water from the north of the Sea of Galilee out of the Jordan basin to the Negev Desert for irrigation. Syrian military actions along the border and international disapproval lead Israel to move its intake to the Sea of Galilee.	Samson & Charrier 1997
1958	Egypt, Sudan	Military tool, Political tool, Control of water resources	Yes	Egypt sends an unsuccessful military expedition into disputed territory amidst pending negotiations over the Nile waters, Sudanese general elections, and an Egyptian vote on Sudan-Egypt unification; Nile Water Treaty signed when pro-Egyptian government elected in Sudan.	Wolf 1997
1960s	North Vietnam, United States	Military target	Yes	Irrigation water supply systems in North Vietnam are bombed during Vietnam War. Some 661 sections of dikes damaged or destroyed.	IWTC 1967, Gleick 1993, Zemmali 1995
1962–1967	Brazil, Paraguay	Military tool, Political tool, Control of water resources	Military maneuvers	Negotiations between Brazil and Paraguay over the development of the Paraná River are interrupted by a unilateral show of military force by Brazil in 1962, which invades the area and claims control over the Guaira Falls site. Military forces were withdrawn in 1967 following an agreement for a joint commission to examine development in the region.	Murphy and Sabadell 1986
1963–1964	Ethiopia, Somalia	Development dispute, Military tool, Political tool	Yes	Creation of boundaries in 1948 leaves Somali nomads under Ethiopian rule; border skirmishes occur over disputed territory in Ogaden desert where critical water and oil resources are located; cease-fire is negotiated only after several hundred are killed.	Wolf 1997
1964	Cuba, United States	Military weapon	No	On February 6, 1964, the Cuban government ordered the water supply to the U.S. Naval Base at Guantanamo Bay cut off.	*Guantanamo Bay Gazette* 1964
1965	Zambia, Rhodesia, Great Britain	Military target	No	President Kenneth Kaunda calls on British government to send troops to Kariba Dam to protect it from possible saboteurs from Rhodesian government.	Chenje 2001

Date	Parties	Type	Violent?	Description	Sources
1965–1966	Israel, Syria	Military tool, Political tool, Control of water resources, Development dispute	Yes	Fire is exchanged over "all-Arab" plan to divert the Jordan River headwaters (Hasbani and Banias) and presumably preempt Israeli National Water Carrier; Syria halts construction of its diversion in July 1966.	Wolf 1995, 1997
1966–1972	Vietnam, United States	Military tool	Yes	United States tries cloud-seeding in Indochina to stop flow of material along Ho Chi Minh trail.	Plant 1995
1967	Israel, Syria	Military target and tool	Yes	Israel destroys the Arab diversion works on the Jordan River headwaters. During Arab-Israeli War Israel occupies Golan Heights, with Banias tributary to the Jordan; Israel occupies West Bank.	Gleick 1993, Wolf 1995, 1997, Wallenstein & Swain 1997
1969	Israel, Jordan	Military target and tool	Yes	Israel, suspicious that Jordan is overdiverting the Yarmouk, leads two raids to destroy the newly-built East Ghor Canal; secret negotiations, mediated by the United States, lead to an agreement in 1970.	Samson & Charrier 1997
1970s	Argentina, Brazil, Paraguay	Control of water resources, Development dispute	No	Brazil and Paraguay announce plans to construct a dam at Itaipu on the Paraná River, causing Argentina concern about downstream environmental repercussions and the efficacy of their own planned dam project downstream. Argentina demands to be consulted during the planning of Itaipu but Brazil refuses. An agreement is reached in 1979 that provides for the construction of both Brazil and Paraguay's dam at Itaipu and Argentina's Yacyreta dam.	Wallenstein & Swain 1997
1972	North Vietnam	Military target	Yes	United States bombs dikes in the Red River delta, rivers, and canals during massive bombing campaign.	Columbia Electronic Encyclopedia 2000
1974	Iraq, Syria	Military target and tool, Political tool, Development dispute	Military maneuvers	Iraq threatens to bomb the al-Thawra dam in Syria and massed troops along the border, alleging that the dam had reduced the flow of Euphrates River water to Iraq.	Gleick 1994

WATER CONFLICT CHRONOLOGY *continued*

Date	Parties Involved	Basis of Conflict	Violent Conflict or in the Context of Violence?	Description	Source
1975	Iraq, Syria	Development dispute, Military tool, Political tool	Military maneuvers	As upstream dams are filled during a low-flow year on the Euphrates, Iraqis claim that flow reaching its territory is "intolerable" and asks the Arab League to intervene. Syrians claim they are receiving less than half the river's normal flow and pull out of an Arab League technical committee formed to mediate the conflict. In May, Syria closes its airspace to Iraqi flights and both Syrian and Iraq reportedly transfer troops to their mutual border. Saudi Arabia successfully mediates the conflict.	Gleick 1993, 1994, Wolf 1997
1975	Angola, South Africa	Military control of water resources	Yes	South African troops move into Angola to occupy and defend the Ruacana hydropower complex, including the Gové Dam on the Kunene River. Goal is to take possession of and defend the water resources of southwestern Africa and Namibia.	Meissner 2000
1978– onwards	Egypt, Ethiopia	Development dispute, Political tool	No	Long-standing tensions over the Nile, especially the Blue Nile, originating in Ethiopia. Ethiopia's proposed construction of dams on the headwaters of the Blue Nile leads Egypt to repeatedly declare the vital importance of water. "The only matter that could take Egypt to war again is water" (Anwar Sadat 1979). "The next war in our region will be over the waters of the Nile, not politics" (Boutrous Ghali 1988).	Gleick 1991, 1994
1978–1984	Sudan	Development dispute, Military target, Terrorism	Yes	Demonstrations in Juba, Sudan in 1978 opposing the construction of the Jonglei Canal led to the deaths of two students. Construction of the Jonglei Canal in the Sudan was forcibly suspended in 1984 following a series of attacks on the construction site.	Suliman 1998; Keluel-Jang 1997
1980s	Mozambique, Rhodesia/ Zimbabwe, South Africa	Military target, Terrorism	Yes	Regular destruction of power lines from Cahora Bassa Dam during fight for independence in the region. Dam targeted by RENAMO.	Chenje 2001
1981	Iran, Iraq	Military target and tool	Yes	Iran claims to have bombed a hydroelectric facility in Kurdistan, thereby blacking out large portions of Iraq, during the Iran-Iraq War.	Gleick 1993
1980–1988	Iran, Iraq	Military tool	Yes	Iran diverts water to flood Iraqi defense positions.	Plant 1995

Year	Parties	Basis of conflict	Violent conflict	Description	Sources
1986	Lesotho, South Africa	Development goal, Access to resources	Yes	Bloodless coup by Lesotho's defense forces, with support from South Africa, lead to immediate agreement with South Africa for water from the Highlands of Lesotho, after 30 previous years of unsuccessful negotiations. There is disagreement over the degree to which water was a motivating factor for either party.	Mohamed 2001
1988	Angola, South Africa, Cuba	Military goal and target	Yes	Cuban and Angolan forces launch an attack on Calueque Dam via land and then air. Considerable damage inflicted on dam wall; power supply to dam cut. Water pipeline to Owamboland cut and destroyed.	Meissner 2000
1982	Israel, Lebanon, Syria	Military tool	Yes	Israel cuts off the water supply of Beirut during siege.	Wolf 1997
1982	Guatemala	Development dispute	Yes	177 civilians killed in Rio Negro over opposition to Chixoy hydroelectric dam.	Levy 2000
1984	United States	Terrorism	No	Members of the Rajneeshee religious cult contaminate a city water supply tank in The Dalles, Oregon, using Salmonella. A community outbreak of over 750 cases occurred in a county that normally reports fewer than five cases per year.	Clark and Deininger 2000
1986	North Korea, South Korea	Military tool	No	North Korea's announcement of its plans to build the Kumgansan hydroelectric dam on a tributary of the Han River upstream of Seoul raises concerns in South Korea that the dam could be used as a tool for ecological destruction or war.	Gleick 1993
1986	Lesotho, South Africa	Military goal, Control of water resources	Yes	South Africa supports coup in Lesotho over support for ANC and anti-apartheid, and water. New government in Lesotho then quickly signs Lesotho Highlands water agreement.	American University 2000b
1990	South Africa	Development dispute, Control of water resources	No	Pro-apartheid council cuts off water to the Wesselton township of 50,000 blacks following protests over miserable sanitation and living conditions.	Gleick 1993
1990	Iraq, Syria, Turkey	Development dispute, Military tool, Political tool	No	The flow of the Euphrates is interrupted for a month as Turkey finishes construction of the Ataturk Dam, part of the Grand Anatolia Project. Syria and Iraq protest that Turkey now has a weapon of war. In mid-1990, Turkish president Turgut Ozal threatens to restrict water flow to Syria to force it to withdraw support for Kurdish rebels operating in southern Turkey.	Gleick 1993 & 1995
1991–present	Karnataka, Tamil Nadu (India)	Development dispute, Control of water resources	Yes	Violence erupts when Karnataka rejects an Interim Order handed down by the Cauvery Waters Tribunal, set up by the Indian Supreme Court. The Tribunal was established in 1990 to settle two decades of dispute between Karnataka and Tamil Nadu over irrigation rights to the Cauvery River.	Gleick 1993, Butts 1997, American University 2000a

WATER CONFLICT CHRONOLOGY *continued*

Date	Parties Involved	Basis of Conflict	Violent Conflict or in the Context of Violence?	Description	Source
1991	Iraq, Kuwait, United States	Military target	Yes	During the Gulf War, Iraq destroys much of Kuwait's desalination capacity during retreat.	Gleick 1993
1991	Iraq, Turkey, United Nations	Military tool	Yes	Discussions are held at the United Nations about using the Ataturk Dam in Turkey to cut off flows of the Euphrates to Iraq.	Gleick 1993
1991	Iraq, Kuwait, United States	Military target	Yes	Baghdad's modern water supply and sanitation system are intentionally and unintentionally damaged by Allied coalition. In the first eight months of 1991, after Iraq's water infrastructure was damaged by the Persian Gulf War, the *New England Journal of Medicine* reported that nearly 47,000 more children than normal died in Iraq and the country's infant mortality rate doubled to 92.7 per 1,000 live births.	Gleick 1993, Barrett 2003
1992	Czechoslovakia, Hungary	Political tool, Development dispute	Military maneuvers	Hungary abrogates a 1977 treaty with Czechoslovakia concerning construction of the Gabcikovo/Nagymaros project based on environmental concerns. Slovakia continues construction unilaterally, completes the dam, and diverts the Danube into a canal inside the Slovakian republic. Massive public protest and movement of military to the border ensue; issue taken to the International Court of Justice.	Gleick 1993
1992	Bosnia, Bosnian Serbs	Military tool	Yes	The Serbian siege of Sarajevo, Bosnia and Herzegovina, includes a cutoff of all electrical power and the water feeding the city from the surrounding mountains. The lack of power cuts the two main pumping stations inside the city despite pledges from Serbian nationalist leaders to United Nations officials that they would not use their control of Sarajevo's utilities as a weapon. Bosnian Serbs take control of water valves regulating flow from wells that provide more than 80 percent of water to Sarajevo; reduced water flow to city is used to "smoke out" Bosnians.	Burns 1992, Husarska 1995
1993–present	Iraq	Military tool	No	To quell opposition to his government, Saddam Hussein reportedly poisons and drains the water supplies of southern Shiite Muslims, the Ma'dan. The European Parliament and UN Human Rights Commission deplore use of water as weapon in region.	Gleick 1993, American University 2000c

Year	Parties	Type	Violent conflict or in the context of violence	Description	Sources
1993	Yugoslavia	Military target and tool	Yes	Peruca Dam intentionally destroyed during war.	Gleick 1993
1995	Ecuador, Peru	Military and political tool	Yes	Armed skirmishes arise in part because of disagreement over the control of the headwaters of Cenepa River. Wolf argues that this is primarily a border dispute simply coinciding with location of a water resource.	Samson and Charrier 1997, Wolf 1997
1997	Singapore, Malaysia	Political tool	No	Malaysia supplies about half of Singapore's water and in 1997 threatened to cut off that supply in retribution for criticisms by Singapore of policy in Malaysia.	Zachary 1997
1998	Tajikistan	Terrorism, Political tool	Potential	On November 6, a guerrilla commander threatened to blow up a dam on the Kairakkhum channel if political demands are not met. Col. Makhmud Khudoberdyev made the threat, reported by the ITAR-Tass News Agency.	WRR 1998
1998	Angola	Military and political tool	Yes	In September 1998, fierce fighting between UNITA and Angolan government forces broke out at Gove Dam on the Kunene River for control of the installation.	Meissner 2001
1998/1994	United States	Cyber-terrorism	No	*The Washington Post* reports a 12-year old computer hacker broke into the SCADA computer system that runs Arizona's Roosevelt Dam, giving him complete control of the dam's massive floodgates. The cities of Mesa, Tempe, and Phoenix, Arizona are downstream of this dam. No damage was done. This report turns out to be incorrect. A hacker did break into the computers of an Arizona water facility, the Salt River Project in the Phoenix area. But he was 27, not 12, and the incident occurred in 1994, not 1998. And while clearly trespassing in critical areas, the hacker never could have had control of any dams—leading investigators to conclude that no lives or property were ever threatened.	Gellman 2002, Lemos 2002
1998	Democratic Republic of Congo	Military target, Terrorism	Yes	Attacks on Inga Dam during efforts to topple President Kabila. Disruption of electricity supplies from Inga Dam and water supplies to Kinshasa.	Chenje 2001, Human Rights Watch 1998
1998–2000	Eritrea and Ethiopia	Military target	Yes	Water pumping plants and pipelines in the border town of Adi Quala were destroyed during the civil war between Eritrea and Ethiopia.	ICRC 2003
1999	Lusaka, Zambia	Terrorism, Political tool	Yes	Bomb blast destroyed the main water pipeline, cutting off water for the city of Lusaka, population 3 million.	FTGWR 1999
1999	Yugoslavia	Military target	Yes	Belgrade reported that NATO planes had targeted a hydroelectric plant during the Kosovo campaign.	Reuters 1999a

247

WATER CONFLICT CHRONOLOGY *continued*

Date	Parties Involved	Basis of Conflict	Violent Conflict or in the Context of Violence?	Description	Source
1999	Bangladesh	Development dispute, Political tool	Yes	50 hurt during strikes called to protest power and water shortages. Protest led by former Prime Minister Begum Khaleda Zia over deterioration of public services and in law and order.	Ahmed 1999
1999	Yugoslavia	Military target	Yes	NATO targets utilities and shuts down water supplies in Belgrade. NATO bombs bridges on Danube, disrupting navigation.	Reuters 1999b
1999	Yugoslavia	Political tool	Yes	Yugoslavia refuses to clear war debris on Danube (downed bridges) unless financial aid for reconstruction is provided; European countries on Danube fear flooding due to winter ice dams will result. Diplomats decry environmental blackmail.	Simons 1999
1999	Kosovo	Political tool	Yes	Serbian engineers shut down water system in Pristina prior to occupation by NATO.	Reuters 1999c
1999	Angola	Terrorism, Political tool	Yes	100 bodies were found in 4 drinking water wells in central Angola.	International Herald Tribune 1999
1999	Puerto Rico, United States	Political tool	No	Protesters blocked water intake to Roosevelt Roads Navy Base in opposition to U.S. military presence and Navy's use of the Blanco River, following chronic water shortages in neighboring towns.	New York Times 1999
1999	China	Development Dispute	Yes	Around Chinese New Years, farmers from Hebei and Henan Provinces fought over limited water resources. Heavy weapons, including mortars and bombs, were used and nearly 100 villagers were injured. Houses and facilities were damaged and the total loss reached $US 1 million. Parties involved: Huanglongkou Village, Shexian County, Hebei Province and Gucheng Village, Linzhou City, Henan Province.	China Water Resources Daily 2002
1999	East Timor	Military and, political tool, Terrorism	Yes	Militia opposing East Timor independence kill pro-independence supporters and throw bodies in water well.	BBC 1999
1999	Kosovo	Terrorism, Political tool	Yes	Contamination of water supplies/wells by Serbs disposing of bodies of Kosovar Albanians in local wells.	CNN 1999

Date	Parties	Basis of Conflict	Violent	Description	Sources
1999–2000	Namibia, Botswana, Zambia	Military goal, Control of water resources	No	Sedudu/Kasikili Island, in the Zambezi/Chobe River. Dispute over border and access to water. Presented to the International Court of Justice.	ICJ 1999
2000	Ethiopia	Development dispute	Yes	One man stabbed to death during fight over clean water during famine in Ethiopia.	Sandrasagra 2000
2000	Central Asia: Kyrgyzstan, Kazakhstan, Uzbekistan	Political tool	No	Kyrgyzstan cuts off water to Kazakhstan until coal is delivered; Uzbekistan cuts off water to Kazakhstan for non-payment of debt.	Pannier 2000
2000	Hazarajat, Afghanistan	Development dispute	Yes	Violent conflicts broke out over water resources in the villages Burna Legan and Taina Legan, and in other parts of the region, as drought depleted local resources.	Cooperation Center for Afghanistan 2000
2000	India, Gujarat	Development dispute	Yes	Water riots reported in some areas of Gujarat to protest against authority's failure to arrange adequate supply of tanker water. Police are reported to have shot into a crowd at Falla village near Jamnagar, resulting in the death of 3 and injuries to 20 following protests against the diversion of water from the Kankavati dam to Jamnagar town.	FTGWR 2000
2000	Australia	Cyber-terrorism	No	In Queensland, Australia, on April 23, 2000, police arrested a man for using a computer and radio transmitter to take control of the Maroochy Shire wastewater system and release sewage into parks, rivers, and property. This is one of the first documented cases of cyber-terrorism (or perhaps just electronic vandalism) in the water industry.	Gellman 2002
2000	China	Development dispute	Yes	Civil unrest erupted over use and allocation of water from Baiyangdian Lake—the largest natural lake in northern China. Several people died in riots by villagers in July 2000 in Shandong after officials cut off water supplies. In August 2000, 6 died when officials in the southern province of Guangdong blew up a water channel to prevent a neighboring county from diverting water.	Pottinger 2000
2001	Israel, Palestine	Terrorism, Control of water resources		Palestinians destroy water supply pipelines to West Bank settlement of Yitzhar and to Kibbutz Kisufim. Agbat Jabar refugee camp near Jericho disconnected from its water supply after Palestinians looted and damaged local water pumps. Palestinians accuse Israel of destroying a water cistern, blocking water tanker deliveries, and attacking materials for a wastewater treatment project.	Israel Line 2001a,b; ENS 2001a

Date	Parties Involved	Basis of Conflict	Violent Conflict or in the Context of Violence?	Description	Source
2001	Pakistan	Development dispute, Terrorism	Yes	Civil unrest over severe water shortages caused by the long-term drought. Protests began in March and April and continued into summer. Riots, 4 bombs in Karachi (June 13), 1 death, 12 injuries, 30 arrests. Ethnic conflicts as some groups "accuse the government of favoring the populous Punjab province [over Sindh province] in water distribution."	Nadeem 2001, Soloman 2001
2001	Macedonia	Terrorism, Control of water resources	Yes	Water flow to Kumanovo (population 100,000) cut off for 12 days in conflict between ethnic Albanians and Macedonian forces. Valves of Glaznja and Lipkovo Lakes damaged.	AFP 2001, Macedonia Information Agency 2001
2001	China	Development dispute	Yes	In an act to protest destruction of fisheries from uncontrolled water pollution, fishermen in northern Jiaxing City, Zhejiang Province, dammed the canal that carries 90 million tons of industrial waste water per year for 23 days. The wastewater discharge into the neighboring Shengze Town, Jiangsu Province, killed fish, and threatened people's health.	China Ministry of Water Resources 2001
2001	Philippines	Terrorism, Political tool	No	Philippine authorities shut off water to 6 remote southern villages yesterday after residents complained of a foul smell from their taps, raising fears Muslim guerrillas had contaminated the supplies. Abu Sayyaf guerrillas, accused of links with Saudi-born militant Osami bin Laden, had threatened to poison the water supply in the mainly Christian town of Isabela on Basilan island if the military did not stop an offensive against them.	*World Environment News* 2001
2001	Afghanistan	Military target	Yes	U.S. forces bombed the hydroelectric facility at Kajaki Dam in Helmand province of Afghanistan, cutting off electricity for the city of Kandahar. The dam itself was apparently not targeted.	BBC 2001, Parry 2001
2002	Nepal	Terrorism, Political Tool	Yes	The Khumbuwan Liberation Front (KLF) blew up a hydroelectric powerhouse of 250 kilowatts in Bhojpur District January 26. The power supply to Bhojpur and adjoining areas was cut off. Estimated repair time was 6 months; repair costs were estimated at 10 million Rupees. By June 2002, Maoist rebels had destroyed more than 7 micro-hydro projects as well as an intake of a drinking water project and pipelines supplying water to Khalanga in western Nepal.	*Kathmandu Post* 2002, *FTGWR* 2002a

Year	Location	Category	Conflict?	Description	Reference
2002	Rome, Italy	Terrorism	No	Italian police arrest four Moroccans allegedly planning to contaminate the water supply system in Rome with a cyanide-based chemical, targeting buildings that included the U.S. embassy. Ties to Al-Queda were suggested.	BBC 2002
2002	Kashmir, India	Development dispute	Yes	Two people were killed and 25 others injured in Kashmir when police fired at a group of villagers clashing over water sharing. The incident took place in Garend village in a dispute over sharing water from an irrigation stream.	*The Japan Times* 2002
2002	United States	Terrorism	No	Among the items seized during the arrest of a Lebanese national who moved to the United States and became an Imam at a Islamist mosque in Seattle were papers by London-based al-Qaida recruiter including "instructions on poisoning water sources."	McDonnell and Meyer 2002
2002	Colombia	Terrorism	Yes	Colombian rebels in January damaged a gate valve in the dam that supplies most of Bogota's drinking water. Revolutionary Armed Forces of Colombia (FARC), detonated an explosive device planted on a German-made gate valve located inside a tunnel in the Chingaza Dam, which provides most of the capital city's water.	*Waterweek* 2002
2002	Karnataka, Tamil Nadu, India	Development dispute	Yes	Continuing violence over the allocation of the Cauvery River between Karnataka and Tamil Nadu. Riots, property destruction, more than 30 injuries, arrests through September and October.	*The Hindu* 2002a,b, *The Times of India* 2002a
2002	United States	Terrorism	No	Earth Liberation Front threatens the water supply for the town of Winter Park. Previously, this group claimed responsibility for the destruction of a ski lodge in Vail, Colorado that threatened lynx habitat.	Crecente 2002, Associated Press 2002
2003	Sudan	Political and military tool	Yes	The ongoing civil war in the Sudan has included violence against water resources. Villagers from around Tina said that bombings had destroyed water wells. In Khasan Basao, they alleged that water wells were poisoned.	*Toronto Daily* 2004
2003	United States	Terrorism	No	Al-Qaida threatens U.S. water systems via call to Saudi Arabian magazine. Al-Qaida does not "rule out…the poisoning of drinking water in American and Western cities."	Associated Press 2003, Waterman 2003, News Max 2003, *US Water News* 2003

WATER CONFLICT CHRONOLOGY *continued*

Date	Parties Involved	Basis of Conflict	Violent Conflict or in the Context of Violence?	Description	Source
2003	Iraq, United States, Others	Military Target	Yes	During the U.S.-led invasion of Iraq, water systems were reportedly damaged or destroyed by different parties, and major dams were military objectives of the U.S. forces. Damage directly attributable to the war includes vast segments of the water distribution system and the Baghdad water system, damaged by a missile.	UNICEF 2003, ARC 2003
2003	Iraq	Terrorism	Yes	Sabotage/bombing of main water pipeline in Baghdad. The sabotage of the water pipeline was the first such strike against Baghdad's water system, city water engineers said. It happened around 7 in the morning when a blue Volkswagen Passat stopped on an overpass near the Nidaa mosque and an explosive was fired at the six-foot-wide water main in the northern part of Baghdad, said Hayder Muhammad, the chief engineer for the city's water treatment plants.	Tierney and Worth 2003

Notes:

1. Conflicts may stem from the drive to possess or control another nation's water resources, thus making water systems and resources a political or military goal. Inequitable distribution and use of water resources, sometimes arising from a water development, may lead to development disputes, heighten the importance of water as a strategic goal, or may lead to a degradation of another's source of water. Conflicts may also arise when water systems are used as instruments of war, either as targets or tools. These distinctions are described in detail in Gleick (1993, 1998). In 2001, the Institute began including incidents involving water and terrorism. We note, however, the difficulty in defining "terrorism" (as opposed to political goal or other category) and caution users to use care with these categories.

2. Thanks to the many people who have contributed to this over time, including William Meyer who sent 9 fascinating items from the 1800s, Patrick Marsh, Hans-Juergen. Liebscher, Robert Halliday, Ma Jun, Marcus Moench, and others I've no doubt forgotten.

Sources: Agence France Press (AFP). 2001. Macedonian troops fight for water supply as president moots amnesty. *AFP,* June 8. http://www.balkanpeace.org/hed/archive/june01/hed3454.shtml

Ahmed, A. 1999. Fifty hurt in Bangladesh strike violence. *Reuters News Service,* Dhaka, April 18.

American Red Cross (ARC). 2003. Baghdad hospitals reopen but health care system strained. Mason Booth, Staff Writer, RedCross.org. April 24. http://www.redcross.org/news/in/iraq/030424baghdad.html

American University (Inventory of Conflict and the Environment ICE). 2000a. Cauvery River dispute. http://www.american.edu/projects/mandala/TED/ice/CAUVERY.HTM

American University (Inventory of Conflict and the Environment ICE). 2000b. Lesotho "watercoup." http://www.american.edu/projects/mandala/TED/ice/LESWATER.HTM

American University (Inventory of Conflict and the Environment ICE). 2000c. Marsh Arabs and Iraq. http://www.american.edu/projects/mandala/TED/ice/MARSH.HTM

Associated Press. 2002. Earth Liberation Front members threaten Colorado town's water. AP, October 15. Associated Press. 2003. Water targeted, magazine reports. AP, May 29.

Barrett, G. 2003. Iraq's bad water brings disease, alarms relief workers. *The Olympian,* Olympia Washington, Gannett News Service, June 29. http://www.theolympian.com/home/news/20030629/frontpage/39442.shtml

Barry J. M. 1997. *Rising tide: The Great Mississippi Flood of 1927 and how it changed America,* 67. New York:Simon and Schuster.

Bingham, G., Wolf, A., and Wohlegenant, T. 1994. Resolving water disputes: Conflict and cooperation in the United States, the Near East, and Asia. United States Agency for International Development (USAID). Bureau for Asia and the Near East. Washington, D.C.

BBC 1999. World: Asia-Pacific Timor atrocities unearthed. September 22. http://news.bbc.co.uk/hi/english/world/asia-pacific/newsid_455000/455030.stm

BBC 2001. U.S. "bombed Afghan power plant." http://news.bbc.co.uk/1/hi/world/south_asia/1632304.stm

BBC 2002. Cyanide attack foiled in Italy. February 20. http://news.bbc.co.uk/hi/english/world/europe/newsid_1831000/1831511.stm

Burns, J. F. 1992. Tactics of the Sarajevo siege: Cut off the power and water, *New York Times*, September 25. p. A1.

Butts, K., editor. 1997. *Environmental change and regional security.* Carlisle, PA:Asia-Pacific Center for Security Studies, Center for Strategic Leadership, United States Army War College.

Cable News Network (CNN). 1999. U.S.:Serbs destroying bodies of Kosovo victims. May 5. www.cnn.com/WORLD/europe/9905/05/kosovo.bodies

Chenje, M. 2001. Hydro-politics and the quest of the Zambezi River Basin Organization. In *International Waters in Southern Africa*, Nakayama, M., editor. Tokyo:United Nations University.

China Ministry of Water Resources. 2001.
http://shuizheng.chinawater.com.cn/ssjf/20021021/20021016087.htm (The web site of the Policy and Regulatory Department).

China Water Resources Daily 2002. Villagers fight over water resources. October 24. Citation provided by Ma Jun, personal communication.

Clark, R. M., and Deininger, R. A., 2000. Protecting the nation's critical infrastructure: The vulnerability of U.S. water supply systems. *Journal of Contingencies and Crisis Management.* 8, no. 2:73–80.

Columbia Electronic Encyclopedia. 2000. Vietnam: History. http://www.infoplease.com/ce6/world/A0861793.html

Columbia Encyclopedia. 2000. Netherlands. 6th Edition. *Columbia Encyclopedia.* http://www.bartleby.com/65/ne/Nethrlds.html

Cooperation Center for Afghanistan. 2000. The social impact of drought in Hazarajat. http://www.ccamata.com/impact.html

Corps of Engineers. 1953. Applications of hydrology in military planning and operations and subject classification index for military hydrology data. Military Hydrology Research and Development Branch, Engineering Division, United States Corps of Engineers, Department of the Army, Washington.

Crecente, B. D. 2002. ELF targets water: Group threatens eco-terror attack on Winter Park tanks. *Rocky Mountain News*, October 15.
http://www.rockymountainnews.com/drmn/state/article/0,1299,DRMN_21_1479883,00.html

Daniel, C., editor. 1995. *Chronicle of the 20th Century.* New York:Dorling Kindersley Publishing, Inc.

Drower, M. S. 1954, Water-supply, irrigation, and agriculture. *In A History of Technology*; Singer, C., Holmyard, E. J., and Hall, A. R., editors. New York:Oxford University Press.

Dutch Water Line. 2002. Information on the historical use of water in defense of Holland.http://www.xs4all.nl/~pho/Dutchwaterline/dutchwaterl.htm

ENS: Environment News Service. 2001a. Environment a weapon in the Israeli-Palestinian conflict. February 5.
http://www.ens-newswire.com/ens/feb2001/2001-02-05-01.asp

Fatout, P. 1972. *Indiana Canals,* 158–162. Purdue University Studies, West Lafayette, Indiana.

Ferguson, R. Brian. 2001. The Birth of War. *Natural History,* 122, no.6:28–35. July–August 2003.

Fickle, J. E. 1983. The "people" versus "progress": Local opposition to the construction of the Wabash and Erie Canal. *Old Northwest,* 8, no. 4:309–328.

Financial Times Global Water Report. 1999. Zambia: Water Cutoff. *FTGWR,* Issue 68, 15. March 19.

Financial Times Global Water Report. 2000. Drought in India comes as no surprise. *FTGWR,* Issue 94, p. 14. April 28.

Financial Times Global Water Report. 2002a. Maoists destroy Nepal's infrastructure. *FTGWR,* Issue 146, pgs. 4–5, May 17.

Forkey, N. S. 1998. Damming the dam: Ecology and community in Ops Township, Upper Canada. *Canadian Historical Review,* 79, no. 1:68–99.

Gellman, B. 2002. Cyber-attacks by Al Qaeda feared. *Washington Post,* A1, June 27.

Gleick, P. H. 1991. Environment and security: The clear connections. *Bulletin of the Atomic Scientists.* April:17–21.

Gleick, P. H. 1993. Water and conflict: Fresh water resources and international security. *International Security,* 18, 1:79–112.

Gleick, P. H. 1994. Water, war, and peace in the Middle East. *Environment,* 36, no. 3:6–on. Washington, D.C.:Heldref Publishers.

Gleick, P. H. 1995. Water and conflict: Critical issues. Presented to the 45th Pugwash Conference on Science and World Affairs. Hiroshima, Japan:July 23–29.

Gleick, P. H. 1998. Water and conflict. In *The world's water 1998–1999.* Washington, D.C., Island Press.

Grant, U. S. 1885. *Personal Memoirs of U.S. Grant.* New York:C. L Webster. ("On the second of February, [1863] this dam, or levee, was cut....The river being high the rush of water through the cut was so great that in a very short time the entire obstruction was washed away.... As a consequence the country was covered with water.")

Green Cross International. The conflict prevention atlas: http://www.greencrossinternational.net/GreenCrossPrograms/waterres/gcwater/report.html

Guantanamo Bay Gazette. 1964. The history of Guantanamo Bay: An online edition. http://www.gtmo.net/gazz/hisidx.htm. Chapter XXI: The 1964 water crisis.
http://www.gtmo.net/gazz/HISCHP21.HTM

Harris, S. H. 1994. Factories of death: Japanese biological warfare 1932–1945 and the American cover-up. Routledge, New York, N.Y.

Hatami, H., and Gleick, P. H. 1994. Chronology of conflict over water in the legends, myths, and history of the ancient Middle East. In Water, war, and peace in the Middle East. *Environment*, 36, no. 3:6–on. Washington:Heldref Publishers.

Hillel, D. 1991. Lash of the Dragon. *Natural History*, August, 28–37.

Hindu, The. 2002a. Ryots on the rampage in Mandya. *The Hindu, India's National Newspaper.* October 31. http://www.hinduonnet.com/thehindu/2002/10/31/stories/2002103106680100.htm

Hindu, The. 2002b. Farmers go berserk; MLA's house attacked. *The Hindu, India's National Newspaper,* October 30. http://www.hinduonnet.com/thehindu/2002/10/30/stories/2002103004870400.htm

Honan, W. H. 1996. Scholar sees Leonardo's influence on Machiavelli. *New York Times*, December 8, p. 18.

Human Rights Watch. 1998. Human rights watch condemns civilian killings by Congo rebels. http://www.hrw.org/press98/aug/congo827.htm

Husarska, A. 1995. Running dry in Sarajevo: Water fight. *The New Republic*, July 17 and 24.

International Committee of the Red Cross. 2003. Eritrea: ICRC repairs war-damaged health centre and water system. Dec 15. ICRC News No. 03/158. http://www.alertnet.org/thenews/fromthefield/107148342038.htm

International Court of Justice. 1999. International Court of Justice Press Communiqué 99/53, Kasikili Island/Sedudu Island (Botswana/Namibia). The Hague, Holland. December 13, p. 2. http://www.icj-cij.org/icjwww/ipresscom/ipress1999/ipresscom9953_ibona_19991213.htm

International Herald Tribune. 1999. 100 bodies found in well. International Herald Tribune, August 14–15, p. 4.

Israel Line. 2001a. Palestinians loot water pumping center, cutting off supply to refugee camp. *Israel Line.* (http://www.israel.org/mfa/go.asp?MFAH0dmp0), January 5. http://www.mfa.gov.il/mfa/go.asp?MFAH0iy50

Israel Line. 2001a. Palestinians vandalize Yitzhar water pipe. *Israel Line*, January 9. http://www.mfa.gov.il/mfa/go.asp?MFAH0izu0

IWCT. 1967. International War Crimes Tribunal Some Facts on Bombing of Dikes. http://www.infotrad.clara.co.uk/antiwar/warcrimes/index.html

Japan Times. 2002. Kashmir water clash. *The Japan Times*, May 27, p. 3

Kathmandu Post. 2002. KLF destroys micro hydro plant. *Kathmandu Post*, January 28. http://www.nepalnews.com.np/contents/englishdaily/ktmpost/2002/jan/jan28/index.htm

Kirschner, O. 1949. Destruction and protection of dams and levees. Military Hydrology, Research and Development Branch, United States Corps of Engineers, Department of the Army, Washington District. From Schweizerische Bauzeitung, March 14. 1949, translated by H. E. Schwarz, Washington.

Keluel-Jang, S. A. 1997. Alier and the Jonglei Canal. *Southern Sudan Bulletin*, 2, no. 3, January. www.sufo.demon.co.uk/poli007.htm

Lemos, R. 2002. Safety: Assessing the infrastructure risk. *CNETnew.com.* August 26. http://news.com.com/2009-1001_3-954780.html

Levy, K. 2000. Guatemalan dam massacre survivors seek reparations from financiers. *World Rivers Review*, 12–13.

International Rivers Network, Berkeley, California. December. Macedonia Information Agency. 2001. Humanitarian catastrophe averted in Kumanovo and Lipkovo. Republic of Macedonia Agency of Information Archive. June 18. http://www.reliefweb.int/w/rwb.nsf/0/dbd4ef105d93da4ac1256a6f005bc328?OpenDocument

McDonnell, P. J., and Meyer, J. 2002. Links to Terrorism Probed in Northwest. *Los Angeles Times*, July 13.

Meissner, R. 2000. "Hydropolitical hotspots in Southern Africa: Will there be a water war? The case of the Kunene river." In *Water wars: Enduring myth or impending reality?* Solomon, H., and Turton, A., editors. Africa Dialogue Monograph Series No. 2. 103–131, Accord, Creda Communications, KwaZulu-Natal, South Africa.

Meissner, R. 2001. Interaction and existing constraints in international river basins: The case of the Kunene River Basin. In *International Waters in Southern Africa*, Nakayama, M., editor. Tokyo:United Nations University.

Mohamed, A.E. 2001. Joint development and cooperation in international water resources: The case of the Limpopo and Orange River Basins in Southern Africa. In *International waters in Southern Africa*, Nakayama, M., editor. Tokyo:United Nations University.

Moorehead, A. 1960. *The White Nile*. England:Penguin Books.

Murphy, I. L., and Sabadell, J. E. 1986. International river basins: A policy model for conflict resolution. *Resources Policy*, 12, no. 1:133–144. United Kingdom:Butterworth and Co. Ltd.

Nadeem, A. 2001. Bombs in Karachi kill one. Associated Press, June 13. http://dailynews.yahoo.com/h/ap/20010613/wl/pakistan_strike_3.html

New York Times. 1999. Puerto Ricans protest Navy's use of water. *New York Times*, October 31, p. 30.

NewsMax. 2003. Al-Qaida threat to U.S. water supply. NewsMax Wires, May 29. http://www.newsmax.com/archives/articles/2003/5/28/202658.shtml

Pannier, B. 2000. Central Asia: Water becomes a political issue. Radio Free Europe. www.rferl.org/nca/features/2000/08/F.RU.0008031222739.html

254

Parry, R. L. 2001. UN fears "disaster" over strikes near huge dam. *The Independent*, London, November 8.

Plant, G. 1995. Water as a weapon in war. *Water and war, symposium on water in armed conflicts*, Montreux, November 21–23 1994, Geneva, ICRC.

Pottinger, M. 2000. Major Chinese lake disappearing in water crisis. *Reuters Science News*. http://us.cnn.com/2000/NATURE/12/20/china.lake.reut/

Rasch, P. J. 1968. The Tularosa Ditch War. *New Mexico Historical Review*, 43, no. 3:229–235.

Reisner, M. 1986, 1993. *Cadillac desert: The American West and its disappearing water.* New York:Penguin Books.

Reuters. 1999a. Serbs say NATO hit refugee convoys. April 14.
    http://www.uia.ac.be/u/carpent/kosovo/messages/397.html

Reuters 1999b. NATO Keeps Up Strikes But Belgrade Quiet. June 5. http://dailynews.yahoo.com/headlines/wl/story.html?s=v/nm/19990605/wl/yugoslavia_strikes_129.html

Reuters 1999c. NATO Builds Evidence Of Kosovo Atrocities. June 17. http://dailynews.yahoo.com/headlines/ts/story.html?s=v/nm/19990617/ts/yugoslavia_leadall_171.html

Rowe, W. T. 1988. Water control and the Qing political process: The Fankou Dam controversy, 1876–1883. *Modern China*, 14, no. 4:353–387.

Samson, P., and Charrier, B. 1997. International freshwater conflict: Issues and prevention strategies. Green Cross International.
    http://www.greencrossinternational.net/GreenCrossPrograms/waterres/gcwater/report.html

Sandrasagra, M. J. 2000. Development Ethiopia: Relief agencies warn of major food crisis. *Inter Press Service.* April 11.

Scheiber, H. N. 1969. *Ohio Canal Era.* 174–175. Athens, Ohio:Ohio University Press.

Semann, D. 1950. Die Kriegsbeschädigungen der Edertalspermauer, die Wiederherstellungsarbeiten und die angestellten Untersuchungen über die Standfestigkeit der Mauer. *Die Wasserwirtschaft*, 41. Jg., Nr. 1 u. 2.

Simons, M. 1999. Serbs refuse to clear bomb-littered river. *New York Times*, October 24.

Soloman, A. 2001. Policeman dies as blasts rock strike-hit Karachi. *Reuters.* June 13.
    http://dailynews.yahoo.com/h/nm/20010613/ts/pakistan_strike_dc_1.html,http://www.labline.de/indernet/partikel/karachi/bombse.htm

Steinberg, T .S. 1990. Dam-breaking in the nineteenth-century Merrimack Valley. *Journal of Social History*, 24, no. 1, 25–45.

Styran, R. M., and Taylor, R. R. 2001. *The great swivel link: Canada's Welland Canal.* Toronto:The Champlain Society.

Suliman, M. 1998. Resource access: A major cause of armed conflict in the Sudan. The case of the Nuba Mountains. Institute for African Alternatives, London, United Kingdom.
    http://srdis.ciesin.org/cases/Sudan-Paper.html

Tierney, J., and Worth, R. F. 2003. Attacks in Iraq may be signals of new tactics. *New York Times*, August 18. 1.
    Also at http://www.nytimes.com/2003/08/18/international/worldspecial/18IRAQ.html?hp

Times of India. 2002a. Cauvery row: Farmers renew stir. October 20. http://timesofindia.indiatimes.com/cms.dll/html/uncomp/articleshow?art_id=26586125

Toronto Daily. 2004. Darfur: Too many people killed for no reason. Amnesty International Index: AFR 54/008/2004, February 3.

UNICEF 2003. Iraq: Cleaning up neglected, damaged water system, clearing away garbage. News Note press release, May 27. http://www.unicef.org/media/media_6998.html

US Water News. 2003. Report suggests al-Qaida could poison U.S. water. *US Water News Online.* June.
    http://www.uswaternews.com/archives/arcquality/3repsug6.html

Vanderwood, P.J. 1969. *Night riders of Reelfoot Lake.* Memphis, Tennessee:Memphis State University Press.

Wallenstein, P., and Swain, A. 1997. International freshwater resources—Conflict or cooperation? Comprehensive Assessment of the Freshwater Resources of the World: Stockholm:Stockholm Environment Institute.

Walters, E. 1948. *Joseph Benson Foraker: Uncompromising Republican*, 44–45. Columbus, Ohio:Ohio History Press.

Waterman, S. 2003. Al-Qaida threat to U.S. water supply. United Press International (UPI), May 28.

Waterweek. 2002. Water facility attacked in Colombia. *Waterweek*, American Water Works Association. January. http://www.awwa.org/advocacy/news/020602.cfm

Wolf, A. T. 1995. *Hydropolitics along the Jordan River: Scarce water and its impact on the Arab-Israeli conflict.* Tokyo:United Nations University Press.

Wolf, A. T. 1997. Water wars and water reality: Conflict and cooperation along international waterways. NATO Advanced Research Workshop on Environmental Change, Adaptation, and Human Security. Budapest, Hungary. October 9–12.

World Environment News. 2001. Philippine rebels suspected of water poisoning. http://www.planetark.org/avantgo/dailynewsstory.cfm?newsid=12807

World Rivers Review (WRR). 1998. Dangerous Dams: Tajikistan, *World Rivers Review*, 13, no. 6:13, December.

Yang Lang. 1989/1994. High Dam: The Sword of Damocles. In *Yangtze! Yangtze!* Qing, Dai, editor. 229–240 Probe International, London:Earthscan Publications.

Zachary G. P. 1997. Water pressure: Nations scramble to defuse fights over supplies. *Wall Street Journal*, December 4, A17.

Zemmali, H. 1995. International humanitarian law and protection of water. *Water and war, symposium on water in armed conflicts*, Montreux, November 21–23, 1994, Geneva, ICRC.

Total Renewable Freshwater Supply by Country

2004 Update

Description

Average annual renewable freshwater resources are listed by country. All quantities are in cubic kilometers per year (km^3/yr). These data represent average freshwater resources in a country—actual annual renewable supply will vary from year to year. The data typically include both renewable surface water and groundwater supplies, including surface inflows from neighboring countries. The United Nations Food and Agriculture Organization (FAO) refers to this as "total natural renewable water resources" while the European Union refers to it as "total freshwater resources." Flows to other countries are not subtracted from these numbers—they represent the water made available by the natural hydrologic cycle, unconstrained by political, institutional, or economic factors. This update includes many significant changes from the last volume of *The World's Water*. New data have become available for most of the countries of Europe (see, especially, the Aquastat reference "u" below). Other changes should be noted for countries like the United States, which now includes estimates for Alaska, countries of Eastern Europe and the former Soviet Union, and parts of Africa.

Limitations

As described in each volume of *The World's Water*, these detailed country data should be viewed, and used, with caution. The data come from different sources and were estimated over different periods. Many countries do not directly measure or report internal water-resources data, so some of these entries were produced using indirect methods. In the past few years, new assessments have begun to standardize definitions and assumptions, particularly the work of the United Nations FAO and the European Union.

Not all of the annual renewable water supply is available for use by the countries to which they are credited here—some flows are committed to downstream users. For example, the Sudan is listed as having 154 km^3/yr, but treaty commitments require them to pass significant flows downstream to Egypt. Similarly, new estimates for Romania, the Slovak Republic, and parts of Central Asia (Kyrgyzstan, Tajikistan, Turkmenistan, and Uzbekistan) have been modified to include new information on external flows into these countries. Other countries such as Turkey, Syria, and France, to name just a few, also pass significant amounts of water to other users. The average annual figures hide large seasonal, inter-annual, and long-term variations. When recent sources "u" and "s" are in substantial disagreement, such as for Slovakia and Romania, both estimates are presented. The value for France includes its Caribbean territories. Estimates for the Netherlands and Latvia exclude underground flows. Finland's estimate includes flows from Russia.

Sources

Compiled by Gleick, P. H., Pacific Institute for Studies in Development, Environment, and Security. Oakland, California.

"nd": no data

a: Total natural renewable surface and groundwater. Typically includes flows from other countries. (FAO: "natural total renewable water resources." European Union: "total freshwater resources.")

b: Estimates from Belyaev, Institute of Geography, USSR.1987.

c: Estimates from FAO. 1995. *Water resources of African countries*. Food and Agriculture Organization, United Nations, Rome.

d: Estimates from WRI.1994. See this source for original data source.

g: Estimates from Goscomstat, USSR, 1989 as cited in Gleick 1993, Table A16.

i: Economic Commission for Europe. 1992. *Environmental statistical database: The Environment in Europe and North America*. United Nations, New York.

j: Estimates from FAO 1997. *Water resources of the Near East Region: A review*. Food and Agriculture Organization, United Nations, Rome.

k: Estimates from FAO 1997. *Irrigation in the countries of the Former Soviet Union in Figures*. Food and Agriculture Organization, United Nations, Rome.

l: UNFAO. 1999. *Irrigation in Asia in Figures*. Food and Agriculture Organization, United Nations, Rome.

m: Nix, H. 1995. *Water/land/life: The eternal triangle*. Water Research Foundation of Australia, Canberra, Australia.

n: UNFAO. 2000. *Irrigation in Latin America and the Caribbean*. Food and Agriculture Organization, United Nations, Rome. AQUASTAT web site. February 2004. www.fao.org

q: Bundesministerium fur Umwelt, Naturschutz, und Reaktorsicherheit. 2001. *Water resources management in Germany, part 1:Fundamentals*. Bonn. October.

r: Margat, J./OSS. 2001. Les ressources en eau des pays de l'OSS. Evaluation, utilisation et gestion. UNESCO/Observatoire du Sahara et du Sahel. (updating of 1995).

s: Estimates from FAO (2003). *Review of world water resources by country*. Food and Agriculture Organization, United Nations, Rome. (See specific references in this document for more information.)

t: United States Geological Survey revised:conterminous U.S. (2071); Alaska (980); Hawaii (18).

u: EUROSTAT, U. Wieland. 2003. *Water resources in the EU and in the candidate countries*. Statistics in Focus, Environment and Energy, European Communities.

v: Margat, J., and Vallée, D. 2000. *Blue Plan—Mediterranean vision on water, population and the environment for the 21st century*. 62 pages. Sophia Antipolis, France.

w: Geres, D. 1998. *Water resources in Croatia*. International Symposium on Water Management and Hydraulic Engineering. September 14–19, Dubrovnic, Croatia.

DATA TABLE 1   Total Renewable Freshwater Supply by Country (2004 Update)

Region and Country	Annual Renewable Water Resources[a] (km³/yr)	Year of Estimate	Source of Estimate
AFRICA			
Algeria	14.3	1997	c, j
Angola	184.0	1987	b
Benin	25.8	2001	r
Botswana	14.7	2001	r
Burkina Faso	17.5	2001	r
Burundi	3.6	1987	b
Cameroon	285.5	2003	s
Cape Verde	0.3	1990	c
Central African Republic	144.4	2003	s
Chad	43.0	1987	b
Comoros	1.2	2003	s
Congo	832.0	1987	b
Congo, Democratic Republic (formerly Zaire)	1,283	2001	r
Cote D'Ivoire	81	2001	r
Djibouti	0.3	1997	j
Egypt	86.8	1997	j
Equatorial Guinea	26	2001	r
Eritrea	6.3	2001	r
Ethiopia	110.0	1987	b
Gabon	164.0	1987	b
Gambia	8.0	1982	c
Ghana	53.2	2001	r
Guinea	226.0	1987	b
Guinea-Bissau	31.0	2003	s
Kenya	30.2	1990	c
Lesotho	5.2	1987	b
Liberia	232.0	1987	b
Libya	0.6	1997	c, j
Madagascar	337.0	1984	c
Malawi	17.3	2001	r
Mali	100.0	2001	p
Mauritania	11.4	1997	c, j
Mauritius	2.2	2001	p
Morocco	29.0	2003	s
Mozambique	216.0	1992	c
Namibia	45.5	1991	c
Niger	33.7	2003	s
Nigeria	286.2	2003	s
Reunion	5.0	1988	s
Rwanda	5.2	2003	s
Senegal	39.4	1987	b
Sierra Leone	160.0	1987	b
Somalia	15.7	1997	j
South Africa	50.0	1990	c
Sudan	154.0	1997	c, j

continues

Data Table 1   *Continued*

Region and Country	Annual Renewable Water Resources[a] (km³/yr)	Year of Estimate	Source of Estimate
AFRICA *(continued)*			
Swaziland	4.5	1987	b
Tanzania	91	2001	r
Togo	14.7	2001	r
Tunisia	4.6	2003	s
Uganda	66.0	1970	c
Zambia	105.2	2001	r
Zimbabwe	20.0	1987	b
NORTH AND CENTRAL AMERICA			
Antigua and Barbuda	0.1	2000	n
Bahamas	nd	nd	
Barbados	0.1	2003	s
Belize	18.6	2000	n
Canada	2,901.0	1980	d
Costa Rica	112.4	2000	n
Cuba	38.1	2000	n
Dominica	nd	nd	
Dominican Republic	21.0	2000	n
El Salvador	25.2	2001	r
Grenada	nd	nd	
Guatemala	111.3	2000	n
Haiti	14.0	2000	n
Honduras	95.9	2000	n
Jamaica	9.4	2000	n
Mexico	457.2	2000	n
Nicaragua	196.7	2000	n
Panama	148.0	2000	n
St. Kitts and Nevis	0.02	2000	n
Trinidad and Tobago	3.8	2000	n
United States of America	3,069.0	1985	t
SOUTH AMERICA			
Argentina	814.0	2000	n
Bolivia	622.5	2000	n
Brazil	8,233.0	2000	n
Chile	922.0	2000	n
Colombia	2,132.0	2000	n
Ecuador	432.0	2000	n
Guyana	241.0	2000	n
Paraguay	336.0	2000	n
Peru	1,913.0	2000	n
Suriname	122.0	2003	s
Uruguay	139.0	2000	n
Venezuela	1,233.2	2000	n
ASIA			
Afghanistan	65.0	1997	j
Bahrain	0.1	1997	j
Bangladesh	1,210.6	1999	l

Region and Country	Annual Renewable Water Resources[a] (km^3/yr)	Year of Estimate	Source of Estimate
ASIA *(continued)*			
Bhutan	95.0	1987	b
Brunei	8.5	1999	l
Cambodia	476.1	1999	l
China	2,829.6	1999	l
India	1,907.8	1999	l
Indonesia	2,838.0	1999	l
Iran	137.5	1997	j
Iraq	96.4	1997	j
Israel	1.7	2001	r, s
Japan	430.0	1999	l
Jordan	0.9	1997	j
Korea DPR	77.1	1999	l
Korea Rep	69.7	1999	l
Kuwait	0.02	1997	j
Laos	333.6	2003	s
Lebanon	4.8	1997	j
Malaysia	580.0	1999	l
Maldives	0.03	1999	l
Mongolia	34.8	1999	l
Myanmar	1,045.6	1999	l
Nepal	210.2	1999	l
Oman	1.0	1997	j
Pakistan	233.8	2003	p
Philippines	479.0	1999	l
Qatar	0.1	1997	j
Saudi Arabia	2.4	1997	j
Singapore	0.6	1975	d
Sri Lanka	50.0	1999	l
Syria	46.1	1997	j
Thailand	409.9	1999	l
Turkey	234.0	2003	p, r, s, u
United Arab Emirates	0.2	1997	j
Vietnam	891.2	1999	l
Yemen	4.1	1997	j
EUROPE			
Albania	41.7	2001	v
Austria	84.0	2003	u
Belgium	16.5	2003	u
Bosnia and Herzegovina	37.5	2003	s
Bulgaria	205.0	1980	d
Croatia	105.5	1998	u, w
Cyprus	0.8	2003	u
Czech Republic	16.0	2003	u
Denmark	6.1	2003	u
Estonia	21.1	2003	u
Finland	110.0	2003	u
France	191.0	2003	u

continues

DATA TABLE 1   *Continued*

Region and Country	Annual Renewable Water Resources[a] (km³/yr)	Year of Estimate	Source of Estimate
EUROPE *(continued)*			
Germany	182.0	2001	q
Greece	72.0	2003	u
Hungary	120.0	1991	i, u
Iceland	170.0	1987	d, u
Ireland	46.8	2003	u
Italy	175.0	2003	u
Luxembourg	1.6	2003	u
Macedonia	6.4	2001	v
Malta	0.02	1997	j
Netherlands	89.7	2003	u
Norway	369.0	2003	u
Poland	63.1	2003	u
Portugal	73.6	2003	u
Romania	42.3/211.90	2003	u, s
Slovak Republic	80.3/50.1	2003	u, s
Slovenia	20.9	2003	u
Spain	111.1	2003	u
Sweden	179.0	2003	u
Switzerland	53.3	2003	u
United Kingdom	160.6	2003	u
Yugoslavia	208.5	2003	s
Russia	4,498.0	1997	g, k
Armenia	10.5	1997	k
Azerbaijan	30.3	1997	k
Belarus	58.0	1997	k
Estonia	12.8	1997	k
Georgia	63.3	1997	k
Kazakhstan	109.6	1997	k
Kyrgyzstan	46.5	1997	s
Latvia	36.2	2003	u
Lithuania	24.5	2003	u
Moldova	11.7	1997	k
Tajikistan	99.7	1997	s
Turkmenistan	60.9	1997	s
Ukraine	139.5	1997	k
Uzbekistan	72.2	2003	s
OCEANIA			
Australia	398.0	1995	m
Fiji	28.6	1987	b
New Zealand	397.0	1995	m
Papua New Guinea	801.0	1987	b
Solomon Islands	44.7	1987	b

Freshwater Withdrawals by Country and Sector

2004 Update

Description

Data on water use by country and economic sector are significantly updated from the previous version of *The World's Water*. These data are among the most sought-after and the most unreliable in the water resources area. The following table presents data on total freshwater withdrawals by country in km^3/yr and m^3 per person per year, using estimated water withdrawals for the year noted and the United Nations population estimates (medium variant) by country for the year 2000. The table also gives the breakdown of that water use for the domestic, agricultural, and industrial sectors, in both percent of total water use and m^3 per person per year. The data sources are identified in the final column.

Since the last volume of this table, the United Nations Food and Agriculture Organization (FAO) and the European Union have published new information on country water use. The Aquastat database of the FAO was updated from previous FAO data by taking into account the change in Gross Domestic Product (GDP) and population growth from the reference year to the year 2000. For poorer countries, the relationship between GDP and water use was considered approximately linear; for richer countries, increases in GDP were not assumed to produce any significant increase in water use (personal communication, Jippe Hoogeveen, FAO, 2004).

The use of water varies greatly from country to country and from region to region. "Withdrawal" refers to water taken from a water source for use. It does not refer to water "consumed" in that use. The domestic sector typically includes household and municipal uses as well as commercial and governmental water use. The industrial sector includes water used for power plant cooling and industrial production. The agricultural sector includes water for irrigation and livestock.

Limitations

Extreme care should be used when applying these data—they are often the least reliable and most inconsistent of all water-resources information. They come from a wide variety of sources and are collected using different approaches with few formal standards. Consistent data collection is needed in this area, using standard methods and assumptions. As a result, this table includes data that are actually measured, estimated, modeled using different assumptions, or derived from other data. The data also come from different years, making direct comparisons difficult, though the effort of FAO to standardize water-use data for 2000 has somewhat reduced this problem. Industrial withdrawals for Panama, St. Lucia, St. Vincent, and the Grenadines are included in the domestic category.

Another major limitation of these data is that they do not include the use of rainfall in agriculture. Many countries use a significant fraction of the rain falling on their territory for agricultural production, but this water use is neither accurately measured nor reported.

Sources

a. New FAO Aquastat estimates from www.fao.org. March 2004. See text for details.

b. World Resources Institute, 1990, *World resources 1990–1991*, New York:Oxford University Press.

c. World Resources Institute, 1994, *World resources 1994–1995*, in collaboration with the United Nations Environment Programme and the United Nations Development Programme, New York: Oxford University Press.

d. Eurostat Yearbook, 1997. *Statistics of the European Union*, EC/C/6/Ser.26GT, Luxembourg.

e. UNFAO 1999. *Irrigation in Asia in figures.*Food and Agriculture Organization, United Nations, Rome.

f. Nix, H. 1995. Water/Land/Life, Water Research Foundation of Australia, Canberra.

g. UNFAO. 2000. *Irrigation in Latin America and the Caribbean*. Food and Agricultural Organization, United Nations, Rome.

h. AQUASTAT web site January 2002. www.fao.org

k. Ministry of Water Resources, China. 2001. *Water resources bulletin of China, 2000*. People's Republic of China, Beijing. September.

m. Hutson, S. S., Barber, N. L., Kenny, J. F., Linsey, K. S., Lumia, D. S., and Maupin, M. A. 2004. *Estimated use of water in the United States in 2000*. United States Geological Survey, Circular 1268. Reston, Virginia.

p. See Wieland, U. 2003. *Water use and waste water treatment in the European Union and in candidate countries*. Eurostat Statistics in Focus, Theme 8. European Communities. And Eurostat. 2004. Statistics in Focus.

http://europa.eu.int/comm/eurostat

http://europa.eu.int/comm/eurostat/newcronos/queen/display.do?screen=detail&language=en&product=THEME8&root=THEME8_copy_151979619462/yearlies_copy_1067300085946/dd_copy_251110364103/dda_copy_649289610368/dda10512_copy_729379227605

DATA TABLE 2 Freshwater Withdrawal, by Country and Sector (2004 Update)

Region and Country	Year	Total Freshwater Withdrawal (km³/yr)	Per-capita Withdrawal (m³/p/yr)	Domestic (%)	Industrial (%)	Agricultural (%)	USE Domestic (m³/p/yr)	Industrial (m³/p/yr)	Agricultural (m³/p/yr)	Source	2000 Population (millions)
AFRICA											
Algeria	2000	6.07	192	22	13	65	42	25	125	a	31.60
Angola	2000	0.34	27	22	16	61	6	4	16	a	12.80
Benin	2000	0.25	40	15	11	74	6	4	30	a	6.20
Botswana	2000	0.14	86	38	19	43	33	16	37	a	1.62
Burkina Faso	2000	0.78	65	11	0	88	7	0	57	a	12.06
Burundi	2000	0.23	33	17	1	82	6	0	27	a	6.97
Cameroon	2000	0.73	48	18	8	74	9	4	36	a	15.13
Cape Verde	2000	0.03	68	15	3	83	10	2	56	a	0.44
Central African Republic	2000	0.02	5	77	19	4	4	1	0	a	3.64
Chad	2000	0.23	32	19	1	80	6	0	25	a	7.27
Comoros	1987	0.01	14	48	5	47	7	1	7	b	0.71
Congo, Democratic Republic (formerly Zaire)	2000	0.36	7	52	16	31	4	1	2	a	51.75
Congo, Republic of	2000	0.04	13	59	30	10	8	4	1	a	2.98
Cote D'Ivoire	2000	0.93	61	23	12	65	14	7	40	a	15.14
Djibouti	2000	0.01	15	11	0	89	2	0	13	a	0.69
Egypt	2000	68.65	1,008	8	14	78	77	140	791	a	68.12
Equatorial Guinea	2000	0.11	243	83	16	1	202	39	2	a	0.45
Eritrea	2000	0.30		4	1	95				a	
Ethiopia	2000	2.65	38	1	6	93	0	2	35	a	69.99
Gabon	2000	0.13	105	48	11	40	51	12	43	a	1.24
Gambia	2000	0.03	24	22	11	67	5	3	16	a	1.24
Ghana	2000	0.52	26	37	15	48	10	4	13	a	19.93
Guinea	2000	1.52	193	8	2	90	15	4	174	a	7.86
Guinea-Bissau	2000	0.11	93	9	1	91	8	1	84	a	1.18

continues

265

266

DATA TABLE 2 Freshwater Withdrawal, by Country and Sector (2004 Update)

Region and Country	Year	Total Freshwater Withdrawal (km³/yr)	Per-capita Withdrawal (m³/p/yr)	USE Domestic (%)	Industrial (%)	Agricultural (%)	Domestic (m³/p/yr)	Industrial (m³/p/yr)	Agricultural (m³/p/yr)	Source	2000 Population (millions)
AFRICA (*continued*)											
Kenya	2000	1.58	52	30	6	64	16	3	33	a	30.34
Lesotho	2000	0.05	22	40	41	19	9	9	4	a	2.29
Liberia	2000	0.11	34	28	15	56	10	5	19	a	3.26
Libya	2000	4.81	753	8	3	89	62	23	668	a	6.39
Madagascar	2000	14.97	861	3	2	96	24	13	823	a	17.40
Malawi	2000	1.01	92	15	5	81	14	4	74	a	10.98
Mali	2000	6.93	552	1	0	99	4	1	547	a	12.56
Mauritania	2000	1.70	659	9	3	88	58	19	582	a	2.58
Mauritius	2000	0.61	517	25	14	60	132	74	312	a	1.18
Morocco	2000	12.76	440	8	2	90	37	7	396	a	28.98
Mozambique	2000	0.64	33	11	2	87	4	1	28	a	19.56
Namibia	2000	0.27	156	33	5	63	51	7	98	a	1.733
Niger	2000	2.19	203	4	1	95	9	1	193	a	10.81
Nigeria	2000	8.00	62	21	10	69	13	6	43	a	128.79
Rwanda	2000	0.08	10	48	14	39	5	1	4	a	7.67
Senegal	2000	1.59	167	6	4	90	10	6	151	a	9.50
Sierra Leone	2000	0.38	78	5	2	93	4	1	73	a	4.87
Somalia	2000	3.30	286	0	0	100	1	0	285	a	11.53
South Africa	2000	15.31	331	17	10	73	56	35	241	a	46.26
Sudan	2000	37.31	1,251	3	1	97	33	9	1,209	a	29.82
Swaziland	2000	0.83	843	3	6	92	23	48	773	a	0.98
Tanzania	2000	2.00	59	6	1	93	4	1	55	a	33.69
Togo	2000	0.17	36	45	8	47	16	3	17	a	4.68
Tunisia	2000	2.73	278	16	2	82	44	7	227	a	9.84
Uganda	2000	0.30	13	45	15	39	6	2	5	a	22.46
Zambia	2000	1.74	191	16	8	76	31	14	145	a	9.13
Zimbabwe	2000	2.61	210	10	5	86	20	10	180	a	12.42

NORTH AND CENTRAL AMERICA

Antigua and Barbuda	1990	0.005	75	60	20	20	45	15	15	g	0.07
Barbados	2000	0.08	308	33	44	23	103	136	69	a	0.26
Belize	2000	0.12	500	11	89	0	56	443	1	a	0.24
Canada	1990	43.89	1,431	20	69	12	280	982	168	d	30.68
Costa Rica	2000	2.68	706	29	17	53	208	120	377	a	3.80
Cuba	2000	8.20	732	19	12	69	139	89	504	a	11.20
Dominica	1996	0.02	239	-	-	-				h	0.07
Dominican Republic	2000	3.39	399	32	2	66	128	7	264	a	8.50
El Salvador	2000	1.27	201	25	16	59	50	32	119	a	6.32
Guatemala	2000	2.00	164	6	13	80	11	22	131	a	12.22
Haiti	2000	0.98	125	5	1	94	6	1	118	a	7.82
Honduras	2000	0.86	133	8	11	81	11	15	107	a	6.49
Jamaica	2000	0.41	158	34	17	49	54	27	77	a	2.59
Mexico	2000	78.22	791	17	5	77	137	43	610	a	98.88
Nicaragua	2000	1.30	277	14	3	83	40	7	230	a	4.69
Panama	2000	0.82	287	66	5	28	191	15	81	a	2.86
St. Lucia	1997	0.01	89	-	-	-				g	0.15
St. Vincent and the Grenadines	1995	0.01	88	-	-	-				g	0.11
Trinidad and Tobago	2000	0.30	221	67	27	6	149	60	12	a	1.34
United States of America	2000	563.00	2,026	13	46	41	257	933	836	m	277.83

SOUTH AMERICA

Argentina	2000	29.07	785	16	9	74	129	74	581	a	37.03
Bolivia	2000	1.39	167	13	3	83	22	6	139	a	8.33
Brazil	2000	59.30	350	20	18	62	71	63	216	a	169.20
Chile	2000	12.54	824	11	25	64	93	208	524	a	15.21
Colombia	2000	10.71	275	50	4	46	138	10	126	a	38.91
Ecuador	2000	16.98	1,343	12	5	82	167	71	1,104	a	12.65
Guyana	2000	1.64	1,876	2	1	97	32	17	1,828	a	0.87
Paraguay	2000	0.49	89	20	9	72	18	8	64	a	5.50
Peru	2000	20.13	784	8	10	82	66	79	640	a	25.66

continues

267

DATA TABLE 2 Freshwater Withdrawal, by Country and Sector (2004 Update)

Region and Country	Year	Total Freshwater Withdrawal (km³/yr)	Per-capita Withdrawal (m³/p/yr)	USE Domestic (%)	Industrial (%)	Agricultural (%)	Domestic (m³/p/yr)	Industrial (m³/p/yr)	Agricultural (m³/p/yr)	Source	2000 Population (millions)
SOUTH AMERICA *(continued)*											
Suriname	2000	0.67	1,482	4	3	93	67	43	1,373	a	0.45
Uruguay	2000	3.15	962	2	1	96	23	11	928	a	3.27
Venezuela	2000	8.37	346	45	7	47	158	24	164	a	24.17
ASIA											
Afghanistan	2000	23.26	909	2	0	98	16	0	893	a	25.59
Armenia	2000	2.95	806	30	4	66	241	35	529	a	3.66
Azerbaijan	2000	17.25	2,204	5	28	68	106	609	1,488	a	7.83
Bahrain	2000	0.30	485	40	4	57	192	17	276	a	0.62
Bangladesh	2000	76.39	595	3	1	96	19	4	572	a	128.31
Bhutan	2000	0.42	207	4	1	95	8	2	197	a	2.03
Brunei	1994	0.92	2,788	nd	nd	nd	8			e	0.33
Cambodia	2000	4.09	365	2	1	98	6	2	357	a	11.21
China	2000	549.76	431	7	26	68	28	111	292	k	1,276.30
Cyprus	2000	0.18	228	27	1	71	62	3	163	p	0.79
Georgia	2000	3.61	666	20	21	59	133	140	393	a	5.42
India	2000	645.84	641	8	5	86	52	35	555	a	1,006.77
Indonesia	2000	82.77	389	8	1	91	31	3	356	a	212.57
Iran	2000	72.88	954	7	2	91	65	22	867	a	76.43
Iraq	2000	42.70	1,848	3	5	92	58	85	1,704	a	23.11
Israel	2000	2.04	336	31	7	63	103	23	210	a	6.08
Japan	2000	88.43	699	20	18	62	138	125	437	a	126.43
Jordan	2000	1.02	161	21	4	75	33	7	121	a	6.33
Kazakhstan	2000	35.01	2,068	2	17	82	35	342	1,691	a	16.93
Korea Democratic People's Republic	2000	9.02	377	20	25	55	75	95	207	a	23.91
Korea Rep	2000	18.59	397	36	16	48	141	65	190	a	46.84

Country	Year									
Kuwait	2000	0.45	229	45	3	52	102	7	120	a	46.88
Kyrgyz Republic	2000	10.08	2,219	3	3	94	70	69	2,081	a	1.97
Laos	2000	2.99	525	4	6	90	22	30	473	a	5.69
Lebanon	2000	1.37	417	33	1	67	136	2	278	a	3.29
Malaysia	2000	9.02	405	17	21	62	68	85	251	a	22.30
Maldives	1987	0.003	10	98	2	0	10	0	0	e	0.29
Mongolia	2000	0.44	161	20	28	52	33	45	83	a	2.74
Myanmar	2000	33.22	673	1	1	98	8	4	661	a	49.34
Nepal	2000	10.18	418	3	1	96	12	3	403	a	24.35
Oman	2000	1.35	497	7	2	91	35	10	452	a	2.72
Pakistan	2000	169.38	1,086	2	2	96	21	22	1,043	a	156.01
Philippines	2000	28.52	380	17	9	74	63	36	281	a	75.04
Qatar	2000	0.29	484	25	3	72	122	14	348	a	0.60
Saudi Arabia	2000	17.32	800	10	1	89	78	9	712	a	21.66
Singapore	1975	0.19	53	45	51	4	24	27	2	c	3.59
Sri Lanka	2000	12.60	669	2	2	95	16	16	637	a	18.82
Syria	2000	19.95	1,237	3	2	95	41	23	1,174	a	16.13
Tajikistan	2000	11.96	1,869	4	5	92	69	87	1,713	a	6.40
Thailand	2000	87.07	1,439	2	2	95	36	35	1,368	a	60.50
Turkey	2001	39.78	605	15	11	74	90	66	449	p	65.73
Turkmenistan	2000	24.64	5,501	2	1	98	93	42	5,366	a	4.48
United Arab Emirates	2000	2.31	945	23	9	68	218	82	645	a	2.44
Uzbekistan	2000	58.33	2,332	5	2	93	111	48	2,173	a	25.02
Vietnam	2000	71.39	886	8	24	68	69	214	604	a	80.55
Yemen	2000	6.63	366	4	1	95	15	2	349	a	18.12
EUROPE										
Albania	2000	1.70	487	27	11	62	130	54	302	a	3.49
Austria	1997	3.56	429	35	64	1	151	274	4	p	8.29
Belarus	2000	2.79	271	23	46	30	64	126	82	a	10.28
Belgium	1998	7.44	725	13	85	1	96	620	9	p	10.26
Bulgaria	2002	6.59	793	3	78	19	24	621	149	p	8.31
Czech Republic	2002	1.91	187	41	57	2	76	107	4	p	10.20

continues

DATA TABLE 2 Freshwater Withdrawal, by Country and Sector (2004 Update)

Region and Country	Year	Total Freshwater Withdrawal (km³/yr)	Per-capita Withdrawal (m³/p/yr)	USE Domestic (%)	Industrial (%)	Agricultural (%)	Domestic (m³/p/yr)	Industrial (m³/p/yr)	Agricultural (m³/p/yr)	Source	2000 Population (millions)
EUROPE (*continued*)											
Denmark	2002	0.67	127	32	26	42	41	32	54	p	5.27
Estonia	2002	1.41	994	56	39	5	554	392	49	p	1.42
Finland	1999	2.33	450	14	84	3	61	377	12	p	5.18
France	2000	30.90	523	16	74	10	82	390	51	p	59.06
Germany	2001	38.00	460	12	68	20	57	312	91	p	82.69
Greece	1997	8.70	821	16	3	81	134	26	661	p	10.60
Hungary	2001	4.55	464	9	59	32	43	272	149	p	9.81
Iceland	2002	0.16	567	34	66	0	193	374	1	p	0.28
Ireland	1980	1.07	299	23	77	0	68	232	0	p	3.57
Italy	1998	42.00	734	18	37	45	134	270	331	p	57.19
Latvia	2001	0.26	108	55	33	12	60	36	13	p	2.40
Lithuania	2002	3.13	848	78	15	7	662	131	56	p	3.69
Luxembourg	1999	0.06	133	42	45	13	56	60	17	p	0.43
Malta	2000	0.02	53	74	1	25	39	0	13	a	0.38
Moldova	2000	2.31	518	9	58	33	49	298	171	p	4.46
Netherlands	2001	8.80	554	6	60	34	34	333	188	p	15.87
Norway	1996	2.40	545	23	67	10	123	364	57	p	4.41
Poland	2002	11.73	303	13	79	8	39	238	25	p	38.73
Portugal	1990	7.29	745	10	12	78	72	90	583	p	9.79
Romania	2002	7.24	322	9	34	57	28	111	183	p	22.51
Russian Federation	2000	76.69	525	19	63	18	98	333	93	a	146.20
Slovak Republic	2002	1.09	203							p	5.37
Slovenia	2001	0.30								p	
Spain	2001	38.60	970	13	19	68	130	180	660	p	39.80
Sweden	2002	2.69	302	37	54	9	111	164	27	p	8.90
				24	74	2	83	253	7	p	7.41

270

Switzerland	2001	2.54	343	12	35	52	90	261	388	a	50.80
Ukraine	2000	37.52	739	22	75	3	44	152	6	d	58.34
United Kingdom	1994	11.75	201	16	72	12	59	265	44	c	23.81
Yugoslavia,[1] former	1980	8.77	368								
OCEANIA											
Australia	1995	17.80	945	15	10	75	139	95	711	f	18.84
Fiji	2000	0.07	83	11	11	78	9	9	65	a	0.85
New Zealand	2000	2.11	561	49	9	42	272	53	236	a	3.76
Papua New Guinea	1987	0.10	21	56	43	1	12	9	0	c	4.81
Solomon Islands	1987			40	20	40	0	0	0	c	0.44

Notes: Figures may not add to totals due to independent rounding. 2000 Population numbers: medium United Nations variant.
1. Includes Bosnia and Herzegovina, Macedonia, Croatia

Deaths and DALYs from Selected Water-Related Diseases

Description

Measuring the scope of water-related diseases has always been a challenge because of the vast extent of the problem, discrepancies in reporting, the quality of health care in different parts of the world, and lack of standard indicators. In 1993, the Harvard School of Public Health in collaboration with the World Health Organization (WHO) and the World Bank began a new assessment of the "global burden of disease" (GBD). This effort introduced a new indicator—the "disability adjusted life year" (DALY)—to quantify the burden of disease. The DALY is a measure of population health that combines in a single indicator years of life lost from premature death and years of life lived with disabilities. Data Table 3 lists deaths and DALYs from selected water-related diseases, as reported by the World Health Organization for 2000. One DALY can be thought of as one lost year of "healthy" life. Current best estimates of water-related deaths appear to fall between 2 and 5 million deaths per year. Of these deaths, the vast majority is of small children struck by virulent but preventable diarrheal diseases, as shown in this table. International policy interest in such indicators is increasing, and the WHO World Health Reports now use deaths and DALYs as basic measures of well-being and health.

Limitations

Deaths and illnesses from water-related diseases are inadequately monitored and reported. A wide range of estimates of deaths is available in the public literature, ranging from 2 million to 12 million deaths per year. The current best estimate of water-related deaths from diarrheal diseases is around 2 million per year (as shown in Data Table 3), but this estimate must be qualified. First of all, huge numbers of cases of diarrheal diseases are not reported at all, suggesting that some—perhaps many—deaths may be misreported as well. Second, the WHO

International Classification of Disease system (ICD) simplifies deaths to a single cause, defined as "the disease or injury which initiated the train of morbid events leading directly to death." But it is well known that diarrhea is a contributing cause of death in many circumstances. Third, other deaths from water-related diseases are also poorly monitored in some places and for some diseases.

This table excludes mortality and DALYs associated with water-related insect vectors, such as malaria, onchocerciasis, and dengue fever. While few deaths from trachoma are reported, approximately 5.9 million cases of blindness or severe complications occur annually.

Source

World Health Organization (WHO). 2001. *World Health Report 2001—Mental health: New understanding, new hope.* Version 2 data tables on the Global Burden of Disease. Geneva. January 2004. http://www.who.int/whr2001/2001/

DATA TABLE 3   Deaths and DALYS from Selected Water-Related Diseases, 2000

	Deaths	**DALYs**
Diarrheal diseases	2,019,585	63,345,722
Childhood cluster diseases		
Poliomyelitis	1,136	188,543
Diphtheria	5,527	187,838
Tropical-cluster diseases		
Trypanosomiasis	49,129	1,570,242
Schistosomiasis	15,335	1,711,522
Trachoma	72	3,892,326
Intestinal nematode infections		
Ascariasis	4,929	1,204,384
Trichuriasis	2,393	1,661,689
Hookworm disease	3,477	1,785,539
Other Intestinal Infections	1,692	53,222
TOTAL	2,103,274	75,601,028

Notes: DALYs: The DALY is a measure of population health that combines in a single indicator years of life lost from premature death and years of life lived with disabilities. One DALY can be thought of as one lost year of "healthy" life.

This table excludes mortality and DALYs associated with water-related insect vectors, such as malaria, onchocerciasis, and dengue fever.

Trachoma: While few deaths from trachoma are reported, approximately 5.9 million cases of blindness or severe complications occur annually.

Official Development Assistance Indicators

Description

Several different measures of "official development assistance" (ODA) are presented in this table for countries of the Organization for Economic Co-operation and Development (OECD). The data in this table are either the 2002 ODA or an average of 2001–2002. The first data column shows total bilateral and multilateral ODA for 2002 (in 2001 US Dollars), followed by total ODA as a percentage of each country's gross national income (GNI). The third column shows multilateral ODA as a fraction of GNI.

Per-capita ODA is also shown for the 1991–1992 and 2001–2002 periods, all reported in normalized 2001 dollars. Over this period, 17 countries out of 22 increased per-capita aid spending. Per-capita spending decreased in 5 countries (Switzerland, Canada, Finland, Japan, and the United States). In 2001–2002, only three countries spent less per-capita than the United States on ODA—Portugal, New Zealand, and Greece, and no country contributed as small a fraction of total gross national income as the United States—0.13 percent of GNI. Denmark, Norway, Sweden, and the Netherlands contribute 7 times as much as a fraction of their GNI, and 3 to 5 times as much as the United States on a per-capita basis. The United States also gives the lowest fraction of ODA as grants (as opposed to loans) of any country of the OECD.

ODA is defined as financial flows to countries and multilateral institutions provided by official agencies, including state and local governments, or by their executive agencies, when those funds are administered with the promotion of economic development and welfare of developing countries as their main objective, and where there are grant elements of at least 25 percent.

Figure DT 4.1 shows Total ODA for 2002 in 2001 US Dollars. The United States contributes the single largest absolute amount, but as shown in Figure DT 4.2, Total 2002 ODA as Percent of GNI, the United States contributes the smallest share as a fraction of total gross national income of all OECD countries. Figure DT 4.3 shows Per-Capita Overseas Development Assistance, 2001 to 2002 and an average of $75 per person for all OECD countries.

Limitations

This table includes data submitted up to December 2003. No details are available here on specific projects funded, priorities within national governments for the types of support given, or the effectiveness of that support.

Source

All data from the OECD Development Co-operation Directorate (DCD), 2003 Statistical Annex Statistical Annex of the 2003 Development Co-operation Report, January 2004. Tables 6a, 7e. http://www.oecd.org/document/9/0,2340,en_2649_34447_1893129_1_1_1_1,00.html

DATA TABLE 4 Official Development Assistance Indicators

Country	Total ODA Bilateral and Multilateral 2002 (USD millions, 2001)	Total ODA 2002 % of GNI	Multilateral ODA % of GNI	ODA Per-capita of Donor Country USD millions, 2001)	
				1991–1992	2001–2002
Australia	916	0.26	0.06	48	92
Austria	488	0.26	0.04	28	63
Belgium	996	0.43	0.07	74	90
Canada	2,011	0.28	0.06	80	57
Denmark	1,540	0.96	0.31	226	296
Finland	434	0.35	0.10	122	79
France	5,125	0.38	0.04	116	158
Germany	4,980	0.27	0.05	77	121
Greece	253	0.21	0.03	–	21
Ireland	360	0.40	0.06	20	168
Italy	2,157	0.20	0.05	51	66
Japan	9,731	0.23	0.06	91	77
Luxembourg	139	0.77	0.10	98	316
Netherlands	3,068	0.81	0.18	161	390
New Zealand	110	0.22	0.06	26	29
Norway	1,517	0.89	0.27	284	316
Portugal	293	0.27	0.04	24	27
Spain	1,559	0.26	0.04	28	82
Sweden	1,848	0.83	0.22	178	197
Switzerland	863	0.32	0.08	136	122
United Kingdom	4,581	0.31	0.05	58	78
United States	13,140	0.13	0.03	55	43

Key:
GNI—Gross National Income
ODA—Overseas Development Assistance
LIC—Low-income Countries, comprise LDCs and all other countries with per-capita income (The World Bank Atlas basis) of $760 or less in 1998. Includes imputed multilateral ODA.
LDC—Least-developed Countries, are countries on the United Nations' list. Includes imputed multilateral ODA.

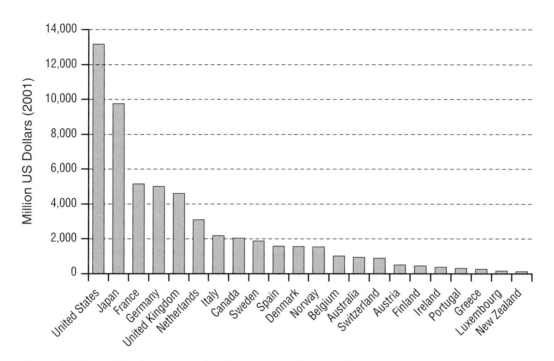

FIGURE DT 4.1 Total overseas development assistance, 2002.

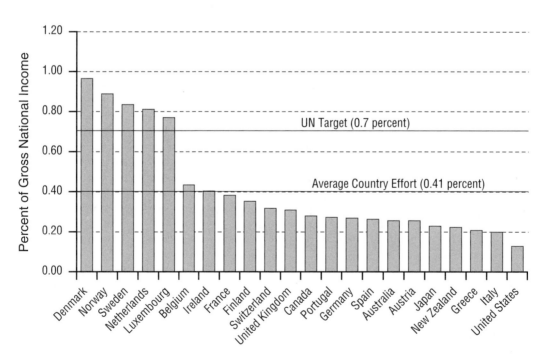

FIGURE DT 4.2 Total 2002 overseas development assistance as percent of Gross National Income.

Note: Also shown is the United Nations recommended level of 0.7 percent of Gross National Income and the average contribution in 2002.

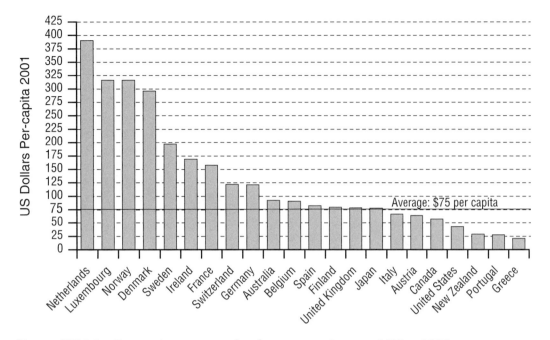

FIGURE DT 4.3   Per-capita overseas development assistance, 2001 to 2002.

Aid to Water Supply and Sanitation by Donor, 1996 to 2001

Description

Official development assistance (ODA) given by organization for economic co-operation and development (OECD) countries and by multilateral organizations is shown for 1996 to 1998 (annual average) and 1999 to 2001 (annual average) for all water supply and sanitation purposes. Data are reported in 2000 US Dollars. The OECD defines aid to water supply and sanitation as being that related to water-resource policy, planning, and programs; water legislation and management; water-resource development and protection; water supply and use; and sanitation. "Sanitation" includes solid-waste management, education and training in water supply and sanitation. The OECD definition excludes dams and reservoirs that are primarily for irrigation and hydropower. Nearly 75 percent of all ODA for water supply and sanitation from 1997 through 2001 went for large-scale water supply and sanitation systems. Japan is by far the largest donor in the sector in value terms, accounting for approximately one-third of all water aid. These data show that total ODA for water supply and sanitation has actually decreased in recent years, despite growing interest and concern about water issues and despite the new emphasis on meeting the Millennium Development Goals (see Chapter 1).

Limitations

These data relate to activities that have water supply and sanitation as their main purpose, but some data related to the water sector may be excluded if the funding is extended through programs focused on "integrated rural or urban development" or "general environmental conservation." Some aid to the water sector delivered through non-governmental organizations may also be excluded. According to the OECD, "from 1996 on the data are close to complete."

Source

Organization for Economic Co-operation and Development. 2003. Creditor Reporting System. *Aid activities in the water sector 1997–2002.* Volume 2003/1. OECD, Paris.

DATA TABLE 5 Aid to Water Supply and Sanitation by Donor, 1996 to 2001

Country / Organization	US Dollars (millions)	
	1996–1998 average	1999–2001 average
Australia	23	40
Austria	34	46
Belgium	12	13
Canada	23	22
Denmark	103	73
Finland	18	12
France	259	148
Germany	435	318
Ireland	6	7
Italy	35	29
Japan	1,442	999
Luxembourg	2	8
Netherlands	103	75
New Zealand	1	1
Norway	16	32
Portugal	0	5
Spain	23	60
Sweden	43	35
Switzerland	25	25
United Kingdom	116	165
United States	186	252
Sub-total Countries	2,906	2,368
Multi-Lateral Organization		
African Development Fund	56	64
Asian Development Bank	150	88
European Community	—	216
International Development Association	323	331
Inter-American Development Bank, Special		
Operations Fund	46	32
Sub-Total Multilateral Organizations	575	730
TOTAL Water Supply/Sanitation Aid	3,482	3,098

Bottled Water Consumption by Country, 1997 to 2002

Description

Total bottled water consumption is reported by region and country for the years from 1997 to 2002. Data for 2002 are preliminary estimates. Units are in thousand cubic meters per year. The greatest consumption occurs in the United States and Mexico. In 1997, China was the ninth largest consumer of bottled water. By 2002, consumption in China had risen to third greatest. Total global consumption exceeds 130 million cubic meters. Data for Fiji show almost no bottled-water consumption, though ironically, Fiji is the source of a premium bottled water sold in the United States. The Beverage Marketing Corporation (BMC) assigns countries to a continental region.

Limitations

Earlier data are not currently available, limiting the significance of observed trends. No distinction among types of bottled water is provided (see Chapter 2 for definitions of types of water), and such definitions may vary from country to country.

Source

Data were provided by the Beverage Marketing Corporation (BMC) to the author in 2003 and are used with permission.

DATA TABLE 6   Bottled Water Consumption by Country, 1997 to 2002

Region	Country	Year (thousand cubic meters)					
		1997	1998	1999	2000	2001	2002(P)
North America	United States	14,361.6	15,634.8	17,348.2	18,563.2	20,534.8	22,893.4
North America	Mexico	10,484.1	10,882.5	11,579.0	12,424.3	13,244.3	14,767.4
Asia	China	2,750.0	3,540.1	4,610.0	5,993.0	7,605.1	9,886.7
Europe	Italy	7,558.5	7,722.5	8,924.6	9,221.5	9,479.7	9,690.1
South America	Brazil	3,931.7	4,741.7	5,658.3	6,816.6	8,166.3	9,628.1
Europe	Germany	8,207.2	8,216.2	8,602.9	8,693.7	8,850.2	8,983.0
Europe	France	6,053.1	6,565.2	6,947.3	7,462.2	7,820.4	8,430.4
Asia	Indonesia	2,261.5	2,735.7	3,435.9	4,300.3	5,121.6	6,145.9
Asia	Thailand	3,567.1	3,842.4	4,063.7	4,286.2	4,539.0	4,837.0
Europe	Spain	3,542.6	3,716.1	3,879.7	4,003.8	4,133.9	4,294.3
Asia	India	1,047.0	1,364.2	1,681.8	2,149.7	2,667.8	3,361.4
Mideast	Saudi Arabia	1,297.7	1,490.1	1,610.0	1,769.9	1,938.0	2,116.3
Europe	Turkey	931.7	1,185.4	1,368.8	1,667.2	1,870.6	2,007.2
Europe	Poland	828.6	943.7	1,106.3	1,279.0	1,460.6	1,722.7
Asia	Japan	646.6	789.5	922.7	1,149.0	1,230.6	1,461.3
Europe	Russian Federation	524.5	610.6	790.7	967.8	1,162.4	1,406.4
Asia	Korea, Republic of	892.5	1,008.8	1,110.3	1,191.5	1,273.7	1,359.0
Europe	United Kingdom	724.1	812.8	967.5	1,071.6	1,195.9	1,339.4
Europe	Belgium-Luxembourg[1]	1,213.0	1,233.6	1,299.0	1,262.4	1,264.9	1,329.7
Asia	Philippines	727.8	837.2	999.1	1,119.0	1,213.0	1,291.8
North America	Canada	541.2	650.5	754.6	847.9	938.6	1,027.3
Europe	Czech Republic	555.1	598.1	639.0	701.5	763.2	820.5
Europe	Portugal	647.1	645.8	705.9	719.3	735.8	761.1
Europe	Romania	406.2	447.6	504.4	553.9	620.9	698.5
Europe	Switzerland	622.7	653.4	651.8	653.6	656.9	668.1
Europe	Austria	569.3	610.0	605.1	609.8	631.5	645.4
South America	Argentina	568.9	575.2	594.2	598.9	600.1	603.1
Oceania	Australia	304.2	354.9	389.6	443.3	488.0	566.5
South America	Colombia	562.8	579.0	560.0	549.0	548.4	556.7
Asia	Pakistan	69.3	108.2	157.6	242.3	360.3	547.7
Europe	Hungary	202.1	245.0	300.4	397.6	467.2	514.8
Europe	Greece	392.8	410.1	436.3	450.3	463.3	483.3
Europe	Ukraine	241.8	274.2	315.6	362.0	420.6	479.1
Mideast	Lebanon	180.7	214.8	239.8	275.3	309.7	346.2
Mideast	United Arab Emirates	229.9	245.3	256.1	269.9	285.2	326.4
Asia	China, Hong Kong SAR	191.0	222.1	245.4	271.1	298.2	331.1
Europe	Netherlands	248.8	240.7	273.3	286.1	296.1	316.8
South America	Venezuela, Republic of Bolivia	201.6	220.9	230.3	247.9	263.3	289.6
Mideast	Israel	100.2	111.7	132.6	170.3	224.9	283.4
Europe	Croatia	135.8	157.7	176.9	199.9	223.7	247.2

continues

DATA TABLE 6 *Continued*

Region	Country	Year (thousand cubic meters)					
		1997	1998	1999	2000	2001	2002(P)
Asia	Malaysia	137.5	157.9	179.6	199.1	217.9	236.8
Mideast	Egypt	133.7	145.7	167.7	188.4	208.6	234.6
Asia	Viet Nam	114.9	139.5	159.3	179.6	199.9	219.4
Europe	Slovakia	158.9	161.8	168.8	170.2	173.5	178.2
Europe	Bulgaria	53.9	67.2	88.1	112.2	142.1	177.8
Europe	Sweden	126.8	127.1	143.7	150.8	158.1	164.2
Mideast	Kuwait	68.2	77.5	95.6	112.5	128.4	144.0
Europe	Slovenia	67.9	80.9	93.2	108.5	124.3	137.9
South America	Peru	75.6	80.9	89.1	103.5	118.0	132.1
South America	Chile	66.8	80.6	88.6	95.8	103.8	113.0
Europe	Ireland	50.5	61.5	75.1	83.5	92.4	99.4
Africa	South Africa	30.9	41.1	57.9	68.8	81.2	96.3
Asia	Singapore	57.2	63.5	69.7	75.6	81.8	88.2
Europe	Norway	67.3	76.3	76.9	77.2	80.5	82.8
Europe	Denmark	71.6	72.0	71.8	72.0	72.3	74.1
Europe	Finland	45.2	51.0	58.3	62.0	65.6	68.4
South America	Paraguay	46.2	50.2	53.5	56.9	60.6	64.5
Europe	Cyprus	48.4	48.7	50.8	54.9	58.3	62.3
Mideast	Qatar	33.0	38.1	42.8	47.2	51.6	56.2
Mideast	Jordan	27.9	31.1	35.7	40.3	44.4	48.5
Mideast	Bahrain	26.7	28.4	31.4	34.1	36.7	39.5
Europe	Lithuania	15.0	17.6	20.4	23.8	28.6	34.6
Europe	Estonia	15.2	18.5	20.9	23.9	26.8	30.2
Mideast	Oman	16.1	18.2	20.3	23.2	26.2	29.5
South America	Uruguay	17.3	18.7	20.5	21.9	23.2	24.5
South America	Nicaragua	13.2	14.5	15.7	17.2	19.0	20.4
North America	Cuba	10.7	12.0	13.2	14.9	16.8	18.7
Asia	Brunei Darussalam	9.9	11.1	12.2	13.6	14.9	16.3
Oceania	Pacific Islands[2]	10.2	10.9	12.0	12.9	14.0	15.1
Europe	Latvia	2.4	3.0	3.7	4.6	5.6	6.9
Asia	Bangladesh[3]	0.0	0.0	0.0	0.0	0.0	0.0
Asia	Fiji Islands[4]	0.0	0.0	0.0	0.0	0.0	0.0
Subtotal		80,141.2	87,243.5	97,110.9	106,490.5	116,843.6	130,178.5
All Others		507.9	595.3	737.1	891.0	1,032.7	1,233.6
TOTAL		80,649.1	87,838.8	97,848.0	107,381.5	117,876.3	131,412.1

Notes:
P = Preliminary
1. Combined.
2. Includes the Caroline Islands (Micronesia excluding Palau), the Marshall Islands, and the Northern Marianas (excluding Guam).
3. Commercial bottled water essentially does not exist in Bangladesh.
4. Consumption in Fiji is virtually nil, immeasurable in terms of thousands of cubic meters.

Global Bottled Water Consumption by Region, 1997 to 2002

Description

Consumption of bottled water by continent is shown from 1997 through 2002. Data for 2002 are preliminary estimates. Units are in thousand cubic meters per year. In 2002, total consumption of bottled water was estimated at 131 million cubic meters, or 131 billion liters—more than 20 liters for everyone on the planet. Regions include Europe, Asia, North America, South America, and Africa/Oceania/Middle East. Bottled water consumption is greatest in Europe, though the rate of increase in North American and Asia is higher. The countries that are assigned to each region are shown in Data Table 6. The Beverage Marketing Corporation (BMC) assigns countries to a continental region.

Limitations

Earlier data are not currently available, limiting the significance of observed trends. No distinctions among types of bottled water are provided (see Chapter 2 for definitions of types of water), and such definitions may vary from country to country.

Source

Data were provided by the Beverage Marketing Corporation (BMC) to the author in 2003 and are used with permission.

DATA TABLE 7   Global Bottled Water Consumption, by Region, 1997 to 2002

Region	Year (thousand cubic meter)						
	1996	1997	1998	1999	2000	2001	2002(P)
Europe		34,328	36,074	39,367	41,507	43,546	45,955
North America		25,398	25,822	29,932	32,338	34,870	38,707
Asia		12,472	14,820	17,647	21,170	24,824	29,783
South America		5,484	6,362	7,310	8,508	9,903	11,432
Africa/Mideast/Oceania		2,459	2,808	3,092	3,456	3,837	3,837
Subtotal		80,141	85,885	97,348	106,978	116,979	129,713
All Others		508	1,953	500	403	898	1,699
TOTAL	72,676	80,649	87,839	97,848	107,381	117,876	131,412

Note: P = Preliminary.

Bottled Water Consumption, Share by Region, 1997 to 2002

Description

Consumption of bottled water by continental region is shown from 1997 through 2002 as a percent of total global consumption. Data for 2002 are preliminary estimates. Regions include Europe, Asia, North America, South America, and Africa/Oceania/Middle East. Europe's share of total consumption has decreased over this period, while Asia's has increased. The countries that are assigned to each region are shown in Data Table 6. The Beverage Marketing Corporation (BMC) assigns countries to a continental region.

Limitations

Earlier data are not currently available, limiting the significance of observed trends. No distinction among types of bottled water are provided (see Chapter 2 for definitions of types of water), and such definitions may vary from country to country.

Source

Data were provided by the Beverage Marketing Corporation (BMC) to the author in 2003 and are used with permission.

DATA TABLE 8   Bottled Water Consumption, Share by Region, 1997 to 2002

	Percent					
Region	**1997**	**1998**	**1999**	**2000**	**2001**	**2002(P)**
Europe	42.6	41.1	40.2	38.7	36.9	35.0
North America	31.5	29.4	30.6	30.1	29.6	29.5
Asia	15.5	16.9	18.0	19.7	21.1	22.7
South America	6.8	7.2	7.5	7.9	8.4	8.7
Africa/Mideast/Oceania	3.0	3.2	3.2	3.2	3.3	2.9
Subtotal	99.4	97.8	99.5	99.6	99.2	98.7
All Others	0.6	2.2	0.5	0.4	0.8	1.3
TOTAL	100	100	100	100	100	100

Note: P = Preliminary

Per-capita Bottled Water Consumption by Region, 1997 to 2002

Description

Per-capita consumption of bottled water by continental region is shown from 1997 through 2002. Units are liters per person per year. Global average consumption has grown from 12.6 liters per person per year in 1996 to over 20 liters per person, with great disparities in regional use. Europeans drink an average of nearly 85 liters per person per year, while average use in Africa, the Middle East, and Oceania is less than 4 liters per person per year. Data for 2002 are preliminary estimates. Regions include Europe, Asia, North America, South America, and Africa/Oceania/Middle East. The countries that are assigned to each region are shown in Data Table 6. The Beverage Marketing Corporation (BMC) assigns countries to a continental region.

Limitations

Earlier data are not currently available, limiting the significance of observed trends. No distinction among types of bottled water are provided (see Chapter 2 for definitions of types of water), and such definitions may vary from country to country.

Source

Data were provided by the Beverage Marketing Corporation (BMC) to the author in 2003 and are used with permission.

DATA TABLE 9 Per-capita Bottled Water Consumption by Region, 1997 to 2002

Region	Liters Per-capita						
	1996	**1997**	**1998**	**1999**	**2000**	**2001**	**2002(P)**
North America		59.4	59.7	68.6	73.3	78.2	84.6
Europe		47.8	50.3	54.8	57.7	60.6	64.0
South America		14.8	16.9	19.2	22.0	25.3	28.5
Asia		3.7	4.3	5.1	6.0	7.0	8.3
Africa/Mideast/Oceania		2.6	2.9	3.1	3.4	3.7	3.6
TOTAL	12.6	13.8	14.8	16.3	17.7	19.1	21.1

Note: P = Preliminary.

United States Bottled Water Sales, 1991 to 2001

Description

The United States is the largest single national market for bottled water. This table shows total bottled water sales in the United States from 1991 to 2001. Data are in million gallons per year, million liters per year, and total annual sales in US Dollars. Total sales volume and revenues have more than doubled since 1991.

Limitations

No separate information on "still" versus "sparkling" sales are provided, nor on the breakdown among spring, processed municipal, or other types of bottled water.

Source

Beverage Marketing Corporation, quoted by K. J. Ransome, 2002. A perspective on the new titans. *Water conditioning and purification*, 42–45, October.

DATA TABLE 10   United States Bottled Water Sales, 1991 to 2001

| Year | Millions Per Year | | |
	Gallons	Liters	US Dollars
1991	2,355.9	8,917	2,512.9
1992	2,486.6	9,411	2,658.7
1993	2,689.4	10,179	2,876.7
1994	2,966.4	11,227	3,164.3
1995	3,226.9	12,213	3,521.9
1996	3,495.1	13,228	3,835.4
1997	3,794.3	14,361	4,222.7
1998	4,130.7	15,634	4,666.1
1999	4,583.4	17,348	5,314.7
2000	4,904.4	18,563	5,809.0
2001	5,425.3	20,534	6,477.0

Types of Packaging Used for Bottled Water in Various Countries, 1999

Description

The type of packaging used for bottled waters is shown for a variety of countries in Europe, plus Brazil, for 1999. Data are shown in percent of sales. Bottled water is sold in a variety of types of packages, but the most common for most countries is the use of plastic or glass bottles. Most bottled water is sold in plastic bottles, though in Germany, bottled water is sold primarily in recyclable glass containers because of national recycling requirements. Plastic is not only more expensive than glass, it is more expensive than the water it contains.

The sum of the plastic and glass data do not always add to 100 percent. The difference for Belgium, Brazil, Hungary, and Switzerland is due to a small number of aluminum cans sold, mostly from vending machines. Figure DT 11.1 shows bottled water packaging in Europe, 1999. Only the fraction of plastic and glass are plotted.

Limitations

These data are for a single year—1999. Over time, the overall trend has been to shift to plastic except where environmental regulations limit this shift.

Source

UNESEM—GISEMES. 2000. Eaux minérales naturelles. Statistiques 1999. Union Européenne et Groupement International des Industries des Eaux Minérales Naturelles et des Eaux de Source.

DATA TABLE 11   Types of Packaging Used for Bottled Water in Various Countries, 1999

Country	Plastic (%)	Glass (%)	Other(%)
Brazil	99.9	–	0.1
Ireland	91.7	8.3	–
Hungary	91.8	7.8	0.4
Spain	88.4	11.6	–
Belgium	83.3	15.7	1.0
Portugal	79.6	20.4	–
Italy	75.0	25.0	–
France	70.0	30.0	–
Switzerland	63.3	36.3	0.4
Slovenia	63.0	37.0	–
Yugoslavia	54.0	46.0	–
Austria	40.0	60.0	–
Germany	3.0	97.0	–

Notes: The sum of the plastic and glass data do not always add to 100 percent. The minor difference for Belgium and Switzerland is due to a small number of aluminum cans sold, mostly from vending machines. Most bottled water sold in Germany is sold in recyclable glass containers. Plastic is not only more expensive than glass, it is more expensive than the water it contains.

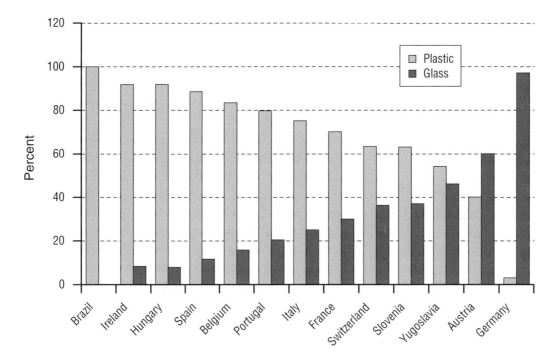

FIGURE DT 11.1   Bottled water packaging, various countries, 1999.

Note: Plastic and glass only.

Irrigated Area by Region, 1961 to 2001

Description

Total irrigated areas by continental region are listed here for 1961, 1965, 1970, 1975, 1980, 1985, 1990, 1995, 1999, 2000, and 2001—the latest year for which reliable data are available. Units are thousands of hectares. After 1990, all irrigated area in the former USSR is split between Europe and Asia. These data have been updated from *The World's Water 2002–2003* to correct misreporting and changes to the United Nations Food and Agriculture Organization (FAOSTAT) database.

Figure DT 12.1 shows total irrigated area plotted over this time period. While irrigated area has continued to increase, it is now increasing at a rate slower than population. As a result, per-capita irrigated area is declining.

Limitations

These data depend on in-country surveys, national reports, and estimates by the United Nations Food and Agriculture Organization. In some regions, multiple cropping may increase the apparent area in production. These data are not reported here. No information is offered about the quality of the land in production. Changes in political borders and the independence of several countries make certain continental time-series comparisons misleading. Data for the Soviet Union, Yugoslavia, and Czechoslovakia are provided through 1990; thereafter, the irrigated areas of the newly independent states are reported. When summing by continental area, however, trends will appear misleading because some of the newly independent states are now included in Asia, while others are in Europe. No meaningful time-series trends by continent can be seen for these areas. The time-series for Africa, North and Central America, South America, and Oceania do not suffer from this problem.

Source

United Nations Food and Agriculture Organization, 2004, FAOSTAT web site. www.fao.org

DATA TABLE 12   Irrigated Area, by Region, 1961 to 2001

Region	Thousand Hectares				
	1961	**1965**	**1970**	**1975**	**1980**
Africa	7,410	7,795	8,483	9,010	9,491
Asia	90,166	97,093	109,666	121,565	132,377
Europe	8,468	9,401	10,583	12,704	14,479
North & Central America	17,950	19,526	20,939	22,833	27,597
Oceania	1,079	1,368	1,588	1,620	1,684
South America	4,661	5,070	5,673	6,403	7,392
USSR	9,400	9,900	11,100	14,500	17,200
World	139,134	150,153	168,032	188,635	210,220

Region	**1985**	**1990**	**1995**	**2000**	**2001**
Africa	10,331	11,235	12,383	12,700	12,813
Asia	141,922	155,009	180,508	190,014	190,385
Europe	16,018	17,414	26,104	25,382	25,347
North & Central America	27,471	28,913	30,473	31,223	31,344
Oceania	1,957	2,114	2,689	2,674	2,674
South America	8,296	9,499	10,086	10,489	10,489
USSR	19,689	20,800			
World	225,684	244,984	262,243	272,482	273,052

Note: After 1990, all irrigated area in the former USSR is split among Europe and Asia.

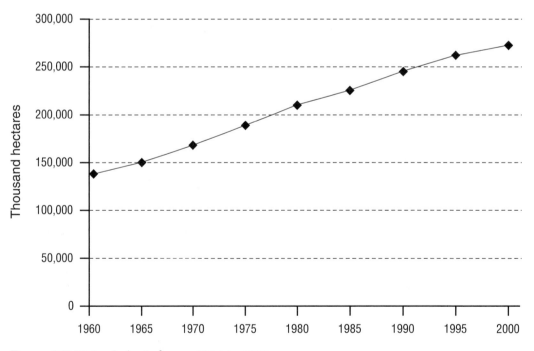

FIGURE DT 12.1   Irrigated area, 1961 to 2001.

Irrigated Area, Developed & Developing Countries, 1961 to 2001

Description

Total irrigated areas by the United Nations Food and Agricultural Organization (FAO) "developed" and "developing" country categories are listed here for 1961, 1965, 1970, 1975, 1980, 1985, 1990, 1995, 2000, and 2001—the latest year for which reliable data are available. Also shown are the annual percentage changes for developed countries, developing countries, and the world. Units of area are thousands of hectares; units for annual changes are percentages. From 1965 to 2000, annual rates are calculated by dividing 5-year differences by 5. While total irrigated area worldwide continues to increase, the average annual rate of change continues to drop. In the developed world, total irrigated area has effectively remained the same for nearly a decade. Overall, the rate of increase in irrigated area worldwide has fallen well below one percent per year, which means that total per-capita irrigated area is actually decreasing.

Figure DT 13.1 shows the area of irrigated agriculture in developed and developing countries every 5 years from 1965 to 2000.

Limitations

These data depend on in-country surveys, national reports, and estimates by the United Nations Food and Agriculture Organization. In some regions, multiple cropping may increase the apparent area in production. These data are not reported here. No differentiation is made about the quality of the land in production. Recent changes in political borders and the independence of several countries make certain time-series comparisons difficult or inappropriate.

Source

United Nations Food and Agriculture Organization, 2004. www.fao.org

DATA TABLE 13   Irrigated Area, Developed and Developing Countries, 1961 to 2001

Region	Thousand Hectare				
	1961	**1965**	**1970**	**1975**	**1980**
Developed Countries[1]	37,180	40,232	44,278	50,381	58,926
Developing Countries	101,954	109,921	123,754	138,254	151,294
World	139,134	150,153	168,032	188,635	210,220
	1985	**1990**	**1995**	**2000**	**2001**
Developed Countries[1]	62,555	66,286	67,964	67,938	67,988
Developing Countries	163,129	178,698	194,279	204,544	205,064
World	225,684	244,984	262,243	272,482	273,052

Region	Annual Rate of Change				
	1961	**1965**	**1970**	**1975**	**1980**
Developed Countries		1.64	2.01	2.76	3.39
Developing Countries		1.56	2.52	2.34	1.89
World		1.58	2.38	2.45	2.29
	1985	**1990**	**1995**	**2000**	**2001**
Developed Countries	1.23	1.19	0.51	−0.01	0.07
Developing Countries	1.56	1.91	1.74	1.06	0.25
World	1.47	1.71	1.41	0.78	0.21

Note:

1. After 1990, all irrigated area in the former USSR is split among Europe and Asia. From 1965 to 2000, annual rates are calculated by dividing 5-year differences by 5.

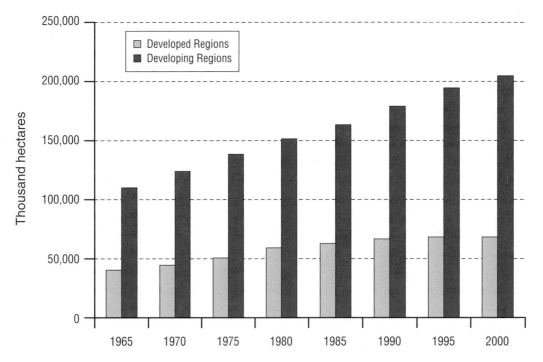

FIGURE DT 13.1   Irrigated area, developed and developing countries, 1961 to 2000.

Global Production and Yields of Major Cereal Crops, 1961 to 2002

Description

As populations and demands for food grow, competition for limited water resources will also grow. This challenge will be met, in part, by improving the efficiency of water use in agriculture, as measured in a variety of ways. This table shows estimates of total global production and yields of major "cereal" crops for every year from 1961 to 2002. "Cereals" include: wheat, rice, corn, barley, oats, millet, rye, buckwheat, quinoa, and others. Units of production are metric tons per year (mt/yr). Units of yield are hectograms per hectare (hg/ha). One hectogram is 100 grams or 0.1 kilograms. Annual percentage changes are also listed. Annual yields for maize, wheat, and rice are also listed separately from 1961 to 2002 in hg/ha.

Figure DT 14.1 shows World Cereal Production, 1961 to 2002 with units of mt/yr. Figure DT 14.2 shows Yields of Maize, Rice, and Wheat, 1961 to 2002.

Limitations

These data come from the United Nations Food and Agriculture Organization (FAOSTAT) data sets and represent data reported by countries. National data depend on in-country surveys and reports, which vary in quality and consistency. Dramatic changes in yield from year to year primarily reflect climatic factors. Over time, better estimates that incorporate water productivity will provide information that is more useful for farmers and water policymakers.

Source

FAOSTAT 2003. http://apps.fao.org/page/collections?subset=agriculture

DATA TABLE 14   Global Production and Yields of Major Cereal Crops, 1961 to 2002

	Total Production mt/Year[1]	Total Yield (Hg/Ha)[2]	Percent Change	Yield (Hg/Ha)[2]		
				Maize	Rice	Wheat
1961	877,026,930	13,532		19,435	18,671	10,889
1962	933,594,928	14,276	5.50	19,808	18,946	12,060
1963	949,458,601	14,400	0.87	20,319	20,548	11,320
1964	1,001,419,730	14,939	3.74	19,961	21,005	12,413
1965	998,775,982	14,966	0.18	21,252	20,329	12,151
1966	1,078,539,510	16,102	7.59	22,096	20,756	14,079
1967	1,124,280,750	16,534	2.68	24,266	21,725	13,392
1968	1,160,884,040	16,964	2.60	22,927	22,303	14,532
1969	1,171,247,790	17,084	0.71	24,226	22,505	14,170
1970	1,192,666,140	17,650	3.31	23,519	23,771	14,939
1971	1,299,920,180	18,919	7.19	26,545	23,584	16,245
1972	1,258,704,660	18,579	-1.80	26,877	23,209	16,048
1973	1,357,231,410	19,402	4.43	27,239	24,483	16,837
1974	1,326,626,710	18,920	-2.48	25,573	24,211	16,154
1975	1,359,985,390	19,121	1.06	28,134	25,146	15,700
1976	1,464,044,420	20,255	5.93	28,380	24,474	17,910
1977	1,456,337,340	20,246	-0.04	29,676	25,678	16,718
1978	1,582,107,620	22,123	9.27	31,566	26,810	19,328
1979	1,537,608,910	21,758	-1.65	33,861	26,561	18,521
1980	1,550,154,170	21,605	-0.70	31,545	27,434	18,554
1981	1,632,692,980	22,463	3.97	34,943	28,224	18,800
1982	1,692,476,220	23,686	5.44	36,101	29,752	19,991
1983	1,626,767,170	23,078	-2.57	29,461	31,304	21,258
1984	1,786,613,300	24,982	8.25	35,261	32,199	22,200
1985	1,820,998,400	25,283	1.20	37,206	32,520	21,718
1986	1,833,605,370	25,559	1.09	36,283	32,405	23,207
1987	1,771,093,640	25,394	-0.65	34,876	32,627	22,900
1988	1,726,968,460	24,596	-3.14	31,012	33,288	22,921
1989	1,870,890,560	26,289	6.88	36,195	34,535	23,733
1990	1,951,587,500	27,547	4.79	36,789	35,259	25,617
1991	1,889,407,070	26,838	-2.57	36,842	35,364	24,488
1992	1,973,254,370	27,827	3.69	38,937	35,869	25,409
1993	1,903,016,860	27,378	-1.61	36,238	36,318	25,321
1994	1,956,755,810	28,133	2.76	41,136	36,618	24,500
1995	1,897,023,470	27,605	-1.88	37,901	36,602	25,077
1996	2,072,162,370	29,428	6.60	42,129	37,898	25,795
1997	2,094,871,530	29,932	1.71	41,400	38,211	27,106
1998	2,084,092,930	30,622	2.31	44,273	38,222	26,957
1999	2,086,012,720	31,182	1.83	43,756	39,874	27,550
2000	2,060,836,280	30,649	-1.71	42,845	39,166	27,193
2001	2,107,953,880	31,141	1.61	44,176	39,528	27,485
2002	2,031,540,320	30,831	-1.00	43,428	39,164	27,202

Notes:
1. Metric tons.
2. Hectograms (100 grams) per hectare.

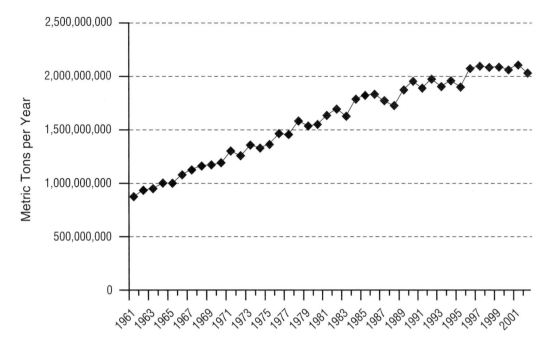

Figure DT 14.1 World cereal production, 1961 to 2002.

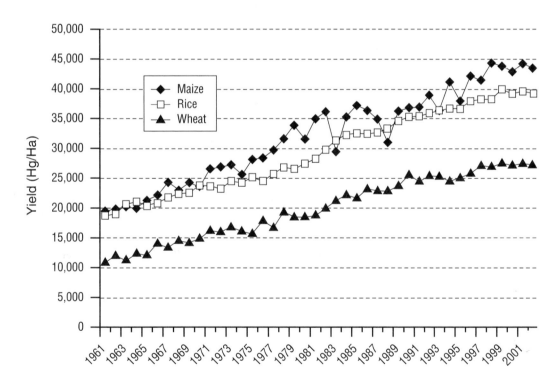

Figure DT 14.2 Yield of maize rice and wheat, 1961 to 2001.

Global Reported Flood Deaths, 1900 to 2002

Description

Total global flood deaths from 1900 to 2002 are reported for the years for which data are available. For years in which very large numbers of deaths are reported, the specific country or event is noted. Most of the truly catastrophic flood events occurred in China. Figure DT 15.1 shows Global Reported Deaths from Floods 1900 to 2002. The scale for the graph stops at 150,000 deaths per year. Higher deaths are reported for 1931, 1939, and 1959.

Limitations

Global data are inconsistently reported prior to 1965, with larger floods reported. Data on minor floods are less reliably available in this earlier period. The scale on the graph is truncated in order to see the more typical level of annual deaths. The three years with deaths greater than the limit on the scale are marked (1931, 1939, 1959).

Sources

EM-DAT: The OFDA/CRED International Disaster database. Université Catholique de Louvain, Brussels, Belgium-C2644. www.cred.be/emdat

DATA TABLE 15   Global Reported Flood Deaths, 1900 to 2002

Year	Reported Deaths	Notes	Year	Reported Deaths	Notes
1900	300		1966	2,006	
1903	250		1967	2,446	
1906	6		1968	7,369	
1909	53		1969	1,544	
1910	1,379		1970	3,246	
1911	100,000	a	1971	2,404	
1913	732		1972	2,548	
1926	1,000		1973	4,574	
1927	3,246		1974	29,431	c
1928	26		1975	848	
1931	3,700,000	a	1976	960	
1933	18,053		1977	2,568	
1935	142,000	a	1978	5,897	
1936	200		1979	1,039	
1937	248		1980	10,442	
1938	954		1981	5,368	
1939	500,010	a	1982	4,691	
1940	125		1983	2,630	
1943	990		1984	3,159	
1947	2,000		1985	4,349	
1948	1,917		1986	2,206	
1949	97,000	a, b	1987	7,482	
1950	3,775		1988	8,892	
1951	5,603		1989	4,734	
1952	199		1990	3,043	
1953	7,037		1991	5,208	
1954	34,894	a	1992	4,967	
1955	2,284		1993	5,835	
1956	3,613		1994	6,403	
1957	2,471		1995	8,145	
1958	405		1996	7,309	
1959	2,003,520	a	1997	6,958	
1960	10,627		1998	9,695	
1961	3,863		1999	34,366	d
1962	1,180		2000	6,330	
1963	1,031		2001	4,678	
1964	1,123		2002	4,354	
1965	1,438				

Notes:
The bulk of the deaths in the respective years occurred in:
a: China
b: Guatamala
c: Bangladesh
d: Venezuela

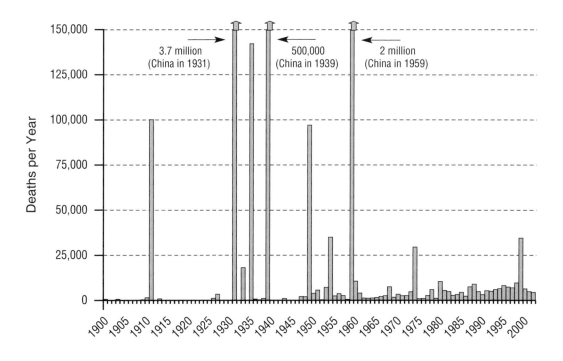

Figure DT 15.1   Global reported deaths from floods, 1900 to 2002.

United States Flood Damage by Fiscal Year, 1926 to 2001

Description

Annual flood damages in the United States from 1926 to 2001 are reported in thousands of dollars, thousands of 1995 dollars, and 1995 dollars per capita. These data use the fiscal year of October to September—thus the 1926 fiscal year runs from October 1925 to September 1926. Even correcting for changes in population and inflation, flood damages appear to be increasing dramatically—by more than a factor of three during the past century. Changes in flood damages can occur for many reasons, including increasing populations and development living in flood-prone areas, changes in climate patterns and flows, and better reporting.

Figure DT 16.1 shows total flood damages in 1995 US Dollars. A linear trend line is included, showing clear increases in total damages over this period.

Limitations

Data are missing for 1980, 1981, and 1982. Corrected dollar damage estimates are missing for 1926 to 28 because of missing inflators. Some of the apparent trend may be due to differences in reporting and data quality.

Sources

ftp://valkyrie.atd.ucar.edu/pub/flood/national.xls
http://www.flooddamagedata.org/national.html

TABLE 16   United States Flood Damage by Fiscal Year, 1926 to 2001

Fiscal Year	Damage— Current USD (thousands)	Damage— 1995 USD (millions)	U.S. Population (millions)	Damage Per-capita (1995 USD)
1926	9,243			
1927	315,187			
1928	88,155			
1929	61,700	480	121.77	3.94
1930	25,832	209	123.08	1.69
1931	2,070	19	124.04	0.15
1932	10,365	106	124.84	0.85
1933	27,366	287	125.58	2.28
1934	18,903	188	126.37	1.49
1935	123,327	1,201	127.25	9.44
1936	287,137	2,767	128.05	21.61
1937	433,339	4,007	128.83	31.10
1938	108,970	1,038	129.83	7.99
1939	13,861	133	130.88	1.02
1940	40,067	381	132.12	2.88
1941	26,092	232	133.40	1.74
1942	91,548	755	134.86	5.60
1943	220,553	1,727	136.74	12.63
1944	99,789	764	138.40	5.52
1945	159,251	1,186	139.93	8.48
1946	68,930	458	141.39	3.24
1947	281,321	1,688	144.13	11.71
1948	213,716	1,213	146.63	8.27
1949	108,586	617	149.19	4.14
1950	129,903	730	152.27	4.80
1951	1,076,687	5,645	154.88	36.45
1952	254,190	1,312	157.55	8.33
1953	121,752	620	160.18	3.87
1954	74,170	374	163.03	2.30
1955	784,672	3,892	165.93	23.45
1956	305,573	1,466	168.90	8.68
1957	352,145	1,635	171.98	9.51
1958	224,939	1,020	174.88	5.83
1959	121,281	544	177.83	3.06
1960	111,168	491	180.67	2.72
1961	147,680	646	183.69	3.51
1962	86,574	373	186.54	2.00
1963	179,496	766	189.24	4.05
1964	194,512	818	191.89	4.26
1965	1,221,903	5,041	194.30	25.94
1966	116,645	468	196.56	2.38
1967	291,823	1,136	198.71	5.71
1968	443,251	1,653	200.71	8.24
1969	889,135	3,161	202.68	15.60
1970	173,803	587	205.05	2.86
1971	323,427	1,040	207.66	5.01
1972	4,442,992	13,698	209.90	65.26
1973	1,805,284	5,271	211.91	24.87
1974	692,832	1,856	213.85	8.68
1975	1,348,834	3,306	215.97	15.31
1976	1,054,790	2,446	218.04	11.22
1977	988,350	2,154	220.24	9.78

Fiscal Year	Damage— Current USD (thousands)	Damage— 1995 USD (millions)	U.S. Population (millions)	Damage Per-capita (1995 USD)
1978	1,028,970	2,093	222.59	9.40
1979	3,626,030	6,808	225.06	30.25
1980	–	–	227.23	–
1981	–	–	229.47	–
1982	–	–	231.66	–
1983	3,693,572	5,260	233.79	22.50
1984	3,540,770	4,862	235.83	20.62
1985	379,303	505	237.92	2.12
1986	5,939,994	7,737	240.13	32.22
1987	1,442,349	1,824	242.29	7.53
1988	214,297	262	244.50	1.07
1989	1,080,814	1,273	246.82	5.16
1990	1,636,366	1,856	249.46	7.44
1991	1,698,765	1,859	252.15	7.37
1992	672,635	718	255.03	2.82
1993	16,364,710	17,069	257.78	66.22
1994	1,120,149	1,145	260.33	4.40
1995	5,110,714	5,111	262.80	19.45
1996	6,121,753	6,005	265.23	22.64
1997	8,934,923	8,597	267.78	32.11
1998	2,465,048	2,343	270.25	8.67
1999	5,450,375	5,107	272.69	18.73
2000	1,336,744	1,227	282.13	4.35
2001	7,158,700	6,418	284.80	22.54

Sources:
ftp://valkyrei.atd.ucar.edu/pub/flood/national/xls
http://www.flooddamagedata.org/national.html

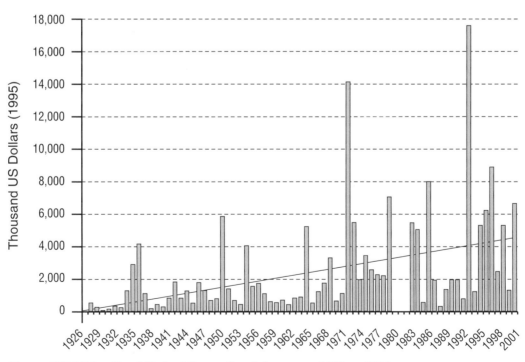

FIGURE DT 16.1   Total United States flood damages, 1929 to 2001.

Total Outbreaks of Drinking Water-Related Disease, United States, 1973 to 2000

Description

The United States has one of the safest water supplies in the world. Nevertheless, waterborne disease outbreaks occur when contaminants find their way into drinking water sources and are not eliminated by treatment. In the United States, local and state public health departments have the primary responsibility for detecting and investigating waterborne disease outbreaks, with voluntary reporting to the United States Centers for Disease Control and Prevention (CDC). CDC data cover all United States and Puerto Rico. The CDC and the United States Environmental Protection Agency (EPA) collaborate to track waterborne disease outbreaks of both microbial and chemical origins. Data on drinking water and recreational water outbreaks and contamination events have been collected and summarized since 1971. From 1991 to 2000, there were 155 outbreaks and 431,846 cases of illness reported to the CDC from public and individual United States' water systems.

This table and the accompanying figures shows three different data sets for the United States with data on the number of outbreaks of water-related diseases and the number of cases reported. Most of the data are those collected and reported by the Centers for Disease Control and Prevention and their program Surveillance for Waterborne-Disease Outbreaks, but two additional datasets that use these data, or modify them using other data, are also included for comparison.

Figure DT 17.1 shows Outbreaks of Waterborne Disease Per Year, 1973 to 2000 for all three data sets. Figure DT 17.2 shows Outbreaks of Waterborne Disease by Etiological Class, 1976 to 1998, as reported by Schneider. This graph breaks the total reported outbreaks into category by type of disease. The categories reported are:

- Bacterial
- Acute Gastrointestinal Illness (AGI)
- Parasitic
- Chemical
- Viral

Of the approximately 500 reported outbreaks during this period, about half were acute gastrointestinal illnesses. Figure DT 17.3 shows Outbreaks of Waterborne Disease by Month, 1973 to 1998, as reported by Schneider. As the graph shows, the majority of outbreaks occur during the summer months.

Limitations

Limited data on the number of outbreaks and cases of water-related diseases are available worldwide because of inadequate and inconsistent reporting. Even in the United States, data are not systematically collected on outbreaks of drinking water-related diseases. It is not possible to say which of the data sets cited here is more accurate, or even if any single set is accurate, given inconsistencies in reporting. They do, however, provide an opportunity to compare reported outbreaks over time.

While useful, statistics derived from surveillance systems may not reflect the true incidence of waterborne disease outbreaks because many people who fall ill from such diseases do not consult medical professionals. For those who do seek medical attention, physicians and laboratory and hospital personnel are required to report diagnosed cases of waterborne illness to state health departments. Further reporting of these illness cases by state health departments to the CDC is voluntary, and more likely to occur for large outbreaks than small ones (http://www.waterandhealth.org/newsletter/new/spring_ 2003/waterborne.html). Indeed, in Canada, one estimate is that there are between 10 and 1,000 times more cases than are officially reported. While most of these are minor, this discrepancy in reporting masks the seriousness of the problem (Edge et al. 2002).

Sources

State of the nation's ecosystem report. Heinz Center.
> http://www.heinzctr.org/ecosystems/fr_water/datasets/freshwater_waterborne_disease_ outbreaks.shtml

Data compiled by Dr. Orren D. Schneider and used by permission.
> http://water.sesep.drexel.edu/outbreaks/US_summaryto1998.htm

Data part of the Centers for Disease Control and Prevention, Surveillance for Waterborne-Disease Outbreaks Program. Waterborne disease outbreaks, 1986–1988.
> http://www.cdc.gov/epo/mmwr/preview/mmwrhtml/00001596.htm

Surveillance for waterborne disease outbreaks—United States, 1991–1992.
> http://www.cdc.gov/epo/mmwr/preview/mmwrhtml/00025893.htm

Surveillance for waterborne disease outbreaks—United States, 1993–1994.
> http://www.cdc.gov/epo/mmwr/preview/mmwrhtml/00040818.htm

Surveillance for waterborne wisease outbreaks—United States, 1995–1996.
> http://www.cdc.gov/epo/mmwr/preview/mmwrhtml/00055820.htm

Surveillance for waterborne disease outbreaks—United States, 1997–1998.
> http://www.cdc.gov/mmwr/preview/mmwrhtml/ss4904a1.htm

Surveillance for waterborne disease outbreaks—United States, 1999–2000.
> http://www.cdc.gov/mmwr/preview/mmwrhtml/ss5108a1.htm

Additional sources of good information and analysis are:

Craun, G. F., Nwachuku, N., Calderon, R. L., and Craun, M. F. 2002. Outbreaks in drinking-water systems, 1991–1998. *Journal of Environmental Health*, 65:16–25.

Edge, T., J., Byrne, M., Johnson, R., Robertson, W., and Stevenson, R. 2002. Waterborne Pathogens. National Water Research Institute, Environment Canada. February 9, 2004.
> http://www.nwri.ca/threatsfull/ch1-1-e.html

Data Table 17   Total Outbreaks of Drinking Water-Related Disease, United States, 1973 to 2000

Year	Heinz Center[1] Outbreaks	Schneider[2] Outbreaks	Schneider[2] Cases	CDC Data[3] Outbreaks	CDC Data[3] Cases
1973	21				
1974	20				
1975	16				
1976	29	22	3,860		
1977	26	17	1,911		
1978	22	33	11,435		
1979	26	41	6,761		
1980	29	49	20,005		
1981	24	32	4,430		
1982	30	40	3,456		
1983	35	40	20,905		
1984	20	26	1,755		
1985	16	25	2,117		
1986	18	22	1,569	22	25,846 (1986–1988)
1987	11	15	22,149	15	
1988	15	15	2,159	13	
1989	12	12	2,540	13	
1990	10	14	1,748	14	
1991	11	15	12,960	15	17,464 (1991–1992)
1992	18	27	4,724	27	
1993	8	17	404,183	18	405,366 (1993–1994)
1994	11	13	1,178	12	
1995	11	16	2,375	16	2,567 (1995–1996)
1996	3	6	192	6	
1997	4	7	304	7	2,038 (1997–1998)
1998	7	10	1,734	10	
1999				15	2,068 (1999–2000)
2000				24	
TOTAL	453	514	534,450	227	455,349

Source:

1. Heinz Center *State of the Nation;s Ecosystem* report.

http://heinzctr.org/ecosystems/fr_water/datasets/freshwater_waterborne_disease_outbreaks.shtm

2. Data compiled by Dr. Orren D. Schneider and used by permission.

http://water.sesep.drexel.edu/outbreaks/US_summaryto1998.htm

3. Data part of the Center for Disease Control.

Surveillance for Waterborne-Disease Outbreaks program.

Waterborne Disease Outbreaks, 1986–1988. http://www.cdc.gov/epo.mmwr/preview/mmwrhtml/00001596.htm
Waterborne-Disease Outbreaks, 1991–1992. http://www.cdc.gov/epo.mmwr/preview/mmwrhtml/000025893.htm
Waterborne-Disease Outbreaks, 1993–1994. http://www.cdc.gov/epo.mmwr/preview/mmwrhtml/0004088.htm
Waterborne-Disease Outbreaks, 1995–1996. http://www.cdc.gov/epo.mmwr/preview/mmwrhtml/00055820.htm
Waterborne-Disease Outbreaks, 1997–1998. http://www.cdc.gov/epo.mmwr/preview/mmwrhtml/ss4904a1.htm
Waterborne-Disease Outbreaks, 1999–2000. http://www.cdc.gov/epo.mmwr/preview/mmwrhtml/ss5108a1.htm

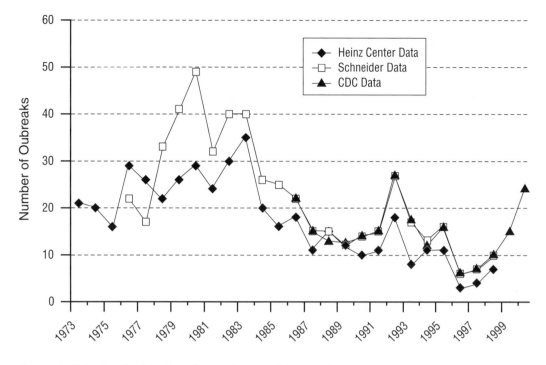

FIGURE DT 17.1 Outbreaks of waterborne disease, 1973 to 2000.

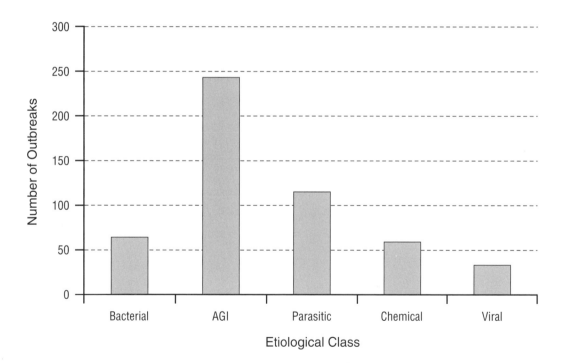

FIGURE DT 17.2 Outbreaks of waterborne disease by etiological class, 1976 to 1998.

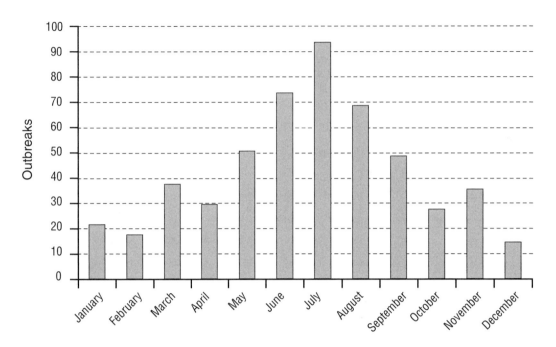

FIGURE DT 17.3   Outbreaks of waterborne disease by month, 1973 to 1998.

a. Extinction Rate Estimates for Continental North American Fauna (percent loss/decade)
b. Imperiled Species for North American Fauna

Description

Two tables from Riccardi and Rasmussen (1999) are shown here, listing recent and projected extinction rates for fauna from continental North America. Data Table 18.a shows extinction rates in percent loss per decade for freshwater, terrestrial, and marine fauna. Data Table 18.b focuses on freshwater species of fish, amphibians, gastropods, and freshwater mussels and includes the total number of species and the number and fraction considered by the authors to be imperiled. A comparable, though different, data set is shown in Data Table 19.

Limitations

While the total number of species of these fauna is relatively well known, as is the recent extinction rate, the projection of future rates of extinction are speculative. These estimates derive from recent accelerations in extinction rates, projections of future activities that are known to imperil different species, and best guesses. They should, therefore, be used with caution. Different definitions of "endangered," "threatened," and "imperiled" also complicate comparing data from different sources. In the United States, there are legal definitions for endangered and threatened, but there are also differing state definitions. Regulations governing actions to protect these threatened species also vary from nation to nation and state to state.

Source

Riccardi, A., and Rasmussen. J. 1999. Extinction rates of North American freshwater fauna. *Conservation Biology*, 13, no. 5. October.

DATA TABLE 18.a Extinction Rate Estimates for Continental North American Fauna
(percent loss per decade)

Freshwater Fauna	Recent	Future
Fishes	0.4	2.4
Crayfishes	0.1	3.9
Mussels	1.2	6.4
Gastropods	0.8	2.6
Amphibians	0.2	3
Mean Rate	0.5	3.7
Terrestrial and Marine Fauna		
Birds	0.3	0.7
Reptiles	–	0.7
Land mammals	–	0.7
Marine mammals	0.2	1.1
Mean Rate	0.1	0.8

DATA TABLE 18.b Imperiled Species for North American Fauna

Freshwater Fauna	Total Number of Species	Number of Imperiled Species	Percent Imperiled
North American freshwater mussels	262	127	49
Freshwater gastropods	474	108	23
Amphibians	243	63	26
Freshwater fishes	1021	217	21
TOTAL	2000	515	26

Proportion of Species at Risk, United States

Description

Stein (2001) reports on the proportion of all species in the United States at risk of extinction, split into 14 categories. Aquatic species are at the highest risk of extinction. Amphibians, fish, mussels, and aquatic insects are especially threatened. Figure DT 19.1,United States Species at Risk, plots these data.

Limitations

These estimates derive from recent trends in extinction rates. Different definitions of "endangered," "threatened," and "imperiled" complicate comparing data from different sources (see Data Table 18.a and 18.b for a different set of data). In the United States, there are legal definitions for endangered and threatened, but there are also differing state definitions. Regulations governing actions to protect these threatened species also vary from nation to nation and state to state.

Source

Stein, B. A. 2001. A fragile cornucopia assessing the status of U.S. biodiversity. *Environment*, 43:11–22.

DATA TABLE 19   Proportion of Species at Risk, United States

Category	Proportion Percent at Risk
Birds	14
Mammals	16
Dragonflies/Damselflies	18
Reptiles	18
Butterflies/Skippers	19
Tiger beetles	19
Ferns/Fern allies	22
Gymnosperms	24
Flowing plants	33
Amphibians	36
Freshwater fish	37
Stoneflies	43
Crayfish	51
Freshwater mussels	69

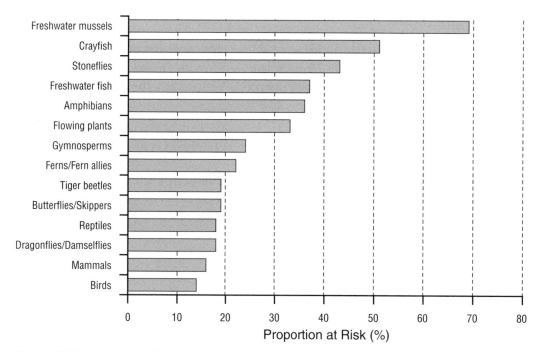

FIGURE DT 19.1   United States species at risk.

United States Population and Water Withdrawals, 1900 to 2000

Description

This table shows the population of the United States and total water withdrawals for all purposes from 1900 to 2000, incorporating the newest information released in March 2004 (for year 2000—the last year for which United States water use data was available). United States population during this period increased from 76 million to 281 million, nearly a four-fold increase. Water withdrawals rose from 56 km^3/yr peaking in 1980 at over 600 km^3/yr— a 10-fold increase—and then dropped nearly 10 percent from that level by 2000. Figure DT 20.1 plots per-capita water withdrawals over the same period, in m^3 per person per year. Per-capita use has dropped 25 percent from its peak.

National population data for the years 1900 to 1949 exclude the population residing in Alaska and Hawaii. National population data for the years 1940 to 1979 cover the resident population plus Armed Forces overseas. National population data for all other years cover only the resident population.

Limitations

The United States is one of the few countries with a consistent system for regularly reporting water use. These data are compiled from state surveys and the quality of water use data varies from location to location. Data from the earliest years should be considered approximations. In March 2004, the data from 2000 were released. The United States Geological Survey (USGS) does not collect annual data. The data released every 5 years were converted to annual data using linear extrapolation. The population data were obtained from the United States Census Bureau. The estimates for the period April 1, 1990 to April 1, 2000 are provided in a monthly time series. These estimates are intercensal, i.e., they are consistent with both the 1990 and 2000 census enumerations.

Sources

Compiled by Gleick, Peter H., Pacific Institute for Studies in Development, Environment, and Security. Oakland, California.

Year 2000 water use data from:

Hudson, S. S., Barber, N. L., Kenny, J. F., Linsey, K. S., Lumia, D. S., and Maupin, M. A. March 2004. Estimated use of water in the United States in 2000. USGS Circular 1268, Reston, Virginia.
   http://water.usgs.gov/pubs/circ/2004/circ1268/

Prior year water data from USGS estimated use of water in the United States report series.
   http://water.usgs.gov/watuse/

Population data are from

Table US-EST90INT-01—Intercensal estimates of the United States population by sex, 1990–2000: Selected months, population civision, United States Census Bureau, Released September 13, 2002.

DATA TABLE 20   United States Population and Water Withdrawals, 1900 to 2000

Year	U.S. Population	Estimates of Annual Total Water Withdrawals (cubic kilometers/year)	Per-capita Water Withdrawals (cubic meters/person/year)
1900	76,094,000	56	736
1901	77,584,000	60	769
1902	79,163,000	63	801
1903	80,632,000	67	832
1904	82,166,000	71	862
1905	83,822,000	75	889
1906	85,450,000	78	915
1907	87,008,000	82	941
1908	88,710,000	86	965
1909	90,490,000	89	987
1910	92,407,000	93	1,006
1911	93,863,000	96	1,025
1912	95,335,000	99	1,043
1913	97,225,000	103	1,055
1914	99,111,000	106	1,067
1915	100,546,000	109	1,084
1916	101,961,000	112	1,100
1917	103,268,000	115	1,117
1918	103,208,000	119	1,149
1919	104,514,000	122	1,165
1920	106,461,000	125	1,174
1921	108,538,000	128	1,177
1922	110,049,000	130	1,185
1923	111,947,000	133	1,189
1924	114,109,000	136	1,190
1925	115,829,000	139	1,196
1926	117,397,000	141	1,203
1927	119,035,000	144	1,209
1928	120,509,000	147	1,217
1929	121,767,000	149	1,226
1930	123,076,741	152	1,235
1931	124,039,648	156	1,254

Year	U.S. Population	Estimates of Annual Total Water Withdrawals (cubic kilometers/year)	Per-capita Water Withdrawals (cubic meters/person/year)
1932	124,840,471	159	1,275
1933	125,578,763	163	1,296
1934	126,373,773	166	1,317
1935	127,250,232	170	1,336
1936	128,053,180	174	1,356
1937	128,824,829	177	1,376
1938	129,824,939	181	1,393
1939	130,879,718	184	1,409
1940	132,122,446	188	1,423
1941	133,402,471	195	1,463
1942	134,859,553	202	1,501
1943	136,739,353	210	1,533
1944	138,397,345	217	1,567
1945	139,928,165	224	1,601
1946	141,388,566	229	1,621
1947	144,126,071	234	1,626
1948	146,631,302	240	1,634
1949	149,188,130	245	1,641
1950	152,271,417	250	1,642
1951	154,877,889	266	1,717
1952	157,552,740	282	1,790
1953	160,184,192	298	1,860
1954	163,025,854	314	1,926
1955	165,931,202	330	1,989
1956	168,903,031	338	2,001
1957	171,984,130	346	2,012
1958	174,881,904	354	2,024
1959	177,829,628	362	2,036
1960	180,671,158	370	2,048
1961	183,691,481	382	2,080
1962	186,537,737	394	2,112
1963	189,241,798	406	2,145
1964	191,888,791	418	2,178
1965	194,302,963	430	2,213
1966	196,560,338	446	2,269
1967	198,712,056	462	2,325
1968	200,706,052	478	2,382
1969	202,676,946	494	2,437
1970	205,052,174	510	2,487
1971	207,660,677	524	2,523
1972	209,896,021	538	2,563
1973	211,908,788	552	2,605
1974	213,853,928	566	2,647
1975	215,973,199	580	2,686
1976	218,035,164	586	2,688

1977	220,239,425	592	2,688

DATA TABLE 20 *Continued*

Year	U.S. Population	Estimates of Annual Total Water Withdrawals (cubic kilometers/year)	Per-capita Water Withdrawals (cubic meters/person/year)
1978	222,584,545	598	2,687
1979	225,055,487	604	2,684
1980	227,224,681	610	2,685
1981	229,465,714	598	2,607
1982	231,664,458	586	2,531
1983	233,791,994	575	2,458
1984	235,824,902	563	2,387
1985	237,923,795	551	2,316
1986	240,132,887	554	2,305
1987	242,288,918	556	2,295
1988	244,498,982	559	2,285
1989	246,819,230	561	2,273
1990	249,622,814	564	2,258
1991	252,980,941	562	2,220
1992	256,514,224	560	2,182
1993	259,918,588	558	2,145
1994	263,125,821	556	2,112
1995	266,278,393	554	2,079
1996	269,394,284	555	2,062
1997	272,646,925	557	2,044
1998	275,854,104	559	2,027
1999	279,040,168	561	2,011
2000	281,421,906	563	2,001

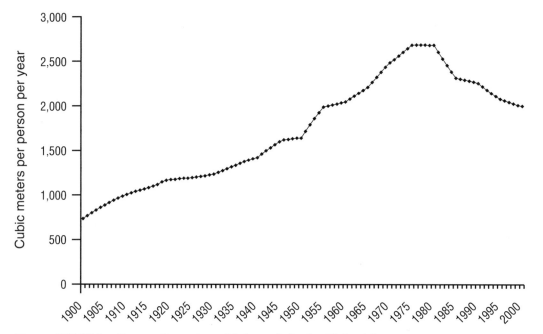

FIGURE DT 20.1 Per-capita water withdrawals in the United States.

United States Economic Productivity of Water, 1900 to 2000

Description

Annual water withdrawals and Gross Domestic Product (GDP) for the United States are presented here. Water data are in km³/yr, and reflect all United States withdrawals. United States GDP data are in corrected 1996 dollars. In the previous volume of *The World's Water*, this table offered data through the mid-1990s. The current table updates these data for 2000. For many years, a fundamental assumption in the water planning community was that increases in water use were necessary to provide for increases in economic activity. The data in this table and accompanying figure (Figure DT 21.1) show that this is no longer the case. Water withdrawals and GDP moved together until the late 1970s at which point gross domestic product continued to rise while water use leveled off and even declined. The change in water use is a result of improving water-use efficiency and productivity in all United States sectors along with a change in the structure of those sectors away from water-intensive activities. Figure DT 21.1 shows GDP per unit of water withdrawn—what we call the "economic productivity of water."

Limitations

The USGS does not collect annual water use data. The data released every five years were converted to annual data here using linear extrapolation. Gross national or domestic product is an imperfect, albeit standard, measure of well-being or economic activity. As a result, GDP data are available for many countries and provide a benchmark against which other activities can be measured. Actual annual water withdrawal data are rarely collected for any country, even the United States. Good data are available every five years going back through most of the twentieth century, though data for earlier decades in the century are less reliable.

Sources

Compiled by Gleick, Peter H., Pacific Institute for Studies in Development, Environment, and Security. Oakland, California.

Year 2000 water use data from:

Hudson, S. S., Barber, N. L., Kenny, J. F., Linsey, K. S., Lumia, D. S., and Maupin, M. A. March 2004. Estimated use of water in the United States in 2000. United States Geologic Survey Circular 1268, Reston, Virginia. http://water.usgs.gov/pubs/circ/2004/circ1268/

Prior year water data from USGS estimated use of water in the United States report series. http://water.usgs.gov/watuse/

United States GDP data come from Johnston, L., and Williamson, S. H., 2002. The annual real and nominal GDP for the United States, 1789–Present. Economic History Services, April 2002. http://www.eh.net/hmit/gdp/

DATA TABLE 21 United States Economic Productivity of Water, 1900 to 2000

Year	Real GDP (billions of 1996 US$)	Estimates of Annual Total Water Withdrawals (cubic kilometers/year)	U.S. Economic Productivity of Water (US$ of GDP/cubic meter)
1900	311	56	5.55
1901	349	60	5.85
1902	355	63	5.60
1903	365	67	5.44
1904	378	71	5.34
1905	413	75	5.54
1906	430	78	5.50
1907	423	82	5.16
1908	400	86	4.67
1909	447	89	5.01
1910	449	93	4.83
1911	463	96	4.81
1912	490	99	4.93
1913	509	103	4.96
1914	470	106	4.44
1915	486	109	4.46
1916	564	112	5.03
1917	563	115	4.88
1918	606	119	5.11
1919	587	122	4.82
1920	576	125	4.61
1921	556	128	4.35
1922	594	130	4.56
1923	677	133	5.09
1924	696	136	5.13
1925	711	139	5.13
1926	754	141	5.34
1927	758	144	5.27
1928	772	147	5.27
1929	822	149	5.51
1930	751	152	4.94
1931	703	156	4.52
1932	611	159	3.84
1933	603	163	3.70

Year	Real GDP (billions of 1996 US$)	Estimates of Annual Total Water Withdrawals (cubic kilometers/year)	U.S. Economic Productivity of Water (US$ of GDP/cubic meter)
1934	668	166	4.01
1935	728	170	4.28
1936	822	174	4.74
1937	865	177	4.88
1938	835	181	4.62
1939	903	184	4.90
1940	980	188	5.21
1941	1140	195	5.84
1942	1360	202	6.72
1943	1580	210	7.54
1944	1710	217	7.89
1945	1690	224	7.54
1946	1500	229	6.54
1947	1490	234	6.36
1948	1560	240	6.51
1949	1550	245	6.33
1950	1680	250	6.72
1951	1810	266	6.80
1952	1880	282	6.67
1953	1970	298	6.61
1954	1960	314	6.24
1955	2090	330	6.33
1956	2140	338	6.33
1957	2180	346	6.30
1958	2160	354	6.10
1959	2310	362	6.38
1960	2370	370	6.41
1961	2430	382	6.36
1962	2570	394	6.52
1963	2690	406	6.63
1964	2840	418	6.79
1965	3020	430	7.02
1966	3220	446	7.22
1967	3300	462	7.14
1968	3460	478	7.24
1969	3570	494	7.23
1970	3570	510	7.00
1971	3690	524	7.04
1972	3890	538	7.23
1973	4120	552	7.46
1974	4090	566	7.23
1975	4080	580	7.03
1976	4310	586	7.35
1977	4510	592	7.62
1978	4760	598	7.96
1979	4910	604	8.13
1980	4900	610	8.03

DATA TABLE 21   *Continued*

Year	Real GDP (billions of 1996 US$)	Estimates of Annual Total Water Withdrawals (cubic kilometers/year)	U.S. Economic Productivity of Water (US$ of GDP/cubic meter)
1981	5020	598	8.39
1982	4910	586	8.37
1983	5130	575	8.93
1984	5500	563	9.77
1985	5710	551	10.36
1986	5910	554	10.68
1987	6110	556	10.99
1988	6360	559	11.39
1989	6590	561	11.74
1990	6700	564	11.89
1991	6670	562	11.88
1992	6880	560	12.29
1993	7060	558	12.66
1994	7340	556	13.21
1995	7540	554	13.62
1996	7810	555	14.06
1997	8150	557	14.62
1998	8500	559	15.20
1999	8850	561	15.77
2000	9190	563	16.32

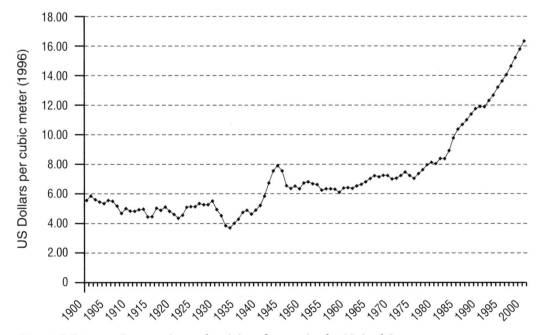

FIGURE DT 21.1   Economic productivity of water in the United States.

Water Units, Data Conversions, and Constants

Water experts, managers, scientists, and educators work with a bewildering array of different units and data. These vary with the field of work: engineers may use different water units than hydrologists; urban water agencies may use different units than reservoir operators; academics may use different units than water managers. But they also vary with regions: water agencies in England may use different units than water agencies in France or Africa; hydrologists in the eastern United States often use different units than hydrologists in the western United States And they vary over time: today's water agency in California may sell water by the acre-foot, but its predecessor a century ago may have sold miner's inches or some other now arcane measure.

These differences are of more than academic interest. Unless a common "language" is used, or a dictionary of translations is available, errors can be made or misunderstandings can ensue. In some disciplines, unit errors can be more than embarrassing; they can be expensive, or deadly. In September 1999, the $125 million Mars Climate Orbiter spacecraft was sent crashing into the face of Mars instead of into its proper safe orbit above the surface because one of the computer programs controlling a portion of the navigational analysis used English units incompatible with the metric units used in all the other systems. The failure to translate English units into metric units was described in the findings of the preliminary investigation as the principal cause of mission failure.

This table is a comprehensive list of water units, data conversions, and constants related to water volumes, flows, pressures, and much more. Most of these units and conversions were compiled by Kent Anderson and initially published in P. H. Gleick, 1993, *Water in Crisis: A Guide to the World's Fresh Water Resources*, Oxford University Press, New York.

Water Units, Data Conversions, and Constants

Prefix (Metric)	Abbreviation	Multiple	Prefix (Metric)	Abbreviation	Multiple
deka-	da	10	deci-	d	0.1
hecto-	h	100	centi-	c	0.01
kilo-	k	1000	milli-	m	0.001
mega-	M	10^6	micro-	μ	10^{-6}
giga-	G	10^9	nano-	n	10^{-9}
tera-	T	10^{12}	pico-	P	10^{-12}
peta-	P	10^{15}	femto-	f	10^{-15}
exa-	E	10^{18}	atto-	a	10^{-18}

LENGTH (L)

1 micron (μ)	$= 1 \times 10^{-3}$ mm	**10 hectometers**	$= 1$ kilometer
	$= 1 \times 10^{-6}$ m	**1 mil**	$= 0.0254$ mm
	$= 3.3937 \times 10^{-5}$ in		$= 1 \times 10^{-3}$ in
1 millimeter (mm)	$= 0.1$ cm	**1 inch (in)**	$= 25.4$ mm
	$= 1 \times 10^{-3}$ m		$= 2.54$ cm
	$= 0.03937$ in		$= 0.08333$ ft
1 centimeter (cm)	$= 10$ mm		$= 0.0278$ yd
	$= 0.01$ m	**1 foot (ft)**	$= 30.48$ cm
	$= 1 \times 10^{-5}$ km		$= 0.3048$ m
	$= 0.3937$ in		$= 3.048 \times 10^{-4}$ km
	$= 0.03281$ ft		$= 12$ in
	$= 0.01094$ yd		$= 0.3333$ yd
1 meter (m)	$= 1000$ mm		$= 1.89 \times 10^{-4}$ mi
	$= 100$ cm	**1 yard (yd)**	$= 91.44$ cm
	$= 1 \times 10^{-3}$ km		$= 0.9144$ m
	$= 39.37$ in		$= 9.144 \times 10^{-4}$ km
	$= 3.281$ ft		$= 36$ in
	$= 1.094$ yd		$= 3$ ft
	$= 6.21 \times 10^{-4}$ mi		$= 5.68 \times 10^{-4}$ mi
1 kilometer (km)	$= 1 \times 10^5$ cm	**1 mile (mi)**	$= 1609.3$ m
	$= 1000$ m		$= 1.609$ km
	$= 3280.8$ ft		$= 5280$ ft
	$= 1093.6$ yd		$= 1760$ yd
	$= 0.621$ mi	**1 fathom (nautical)**	$= 6$ ft
10 millimeters	$= 1$ centimeter	**1 league (nautical)**	$= 5.556$ km
10 centimeters	$= 1$ decimeter		$= 3$ nautical miles
10 decimeters (dm)	$= 1$ meter	**1 league (land)**	$= 4.828$ km
			$= 5280$ yd
10 meters	$= 1$ dekameter		$= 3$ mi
10 dekameters (dam)	$= 1$ hectometer	**1 international nautical mile**	$= 1.852$ km
			$= 6076.1$ ft
			$= 1.151$ mi

Water Units, Data Conversions, and Constants *(continued)*

AREA (L^2)

1 square centimeter (cm²)	$= 1 \times 10^{-4} m^2$	**1 square foot (ft²)**	$= 929.0\ cm^2$
	$= 0.1550\ in^2$		$= 0.0929\ m^2$
	$= 1.076 \times 10^{-3}\ ft^2$		$= 144\ in^2$
	$= 1.196 \times 10^{-4}\ yd^2$		$= 0.1111\ yd^2$
1 square meter (m²)	$= 1 \times 10^{-4}\ hectare$		$= 2.296 \times 10^{-5}\ acre$
	$= 1 \times 10^{-6}\ km^2$		$= 3.587 \times 10^{-8}\ mi^2$
	$= 1\ centare$ (French)	**1 square yard (yd²)**	$= 0.8361\ m^2$
	$= 0.01\ are$		$= 8.361 \times 10^{-5}\ hectare$
	$= 1550.0\ in^2$		$= 1296\ in^2$
	$= 10.76\ ft^2$		$= 9\ ft^2$
	$= 1.196\ yd^2$		$= 2.066 \times 10^{-4}\ acres$
	$= 2.471 \times 10^{-4}\ acre$		$= 3.228 \times 10^{-7}\ mi^2$
1 are	$= 100\ m^2$	**1 acre**	$= 4046.9\ m^2$
1 hectare (ha)	$= 1 \times 10^4\ m^2$		$= 0.40469\ ha$
	$= 100\ are$		$= 4.0469 \times 10^{-3}\ km^2$
	$= 0.01\ km^2$		$= 43,560\ ft^2$
	$= 1.076 \times 10^5\ ft^2$		$= 4840\ yd^2$
	$= 1.196 \times 10^4\ yd^2$		$= 1.5625 \times 10^{-3}\ mi^2$
	$= 2.471\ acres$	**1 square mile (mi²)**	$= 2.590 \times 10^6\ m^2$
	$= 3.861 \times 10^{-3}\ mi^2$		$= 259.0\ hectares$
1 square kilometer (km²)	$= 1 \times 10^6\ m^2$		$= 2.590\ km^2$
	$= 100\ hectares$		$= 2.788 \times 10^7\ ft^2$
	$= 1.076 \times 10^7\ ft^2$		$= 3.098 \times 10^6\ yd^2$
	$= 1.196 \times 10^6\ yd^2$		$= 640\ acres$
	$= 247.1\ acres$		$= 1\ section\ (of\ land)$
	$= 0.3861\ mi^2$	**1 feddan (Egyptian)**	$= 4200\ m^2$
1 square inch (in²)	$= 6.452\ cm^2$		$= 0.42\ ha$
	$= 6.452 \times 10^{-4}\ m^2$		$= 1.038\ acres$
	$= 6.944 \times 10^{-3}\ ft^2$		
	$= 7.716 \times 10^{-4}\ yd^2$		

(continues)

Water Units, Data Conversions, and Constants *(continued)*

VOLUME (L^3)

1 cubic centimeter (cm^3)	$= 1 \times 10^{-3}$ liter	**1 cubic foot (ft^3)**	$= 2.832 \times 10^4$ cm^3
	$= 1 \times 10^{-6}$ m^3		$= 28.32$ liters
	$= 0.06102$ in^3		$= 0.02832$ m^3
	$= 2.642 \times 10^{-4}$ gal		$= 1728$ in^3
	$= 3.531 \times 10^{-3}$ ft^3		$= 7.481$ gal
1 liter (1)	$= 1000$ cm^3		$= 0.03704$ yd^3
	$= 1 \times 10^{-3}$ m^3	**1 cubic yard (yd^3)**	$= 0.7646$ m^3
	$= 61.02$ in^3		$= 6.198 \times 10^{-4}$
	$= 0.2642$ gal		acre-ft
	$= 0.03531$ ft^3		$= 46656$ in^3
1 cubic meter (m^3)	$= 1 \times 10^6$ cm^3		$= 27$ ft^3
	$= 1000$ liter	**1 acre-foot**	$= 1233.48$ m^3
	$= 1 \times 10^{-9}$ km^3	(acre-ft or AF)	$= 3.259 \times 10^5$ gal
	$= 264.2$ gal		$= 43560$ ft^3
	$= 35.31$ ft^3	**1 Imperial gallon**	$= 4.546$ liters
	$= 6.29$ bbl		$= 277.4$ in^3
	$= 1.3078$ yd^3		$= 1.201$ gal
	$= 8.107 \times 10^{-4}$		$= 0.16055$ ft^3
	acre-ft	**1 cfs-day**	$= 1.98$ acre-feet
1 cubic decameter (dam^3)	$= 1000$ m^3		$= 0.0372$ in-mi^2
	$= 1 \times 10^6$ liter	**1 inch-mi^2**	$= 1.738 \times 10^7$ gal
	$= 1 \times 10^{-6}$ km^3		$= 2.323 \times 10^6$ ft^3
	$= 2.642 \times 10^5$ gal		$= 53.3$ acre-ft
	$= 3.531 \times 10^4$ ft^3		$= 26.9$ cfs-days
	$= 1.3078 \times 10^3$ yd^3	**1 barrel (of oil)**	$= 159$ liter
	$= 0.8107$ acre-ft	(bbl)	$= 0.159$ m^3
1 cubic hectometer (ha^3)	$= 1 \times 10^6$ m^3		$= 42$ gal
	$= 1 \times 10^3$ dam^3		$= 5.6$ ft^3
	$= 1 \times 10^9$ liter	**1 million gallons**	$= 3.069$ acre-ft
	$= 2.642 \times 10^8$ gal	**1 pint (pt)**	$= 0.473$ liter
	$= 3.531 \times 10^7$ ft^3		$= 28.875$ in^3
	$= 1.3078 \times 10^6$ yd^3		$= 0.5$ qt
	$= 810.7$ acre-ft		$= 16$ fluid ounces
1 cubic kilometer (km^3)	$= 1 \times 10^{12}$ liter		$= 32$ tablespoons
	$= 1 \times 10^9$ m^3		$= 96$ teaspoons
	$= 1 \times 10^6$ dam^3	**1 quart (qt)**	$= 0.946$ liter
	$= 1000$ ha^3		$= 57.75$ in^3
	$= 8.107 \times 10^5$		$= 2$ pt
	acre-ft		$= 0.25$ gal
	$= 0.24$ mi^3	**1 morgen-foot**	$= 2610.7$ m^3
1 cubic inch (in^3)	$= 16.39$ cm^3	(S. Africa)	
	$= 0.01639$ liter	**1 board-foot**	$= 2359.8$ cm^3
	$= 4.329 \times 10^{-3}$ gal		$= 144$ in^3
	$= 5.787 \times 10^{-4}$ ft^2		$= 0.0833$ ft^3
1 gallon (gal)	$= 3.785$ liters	**1 cord**	$= 128$ ft^3
	$= 3.785 \times 10^{-3}$ m^3		$= 0.453$ m^3
	$= 231$ in^3		
	$= 0.1337$ ft^3		
	$= 4.951 \times 10^{-3}$ yd^3		

Water Units, Data Conversions, and Constants *(continued)*

VOLUME/AREA (L^3/L^2)

1 inch of rain	= 5.610 gal/yd^2	**1 box of rain**	= 3,154.0 lesh
	= 2.715×10^4 gal/acre		

MASS (M)

1 gram (g or gm)	= 0.001 kg	**1 ounce (oz)**	= 28.35 g
	= 15.43 gr		= 437.5 gr
	= 0.03527 oz		= 0.0625 lb
	= 2.205×10^{-3} lb	**1 pound (lb)**	= 453.6 g
1 kilogram (kg)	= 1000 g		= 0.45359237 kg
	= 0.001 tonne		= 7000 gr
	= 35.27 oz		= 16 oz
	= 2.205 lb	**1 short ton (ton)**	= 907.2 kg
1 hectogram (hg)	= 100 gm		= 0.9072 tonne
	= 0.1 kg		= 2000 lb
1 metric ton (tonne or te or MT)	= 1000 kg	**1 long ton**	= 1016.0 kg
	= 2204.6 lb		= 1.016 tonne
	= 1. 102 ton	**1 long ton**	= 2240 lb
	= 0.9842 long ton		= 1.12 ton
1 dalton (atomic mass unit)	= 1.6604×10^{-24} g	**1 stone (British)**	= 6.35 kg
			= 14 lb
1 grain (gr)	= 2.286×10^{-3} oz		
	= 1.429×10^{-4} lb		

TIME (T)

1 second (s or sec)	= 0.01667 min	**1 day (d)**	= 24 hr
	= 2.7778×10^{-4} hr		= 86400 s
1 minute (min)	= 60 s	**1 year (yr or y)**	= 365 d
	= 0.01667 hr		= 8760 hr
1 hour (hr or h)	= 60 min		= 3.15×10^7 s
	= 3600 s		

DENSITY (M/L^3)

1 kilogram per cubic meter (kg/m^3)	= 10^{-3} g/cm^3	**1 metric ton per cubic meter (te/m^3)**	= 1.0 specific gravity
	= 0.062 lb/ft^3		= density of H$_2$O at 4°C
1 gram per cubic centimeter (g/cm^3)	= 1000 kg/m^3		= 8.35 lb/gal
	= 62.43 lb/ft^3	**1 pound per cubic foot (lb/ft^3)**	= 16.02 kg/m^3

(continues)

Water Units, Data Conversions, and Constants *(continued)*

VELOCITY (L/T)

1 meter per	= 3.6 km/hr	**1 foot per second**	= 0.68 mph
second (m/s)	= 2.237 mph	**(ft/s)**	= 0.3048 m/s
	= 3.28 ft/s	**velocity of light in**	= 2.9979×10^8 m/s
1 kilometer per	= 0.62 mph	**vacuum (c)**	= 186,000 mi/s
hour (km/h	= 0.278 m/s	**1 knot**	= 1.852 km/h
or kph)			= 1 nautical
1 mile per hour	= 1.609 km/h		mile/hour
(mph or mi/h)	= 0.45 m/s		= 1.151 mph
	= 1.47 ft/s		= 1.688 ft/s

VELOCITY OF SOUND IN WATER AND SEAWATER
(assuming atmospheric pressure and sea water salinity of 35,000 ppm)

Temp, °C	Pure water, (meters/sec)	Sea water, (meters/sec)
0	1,400	1,445
10	1,445	1,485
20	1,480	1,520
30	1,505	1,545

FLOW RATE (L^3/T)

1 liter per second	= 0.001 m³/sec	**1 cubic decameters**	= 11.57 l/sec
(1/sec)	= 86.4 m³/day	**per day (dam³/day)**	= 1.157×10^{-2}
	= 15.9 gpm		m³/sec
	= 0.0228 mgd		= 1000 m³/day
	= 0.0353 cfs		= 1.83×10^6 gpm
	= 0.0700 AF/day		= 0.264 mgd
1 cubic meter per	= 1000 l/sec		= 0.409 cfs
second (m³/sec)	= 8.64×10^4 m³/day		= 0.811 AF/day
	= 1.59×10^4 gpm	**1 gallon per minute**	= 0.0631 l/sec
	= 22.8 mgd	**(gpm)**	= 6.31×10^{-5}
	= 35.3 cfs		m³/sec
	= 70.0 AF/day		= 1.44×10^{-3} mgd
1 cubic meter per	= 0.01157 l/sec		= 2.23×10^{-3} cfs
day (m³/day)	= 1.157×10^{-5}		= 4.42×10^{-3}
	m³/sec		AF/day
	= 0.183 gpm	**1 million gallons**	= 43.8 l/sec
	= 2.64×10^{-4} mgd	**per day (mgd)**	= 0.0438 m³/sec
	= 4.09×10^{-4} cfs		= 3785 m³/day
	= 8.11×10^{-4}		= 694 gpm
	AF/day		= 1.55 cfs
			= 3.07 AF/day

(continues)

Water Units, Data Conversions, and Constants *(continued)*

FLOW RATE (L³/T) (continued)

1 cubic foot per second (cfs)	= 28.3 l/sec	**1 miner's inch**	= 0.02 cfs (in Idaho, Kansas, Nebraska, New Mexico, North Dakota, South Dakota, and Utah)
	= 0.0283 m³/ sec		
	= 2447 m³/day		
	= 449 gpm		
	= 0.646 mgd		= 0.026 cfs (in Colorado)
	= 1.98 AF/day		
1 acre-foot per day (AF/day)	= 14.3 l/sec		= 0.028 cfs (in British Columbia)
	= 0.0143 m³/sec		
	= 1233.48 m³/day	**1 weir**	= 0.02 garcia
	= 226 gpm	**1 quinaria**	= 0.47–0.48 l/sec
	= 0.326 mgd	**(ancient Rome)**	
	= 0.504 cfs		
1 miner's inch	= 0.025 cfs (in Arizona, California, Montana, and Oregon: flow of water through 1 in² aperture under 6-inch head)		

ACCELERATION (L/T²)

standard acceleration of gravity	= 9.8 m/s²
	= 32 ft/s²

FORCE (ML/T² = Mass × Acceleration)

1 newton (N)	= kg-m/s²	**1 dyne**	= g·cm/s²
	= 10⁵ dynes		= 10⁻⁵ N
	= 0.1020 kg force	**1 pound force**	= lb mass × acceleration of gravity
	= 0.2248 lb force		
			= 4.448 N

(continues)

Water Units, Data Conversions, and Constants *(continued)*

PRESSURE (M/L^2 = Force/Area)

		1 kilogram per sq.	= 14.22 lb/in^2
1 pascal (Pa)	= N/m^2	**centimeter**	
1 bar	= 1 × 10^5 Pa	**(kg/cm^2)**	
	= 1 × 10^6 dyne/cm^2	**1 inch of water**	= 0.0361 lb/in^2
	= 1019.7 g/cm^2	**at 62°F**	= 5.196 lb/ft^3
	= 10.197 te/m^2		= 0.0735 inch of
	= 0.9869 atmos-		mercury at 62°F
	phere	**1 foot of water**	= 0.433 lb/in^2
	= 14.50 lb/in^2	**at 62°F**	= 62.36 lb/ft^2
	= 1000 millibars		= 0.833 inch of
1 atmosphere (atm)	= standard		mercury at 62°F
	pressure		= 2.950 × 10^{-2}
	= 760 mm of		atmosphere
	mercury at 0°C	**1 pound per sq.**	= 2.309 feet of
	= 1013.25 millibars	**inch (psi or**	water at 62°F
	= 1033 g/cm^2	**lb/in^2)**	= 2.036 inches of
	= 1.033 kg/cm^2		mercury at 32°F
	= 14.7 lb/in^2		= 0.06804
	= 2116 1b/ft^2		atmosphere
	= 33.95 feet of		= 0.07031 kg/cm^2
	water at 62°F	**1 inch of mercury**	= 0.4192 lb/in^2
	= 29.92 inches of	**at 32°F**	= 1.133 feet of
	mercury at 32°F		water at 32°F

TEMPERATURE

degrees Celsius or	= (°F–32) × 5/9	**degrees Fahrenheit**	= 32 + (°C x 1.8)
Centigrade (°C)	= K–273.16	**(°F)**	= 32 + ((°K–273.16)
Kelvins (K)	= 273.16 + °C		× 1.8)
	= 273.16 + ((°F- 32)		
	× 5/9)		

Water Units, Data Conversions, and Constants *(continued)*

ENERGY(ML^2/T^2 = Force $\times$ Distance)

1 joule (J)	$= 10^7$ ergs	**1 kilowatt-hour**	$= 3.6 \times 10^6$ J
	$=$ N·m	**(kWh)**	$= 3412$ Btu
	$=$ W·s		$= 859.1$ kcal
	$=$ kg·m^2/s^2	**l quad**	$= 10^{15}$ Btu
	$= 0.239$ calories		$= 1.055 \times 10^{18}$J
	$= 9.48 \times 10^{-4}$ Btu		$= 293 \times 10^9$ kWh
1 calorie (cal)	$= 4.184$ J		$= 0.001$ Q
	$= 3.97 \times 10^{-3}$ Btu		$= 33.45$ GWy
	(raises 1 g H$_2$O	**1 Q**	$= 1000$ quads
	1°C)		$\approx 10^{21}$ J
1 British thermal	$= 1055$ J	**1 foot-pound (ft-lb)**	$= 1.356$ J
unit (Btu)	$= 252$ cal (raises		$= 0.324$ cal
	1 lb H$_2$O 1°F)	**1 therm**	$= 10^5$ Btu
	$= 2.93 \times 10^{-4}$ kWh	**1 electron-volt (eV)**	$= 1.602 \times 10^{-19}$ J
1 erg	$= 10^{-7}$ J	**1 kiloton of TNT**	$= 4.2 \times 10^{12}$ J
	$=$ g·cm^2/s^2	**1 10^6 te oil equiv.**	$= 7.33 \times 10^6$ bbl oil
	$=$ dyne·cm	**(Mtoe)**	$= 45 \times 10^{15}$ J
1 kilocalorie (kcal)	$= 1000$ cal		$= 0.0425$ quad
	$= 1$ Calorie (food)		

POWER (ML^2/T^3 = rate of flow of energy)

1 watt (W)	$=$ J/s	**1 horsepower**	$= 0.178$ kcal/s
	$= 3600$ J/hr	**(H.P. or hp)**	$= 6535$ kWh/yr
	$= 3.412$ Btu/hr		$= 33,000$ ft-lb/min
1 TW	$= 10^{12}$ W		$= 550$ ft-lb/sec
	$= 31.5 \times 10^{18}$ J		$= 8760$ H.P.-hr/yr
	$= 30$ quad/yr	**H.P. input**	$= 1.34 \times$ kW input
1 kilowatt (kW)	$= 1000$W		to motor
	$= 1.341$ horsepower		$=$ horsepower
	$= 0.239$ kcal/s		input to motor
	$= 3412$ Btu/hr	**Water H.P.**	$=$ H.P. required to
10^6 bbl (oil) /day	≈ 2 quads/yr		lift water at a
(Mb/d)	≈ 70 GW		definite rate to
1 quad/yr	$= 33.45$ GW		a given distance
	≈ 0.5 Mb/d		assuming 100%
1 horsepower	$= 745.7$W		efficiency
(H.R or hp)	$= 0.7457$ kW		$=$ gpm $\times$ total head
			(in feet)/3960

(continues)

Water Units, Data Conversions, and Constants *(continued)*

EXPRESSIONS OF HARDNESS[a]

1 grain per gallon	= 1 grain $CaCO_3$ per U.S. gallon	**1 French degree**	= 1 part $CaCO_3$ per 100,000 parts water
1 part per million	= 1 part $CaCO_3$ per 1,000,000 parts water	**1 German degree**	= 1 part CaO per 100,000 parts water
1 English, or Clark, degree	= 1 grain $CaCO_3$ per Imperial gallon		

CONVERSIONS OF HARDNESS

1 grain per U.S. gallon	= 17.1 ppm, as $CaCO_3$	**1 French degree**	= 10 ppm, as $CaCO_3$
1 English degree	= 14.3 ppm, as $CaCO_3$	**1 German degree**	= 17.9 ppm, as $CaCO_3$

WEIGHT OF WATER

1 cubic inch	= 0.0361 lb	**1 imperial gallon**	= 10.0 lb
1 cubic foot	= 62.4 lb	**1 cubic meter**	= 1 tonne
1 gallon	= 8.34 lb		

DENSITY OF WATER[a]

Temperature		Density
°C	°F	gm/cm³
0	32	0.99987
1.667	35	0.99996
4.000	39.2	1.00000
4.444	40	0.99999
10.000	50	0.99975
15.556	60	0.99907
21.111	70	0.99802
26.667	80	0.99669
32.222	90	0.99510
37.778	100	0.99318
48.889	120	0.98870
60.000	140	0.98338
71.111	160	0.97729
82.222	180	0.97056
93.333	200	0.96333
100.000	212	0.95865

Note: Density of Sea Water: approximately 1.025 gm/cm³ at 15°C.

[a]*Source:* van der Leeden, F., Troise, F. L., and Todd, D. K., 1990. *The Water Encyclopedia,* 2d edition. Lewis Publishers, Inc., Chelsea, Michigan.

For the first time, we include a comprehensive Table of Contents that allows readers to find information across previous editions.

Comprehensive Table of Contents

Volume 1
World's Water 1998-1999: The Biennial Report on Freshwater Resources

			Conflicts Over Shared Water Resources 107
			Reducing the Risk of Water-Related Conflict 113
			The Israel-Jordan Peace Treaty of 1994 115
			The Ganges-Brahmaputra Rivers: Conflict and Agreement 118
			Water Disputes in Southern Africa 119
			Summary 124
			Appendix A
				Chronology of Conflict Over Waters in the Legends,
				Myths, and History of the Ancient Middle East 125
			Appendix B
				Chronology of Conflict Over Water:
				1500 to the Present 128
			References 132

FIVE		Climate Change and Water Resources:
			What Does the Future Hold? 137

			What Do We Know? 138
			Hydrologic Effects of Climate Change 139
			Societal Impacts of Changes in Water Resources 144
			Is the Hydrologic System Showing Signs of Change? 145
			Recommendations and Conclusions 148
			References 150

SIX			New Water Laws, New Water Institutions 155

			Water Law and Policy in New South Africa: A Move Toward
				Equity 156
			The Global Water Partnership 165
			The World Water Council 172
			The World Commission on Dams 175
			References 180

SEVEN		Moving Toward a Sustainable Vision for the Earth's
			Fresh Water 183

			Introduction 183
			A Vision for 2050: Sustaining Our Waters 185

WATER BRIEFS

			The Best and Worst of Science: Small Comets and the New
				Debate Over the Origin of Water on Earth 193
			Water Bag Technology 200
			Treaty Between the Government of the Republic of India and
				the Government of the People's Republic of Bangladesh
				on Sharing of the Ganga/Ganges Waters at Farakka 206

DATA SECTION

Volume 2
World's Water 2000-2001: The Biennial Report on Freshwater Resources

Index   311

Volume 3
World's Water 2002-2003: The Biennial Report on Freshwater Resources

DATA SECTION   237

Water Units, Data Conversions, and Constants   318

Index   329

For the first time, we include a comprehensive Index that allows readers to find information across previous volumes.

Comprehensive Index

KEY (book volume in boldface numerals)

1: The World's Water 1998–1999: The Biennial Report on Freshwater Resources

2: The World's Water 2000–2001: The Biennial Report on Freshwater Resources

3: The World's Water 2002–2003: The Biennial Report on Freshwater Resources

4: The World's Water 2004–2005: The Biennial Report on Freshwater Resources